FLUID STRUCTURE INTERACTION

Applied Numerical Methods

FLUID STRUCTURE INTERACTION

Applied Numerical Methods

Henri J.-P. MORAND

Université des Antilles et de la Guyane
Centre national d'études spatiales (CNES)
Centre spatial guyanais (Kourou)

Roger OHAYON

Conservatoire national des arts et métiers
(CNAM-Paris)
Office national d'études et de recherches
aérospatiales (ONERA-Châtillon)

Translated by:

Claude Andrew JAMES

INSET, Saint-Quentin, France

JOHN WILEY & SONS
Chichester • New York • Brisbane • Toronto • Singapore

1995

MASSON
Paris • Milan • Barcelona

Published with the support of the ministère de l'Enseignement supérieur et de la Recherche (France) : Direction de l'information scientifique et technique et des bibliothèques.

The French edition of this book is published in the series *Recherches en Mathématiques Appliquées* edited by P.G. Ciarlet and J.L. Lions.

Wiley Editorial Offices

John Wiley & Sons Ltd
Baffins Lane, Chichester
West Sussex P019 1UD, England

John Wiley & Sons, Inc., 605 Third Avenue,
New York, NY 10158-0012, USA

Jacaranda Wiley Ltd, G.P.O. Box 859, Brisbane,
Queensland 4001, Australia

John Wiley & Sons (Canada) Ltd, 22 Worceter Road,
Rexdale, Ontario M9W 1L1, Canada

John Wiley & Sons (SEA) Pte Ltd, 37 Jalan Pemimpin 05-04,
Block B, Union Industrial Building, Singapore 2057

ISBN : 0-471-94459-9 (Wiley)
ISBN : 2-225-84682-0 (Masson)

A catalogue record for this book is available from the British Library

Printed in France

Contents

Preface

The purpose of this book is to analyze the methods leading to the numerical modeling of the linear vibrations of elastic structures coupled to internal fluids (liquid or gas), for wavelength excluding high modal densities cases. The numerical methods presented throughout this book have proven their efficiency, accuracy and versatility in the aerospace field.

The proposed methods correspond to the research work carried out by the authors at ONERA (Office National d'Etudes et de Recherches Aérospatiales) and CNES (Centre National d'Etudes Spatiales), as well as from teaching activities at CNAM (Conservatoire National des Arts et Métiers), University Pierre et Marie Curie (Paris), Ecole Centrale de Paris, Ecole Nationale de Techniques Avancées, and Ecole Nationale Supérieure de l'Aéronautique et de l'Espace.

Among the various applications, let us cite:
— the hydroelastic and sloshing vibrations of liquid propelled launch vehicles taking into account if necessary surface tension effects — which occur in stability studies ("*pogo*" effect, attitude control of satellites and launchers, etc.) and in vibration studies of satellites, spacecrafts, hypersonic planes, space stations, etc.
— and the structural-acoustic vibrations occurring in the payload of launchers (due to engine noise), in aircraft, in automobiles, etc.

The particularity of the problems under consideration lies in their multidisciplinary aspects, involving structural and fluid representation and related numerical aspects.

The numerical methods considered herein are of interest for various industrial domains, such as automotive industry engineering, nuclear engineering, civil engineering, naval engineering, and biomechanics, etc.

Various symmetric matrix equations derived from appropriate *variational formulations* by means of the finite element method are given.
The following two objectives are pursued:

1. Direct resolution of the coupled systems by the *finite element method*;

2. *Modal reduction procedures* using the eigenmodes of appropriately defined "elementary subsystems" (structural modes, sloshing modes, acoustic modes).

In this book, we do not consider the presence of an external environment, such as, for instance, an external unbounded gas or liquid medium. The effects of such an external environment involve, at low frequency range, slowly varying operators.

The coupled system constituted by the structure and its internal fluid is relevant to the numerical methods presented here. One may then couple this system (with its numerical description) to an external fluid (described for example by boundary integral procedures).

For very low frequency, sloshing free surface effects due to gravity are predominant, while for higher frequencies, neglecting gravity, hydroelastic incompressible effects and acoustic effects may be predominant. That is why we describe various physical subsystems which may interact whenever their frequencies are close.

The book is divided into nine chapters, starting from a synthesis of structural vibrations in chapter one, and a summary of the main linearized fluid equations in chapter two.

Then, from chapter three to chapter eight, we describe the various physical subsystems: liquid sloshing modes (taking into account surface tension effects), hydroelastic incompressible vibrations (plus possible gravity effects), structural-acoustic vibrations (for gas and compressible liquids).

Finally, in chapter nine, modal reduction methods (dynamic substructuring) allow the analysis of the interactions between those subsystems.

The authors express their gratitude to Professor J.L. Lions and to Professor P.G. Ciarlet, Members of the Academy of Sciences, for their continuing support and advice concerning the manuscript and for having accepted the original French version in their series on Research in Applied Mathematics.

The authors express also their gratitude to Professor O.C. Zienkiewicz, Fellow of the Royal Society, for all the continuing support concerning the overall research works of the authors and for having introduced them to the international community of computational mechanics.

The authors deeply acknowledge the contributions of their colleagues of the Launcher Directorate of CNES and of the Structures Department of ONERA.

Finally, the authors wish to thank Ingrid Ohayon, Yves Gorge and Tran Duc Minh for their contributions to the present LaTeX edition of the English manuscript.

CHAPTER 1

Vibrations of elastic structures

1.1 Introduction

We shall consider here oscillations of "low amplitude" about a "mean" position. In practice, we take a *mean position* to be either:

- an equilibrium position, or
- a "reference configuration". For example, the vibrations of a launch vehicle are described in relation to a rigid body motion of the vehicle on its trajectory.

This chapter is devoted to a review of the modeling of elastic structure vibrations, in terms of a *displacement* u, in a variational frame of reference suitable for application of the finite element method and the use of modal projection techniques.

As a first step, we use the test-function method to derive the variational formulation in displacements u of the linearized elastodynamic problem, together with the matrix structure of the problem discretized by finite elements.

We then review the general properties of the vibrational eigenmodes u_α and we discuss particularly the phenomenon of mode crossing, often called "curve veering", in a system whose characteristics are continuously dependent on one parameter.

We next discuss the harmonic problem with given forces and displacements and the variational properties of reaction forces.

Finally, we undertake the modal analysis of the vibratory response of a structure leading to a dynamic superelement of substructure in $(u_\Sigma, \{q_\alpha\})$, where u_Σ is the displacement of a coupling surface Σ and q_α the generalized coordinates associated with the eigenmodes u_α of the structure fixed on Σ.

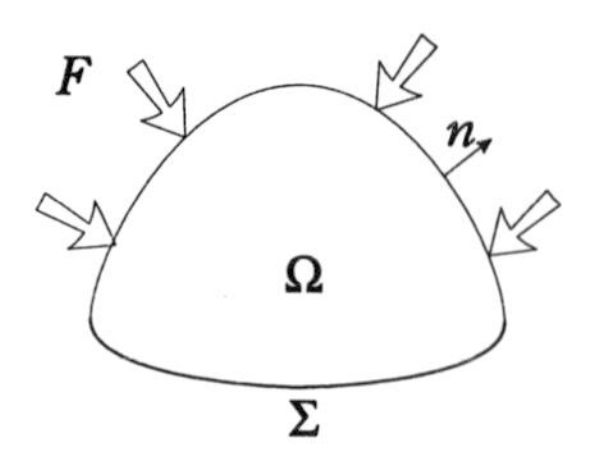

Figure 1.1: Geometrical configuration

1.2 Elastodynamic linearized equations

Let Ω designate the bounded domain occupied by the structure at equilibrium, and n, the unitary normal external to the boundary $\partial\Omega$ of Ω (Fig. 1.1).
Let $u(M,t)$ designate the displacement, at an instant t, of a particle situated in $M \in \Omega$ at equilibrium. Let x_1, x_2, x_3 designate the cartesian coordinates of M and u_1, u_2, u_3 the components of u. We shall be using the notation $A,j = \partial A/\partial x_j$, together with the classical convention for summations over repeated indices.
We operate within the framework of the linearized elasticity theory, and refer the reader to the basic works of Germain [83], Fung [76], Malvern [127], Salençon [194].
Stresses in the equilibrium state are neglected, and the displacements u are assumed infinitesimal. Under these conditions, the Cauchy stress tensor σ_{ij} is related to the linearized strain tensor ϵ_{ij} by the constitutive law:

$$\sigma_{ij} = a_{ijkh}\epsilon_{kh} \tag{1.1}$$

where a_{ijkh} are the coefficients of elasticity endowed with the usual properties of positivity ($a_{ijkh}\epsilon_{ij}\epsilon_{kh} > 0 \quad \forall \epsilon_{ij}$ symmetric $\neq 0$) and symmetry (with respect to the index exchange $i \leftrightarrow j, k \leftrightarrow h$ and $[ij] \leftrightarrow [kh]$). ϵ_{kh} is related to u by

$$\epsilon_{kh} = \frac{1}{2}(u_{k,h} + u_{h,k}) \tag{1.2}$$

In what follows, $\sigma_{ij}(u)$ will designate the stress tensor σ_{ij} as a function of u, obtained on replacing ϵ_{kh} by its expression (1.2) in terms of u in the constitutive law (1.2):

$$\sigma_{ij}(u) = a_{ijkh}\,\epsilon_{kh}(u) \;=\; \frac{1}{2}a_{ijkh}(u_{k,h} + u_{h,k}) \tag{1.3}$$

Neglecting body forces, u then satisfies the *elastodynamic equation*:

$$\boxed{\sigma_{ij,j}(u) \;-\; \rho\frac{\partial^2 u_i}{\partial t^2} = 0 \quad \text{in } \Omega} \tag{1.4}$$

where ρ is the density at equilibrium. For prescribed displacements $u_\Sigma(M,t)$ on a part Σ of $\partial\Omega$, we have the following boundary condition:

$$\boxed{u = u_\Sigma \quad \text{on } \Sigma} \tag{1.5}$$

For a given surface force density $F(M,t)$ on $\partial\Omega\backslash\Sigma$, we have the following boundary condition:

$$\boxed{\sigma_{ij}(u)n_j = F_i \quad \text{on } \partial\Omega\setminus\Sigma} \tag{1.6}$$

Finally, the above equations must be completed by the initial conditions of Cauchy on $u(M,0)$ and $\left.\dfrac{\partial u(M,t)}{\partial t}\right|_{t=0}$ in Ω.

1.3 Variational formulation of the response to given forces

Elastodynamic boundary value problem

We consider a structure Ω, fixed on a part Σ of its boundary, subject to forces of surface density $F(M,t)$ on $\partial\Omega\setminus\Sigma$. Under these conditions, u satisfies the boundary value problem:

$$\begin{array}{rcllr} \sigma_{ij,j}(u) - \rho\dfrac{\partial^2 u_i}{\partial t^2} & = & 0 & \text{in } \Omega & (a) \\ \sigma_{ij}(u)n_j & = & F_i & \text{on } \partial\Omega\setminus\Sigma & (b) \\ u_i & = & 0 & \text{on } \Sigma & (c) \\ & & \multicolumn{2}{l}{+ \text{ initial conditions}} & \end{array} \tag{1.7}$$

Variational formulation

We proceed formally by the test-functions method. We consider a solution of (1.7).

1. First, we introduce the space $\mathcal{C}$ of the "sufficiently" smooth functions $v(M)$, $M \in \Omega$,

   In a second step, we multiply (1.7a) by an arbitrary *time-independent* test-function $v \in \mathcal{C}$. We then integrate in the domain Ω. Finally we proceed by an integration by part and we use Stoke's formula (the result is also known as Green's formula):

$$\int_\Omega \sigma_{ij,j}v_i\,dx = \int_{\partial\Omega} \sigma_{ij}n_jv_i\,d\sigma - \int_\Omega \sigma_{ij}v_{i,j}\,dx \tag{1.8}$$

we obtain:

$$\int_{\partial\Omega} \sigma_{ij}(u) n_j v_i \, d\sigma \; - \; \int_\Omega \sigma_{ij}(u) v_{i,j} \, dx \; - \; \int_\Omega \rho \frac{\partial^2 u_i}{\partial t^2} v_i \, dx \; = 0 \qquad (1.9)$$

2. Secondly, we use boundary condition (1.2), together with the symmetry of σ_{ij} which enables $\sigma_{ij} v_{i,j}$ to be replaced by $\sigma_{ij}(1/2)(v_{i,j} + v_{j,i}) = \sigma_{ij}\epsilon_{ij}(v)$, giving:

$$\int_\Omega \sigma_{ij}(u)\epsilon_{ij}(v) \, dx + \int_\Omega \rho \frac{\partial^2 u_i}{\partial t^2} v_i \, dx \; = \; \int_{\partial\Omega\backslash\Sigma} F_i v_i \, d\sigma + \int_\Sigma \sigma_{ij} n_j v_i \, d\sigma \quad (1.10)$$

3. Finally we introduce the space $\mathcal{C}^0 \subset \mathcal{C}$ of the $v \in \mathcal{C}$ vanishing on Σ:

$$\mathcal{C}^0 \; = \; \{ v \in \mathcal{C} \mid v = 0 \text{ on } \Sigma \} \qquad (1.11)$$

For $v \in \mathcal{C}^0$, the variational property (1.10) becomes:

$$\boxed{\int_\Omega \sigma_{ij}(u)\epsilon_{ij}(v) \, dx + \int_\Omega \rho \frac{\partial^2 u_i}{\partial t^2} v_i \, dx \; = \; \int_{\partial\Omega\backslash\Sigma} F_i v_i \, d\sigma \qquad u \in \mathcal{C}^0, \; \forall v \in \mathcal{C}^0} \quad (1.12)$$

The variational formulation of (1.7) can then be stated as follows:
Find $u \in \mathcal{C}^0$ such that the property of (1.12) is satisfied $\forall v \in \mathcal{C}^0$.

Converse — Let us demonstrate conversely that if $u(M,t)$ satisfies the property (1.12), $u(M,t)$ satisfies equations (1.7a-c).
On going through the calculations formally using Green's formula (1.8), and noting that $v = 0$ on Σ, we find:

$$\int_\Omega (\sigma_{ij,j}(u) \; - \; \rho \frac{\partial^2 u_i}{\partial t^2}) v_i \, dx \; - \; \int_{\partial\Omega\backslash\Sigma} (\sigma_{ij} n_j - F_i) v_i \, d\sigma \; = 0 \qquad \forall v \in \mathcal{C} \qquad (1.13)$$

To retrieve the local equations, we proceed in two steps:

1. By choosing test-functions vanishing on $\partial\Omega\backslash\Sigma$, (1.13) is reduced to its first term, whose nullity leads to (1.7a) being satisfied.

2. We refer to (1.13). The first term is identically null, since we have just established that u satisfies (1.7a). The nullity of the remaining integral over $\partial\Omega$ ($\forall v \in \mathcal{C}^0, v \neq 0$ on $\partial\Omega \setminus \Sigma$) then leads to (1.7b).

Case of a free structure ($\Sigma = \emptyset$) — The equations and variational formulation are written:

$$\boxed{\begin{array}{lll} \sigma_{ij,j}(u) \; - \; \rho \dfrac{\partial^2 u_i}{\partial t^2} = 0 & \text{in } \Omega & (a) \\ \sigma_{ij}(u) n_j = F_i & \text{on } \partial\Omega & (b) \end{array}} \qquad (1.14)$$

$$\boxed{\int_\Omega \sigma_{ij}(u)\epsilon_{ij}(v)\,dx + \int_\Omega \rho \frac{\partial^2 u_i}{\partial t^2} v_i\,dx = \int_{\partial\Omega} F_i v_i\,d\sigma \qquad u \in \mathcal{C},\ \forall v \in \mathcal{C}} \tag{1.15}$$

Discussion — A rigorous presentation of variational methods in elastodynamics can be found in Dautray & Lions [54] (vol. 4 and 7), Duvaut & Lions [61], Raviart & Thomas [191].

- The space $\mathcal{C}$ of the test-functions is Sobolev space $(H^1(\Omega))^3$.
- t is considered as a parameter and $u_i(\cdot, t)$ as a mapping $]0, T[\to u_i(t) \in H^1(\Omega)$.
- We note that the variational formulation (1.12) of the problem relating to the structure fixed on Σ, is "deducible" from that relating to the free structure (1.15) by restricting the admissible class to the subspace $\mathcal{C}^0 = \{v \in \mathcal{C} \mid v|_\Gamma = 0\}$.
- In the case of *slender* structures such as beams, plates and shells, the bilinear form $\int_\Omega \sigma_{ij}(u)\epsilon_{ij}(v)\,dx$ is simply replaced by the corresponding variational expressions (*cf.* Novozhilov [163], Soedel [201], Ciarlet [39]).

1.4 Discretization: Ritz-Galerkin and finite element methods

We use the method of Ritz-Galerkin, which consists of seeking an approximate solution of (1.15) in a subspace $\mathcal{C}^h$ of finite dimension N of $\mathcal{C}$.
If $e_j (j = 1, \ldots, N)$ denotes a basis in this vector subspace, an element $u^h \in \mathcal{C}^h$ can be written in the form $u^h = \sum_{i=1}^N \xi_j e_j$.
Satisfaction of a variational property involving bilinear and linear forms in this subspace then yields matrix equations.
In the particular case of the *finite element method*, we recall that ξ_j coincides with the value of u^h at the nodes of a mesh (in the case of a Lagrange interpolation) —u^h being defined by interpolation from the *nodal values* U_j called *degrees of freedom*.
Presentations of the finite element method and its application in science and technology can be found in Zienkiewicz & Taylor [217], Ciarlet [35], Hughes [99], Dautray & Lions [54] (vol. 6), Raviart & Thomas [191], Onate, Périaux, & Samuelsson [177].

Mass and stiffness matrices, force vector — In what follows, $\boldsymbol{U}$ (*resp.* $\boldsymbol{V}$) denotes the vector of $\mathbb{R}^N$ of components U_j (*resp.* V_j) and $\boldsymbol{V}^T$ the transpose of $\boldsymbol{V}$. The discretized expressions of bilinear and linear forms involved in the variational formulation (1.15) yield $N \times N$ matrices of *mass* $\boldsymbol{M}$ (*symmetric, nonsingular*), of *stiffness* $\boldsymbol{K}$ (*symmetric, singular*), together with the force vector $\boldsymbol{F}$, defined respectively by:

$$
\begin{aligned}
\int_\Omega \rho u\cdot v\, dx &\implies \boldsymbol{V}^T\boldsymbol{M}\boldsymbol{U} && (a)\\
\int_\Omega \sigma_{ij}(u)\epsilon_{ij}(v)\, dx &\implies \boldsymbol{V}^T\boldsymbol{K}\boldsymbol{U} && (b)\\
\int_{\partial\Omega} F\cdot v\, d\sigma &\implies \boldsymbol{V}^T\boldsymbol{F} && (c)
\end{aligned}
\tag{1.16}
$$

In discretized form (1.15) can be written:

$$
\begin{aligned}
\boldsymbol{V}^T\boldsymbol{M}\ddot{\boldsymbol{U}} + \boldsymbol{V}^T\boldsymbol{K}\boldsymbol{U} &= \boldsymbol{V}^T\boldsymbol{F} \quad \forall \boldsymbol{V} && (a)\\
&\downarrow \\
\boldsymbol{M}\ddot{\boldsymbol{U}} + \boldsymbol{K}\boldsymbol{U} &= \boldsymbol{F} && (b)
\end{aligned}
\tag{1.17}
$$

We are thus led to solving a system of large coupled differential equations (*cf.* e.g. Bathe [9], Belytschko & Hughes [13]).
For a treatment of the finite elements used in the construction of stiffness and mass matrices of structures modelled by beams, plates or shells, see e.g. Batoz & Dhatt [10], Ciarlet [35], Zienkiewicz & Taylor [217].

Case of a structure fixed on Σ — The components of $\boldsymbol{U}$ and $\boldsymbol{V}$ corresponding to nodes situated on Σ are null.
The matrices $\boldsymbol{K}$ and $\boldsymbol{M}$ of the structure fixed on Σ, are therefore deducible from those of the free structure by removing the rows and columns corresponding to the constrained degrees of freedom (in this case, K is nonsingular).

Matrix reduction by projection on Ritz vectors

An approximate solution to (1.17) can be obtained by projecting these equations by the Ritz method, over the subspace $\mathcal{C}_n$ of $\mathbb{R}^N$ spanned by n linearly independent vectors $\boldsymbol{U}_\alpha$, $1 \leq \alpha \leq n \leq N$, also called "Ritz vectors".
We thus write:

$$
\boldsymbol{U} = \sum_{\alpha=1}^{n} q_\alpha \boldsymbol{U}_\alpha \iff \boldsymbol{U} = \boldsymbol{H}\boldsymbol{q} \tag{1.18}
$$

where $\boldsymbol{H} = [\boldsymbol{U}_1, \ldots, \boldsymbol{U}_n]$ is a $N \times n$ matrix and $\boldsymbol{q}^T = [q_1, \ldots, q_n]$.
Restricting (1.17a) to $\mathcal{C}_n$, is equivalent to satisfying this property for $\boldsymbol{U}$ and $\boldsymbol{V}$ of the form (1.18).
By putting $\boldsymbol{U} = \boldsymbol{H}\boldsymbol{q}$ and $\boldsymbol{V} = \boldsymbol{H}\boldsymbol{r}$ in (1.17a), we obtain a variational equation whose satisfaction $\forall \boldsymbol{r} \in \mathbb{R}^n$ leads to the following "reduced problem":

$$
(\boldsymbol{H}^T\boldsymbol{K}\boldsymbol{H})\boldsymbol{q} + (\boldsymbol{H}^T\boldsymbol{M}\boldsymbol{H})\ddot{\boldsymbol{q}} = \boldsymbol{H}^T\boldsymbol{F} \tag{1.19}
$$

Note that this last equation can be obtained by substituting (1.18) into (1.17a) and putting successively $\boldsymbol{V} = \boldsymbol{U}_1, \boldsymbol{V} = \boldsymbol{U}_2, \ldots, \boldsymbol{V} = \boldsymbol{U}_n$.
To summarize, the following matrix *projection rule* is obtained (where $\boldsymbol{H}$ is a matrix whose n column-vectors are linearly independent):

$$
\{\boldsymbol{K}, \boldsymbol{M}, \boldsymbol{F}\} \implies \left\{(\boldsymbol{H}^T\boldsymbol{K}\boldsymbol{H}), (\boldsymbol{H}^T\boldsymbol{M}\boldsymbol{H}), (\boldsymbol{H}^T\boldsymbol{F})\right\} \tag{1.20}
$$

1.5 Harmonic response to forces

We assume $F(M,t) = F(M)\cos\omega t$ and seek solutions $u(M,t) = u(M)\cos\omega t$ of (1.14) (or (1.15), (1.17)).
The boundary value problem, the variational formulation and the matrix equations are then written, for a given ω and F:

$$\boxed{\begin{aligned} \sigma_{ij,j}(u) + \rho\omega^2 u_i &= 0 \qquad \text{in } \Omega \\ \sigma_{ij}(u)n_j &= F_i \qquad \text{on } \partial\Omega \end{aligned}} \tag{1.21}$$

$$\boxed{\int_\Omega \sigma_{ij}(u)\epsilon_{ij}(v)\,dx - \omega^2\int_\Omega \rho u\cdot v\,dx = \int_{\partial\Omega} F\cdot v\,d\sigma \quad u \in \mathcal{C},\ \forall v \in \mathcal{C}} \tag{1.22}$$

$$\boxed{\boldsymbol{KU} - \omega^2\boldsymbol{MU} = \boldsymbol{F}} \tag{1.23}$$

Fredholm alternative:

1. either ω^2 does not coincide with any of the eigenvalues λ_α of the associated spectral problem obtained on setting $F = 0$ — and the problem then has a unique solution,

2. or $\omega^2 = \lambda_\alpha$ — and in that case the problem has a solution if $\int_\Sigma F\cdot u_\alpha\,d\sigma = 0$ (i.e. if F is orthogonal to the eigenspace associated with λ_α). This solution is defined up to an additive vector contained in this eigenspace.

Remarks

- In the case of a structure fixed on Σ, we note that the corresponding variational formulation is obtained by restricting (1.15) to $\mathcal{C}^0$ as defined by (1.11).

- In practice, (1.23) has to be solved for a given ω and $\boldsymbol{F}$, which presupposes that the value of ω is "not too high" (the mesh used must be compatible with the spatial scale of the phenomena being studied, for the value of ω used).

1.6 Vibrational eigenmodes

Free structure —Solutions of (1.21) are sought for $F = 0$. Putting $\lambda = \omega^2$, one seeks the solutions of the following spectral problem:
find λ such that there exists $u \neq 0$ satisfying:

$$\boxed{\begin{aligned} \sigma_{ij,j}(u) + \rho\lambda u_i &= 0 \qquad \text{in } \Omega \qquad (a) \\ \sigma_{ij}(u)n_j &= 0 \qquad \text{on } \partial\Omega \qquad (b) \end{aligned}} \tag{1.24}$$

The solutions of this eigenvalue problem are called the eigenmodes. From (1.22) and (1.23), the variational formulation and the corresponding matrix equations are written:

$$\boxed{\begin{array}{l}\text{Find } \lambda,\ u \in \mathcal{C}(u \neq 0), \text{ such that } \forall v \in \mathcal{C}: \\ \displaystyle\int_\Omega \sigma_{ij}(u)\epsilon_{ij}(v)\,dx - \lambda \int_\Omega \rho u \cdot v\,dx = 0\end{array}} \Rightarrow \boxed{\boldsymbol{KU} = \lambda \boldsymbol{MU}} \qquad (1.25)$$

The solution of (1.25) involves specific algorithms (*cf.* Ciarlet [36], Bathe [9], Parlett [180]).

Case of a fixed structure — The variational formulation of the problem of eigenmodes of a structure fixed on Σ are obtained by restricting (1.17) to the class $\mathcal{C}^0 \subset \mathcal{C}$ (*cf.* (1.11)):
find λ and $u \in \mathcal{C}^0, u \neq 0$ satisfying $\forall v \in \mathcal{C}^0$:

$$\boxed{\int_\Omega \sigma_{ij}(u)\epsilon_{ij}(v)\,dx \;-\; \lambda \int_\Omega \rho u \cdot v\,dx \;=\; 0} \qquad (1.26)$$

Properties of the eigenmodes

These properties, established below for a fixed structure, are valid in the "special case" of a free structure ($\Sigma = \emptyset$), by means of several modifications which will be pointed out at the appropriate times.
Let $(\lambda_\alpha, u_\alpha)$ to be a solution of (1.26), and therefore satisfying:

$$\int_\Omega \sigma_{ij}(u_\alpha)\epsilon_{ij}(v)\,dx \;=\; \lambda_\alpha \int_\Omega \rho u_\alpha \cdot v\,dx \qquad u_\alpha \in \mathcal{C}^0,\ \forall v \in \mathcal{C}^0. \qquad (1.27)$$

- λ_α is positive. Indeed, on setting $v = u_\alpha$ in (1.27), we obtain:

  $$\int_\Omega \sigma_{ij}(u_\alpha)\epsilon_{ij}(u_\alpha)\,dx \;=\; \lambda_\alpha \int_\Omega \rho |u_\alpha|^2\,dx \qquad (1.28)$$

  which shows that λ_α is the quotient of two positive quantities, since the stiffness and mass operators are positive.

- $\mu_\alpha = \int_\Omega \rho |u_\alpha|^2\,dx$ is called the *generalized mass* of the mode under consideration.

- $\gamma_\alpha = \int_\Omega \sigma_{ij}(u_\alpha)\epsilon_{ij}(u_\alpha)\,dx$ is called the *generalized rigidity* of the mode under consideration. (1.28) can thus be written:

  $$\gamma_\alpha \;=\; \lambda_\alpha \mu_\alpha \qquad (1.29)$$

- u_α is defined to within a factor of one. In practice, this indeterminacy is removed by choice of a normalization convention. One such convention consists, for example, of choosing a scale factor such that $\mu_\alpha = 1$, giving $\gamma_\alpha = \lambda_\alpha = \omega_\alpha^2$.

- *Energy interpretation*: consider the oscillation $u(M,t) = u_\alpha(M)\cos\omega_\alpha t$ corresponding to this eigenmode. $(1/2)\gamma_\alpha \cos^2\omega_\alpha t$ represents the potential energy of deformation of the structure and $(1/2)\omega_\alpha^2\mu_\alpha \sin^2\omega_\alpha t$ the kinetic energy of the structure. Consequently, (1.29) represents the conservation of total mechanical energy.

- By taking Sobolev space $(H^1(\Omega))^3$ as admissible space $\mathcal{C}$, it can be shown, by using the properties of elasticity operators, that there exists an increasing sequence of positive eigenvalues tending to infinity.

- The set of eigenvectors u_α corresponding to a given eigenvalue spanned a subspace $\mathcal{C}_\alpha$ of finite dimension N_α called *degeneracy*, or *multiplicity* of the eigenvalue under consideration. In the case of structures showing symmetries (geometrical domain and mechanical characteristics), the eigenvalues may be degenerate. For example, in the case of spherical symmetry, it can be shown that each eigenvalue shows a degeneracy of $2j+1$, where j is an integer. We shall see later that axisymmetric structures have eigenvalues of degeneracy 2. For systems with no symmetry, the eigenvalues are in general simple. Any degeneracy not attributable to the symmetries of a structure is qualified as "accidental".

  In reality, natural imperfections in a symmetrical structure have the effect of removing these degeneracies, and this can complicate the experimental identification of calculated modes when the frequencies of different types of modes are close to each other.

  In general, the classification of eigenmodes of systems presenting symmetries involves the theory of irreducible representations of groups (*cf.* e.g. Ludwig & Falter [124], Bossavit [22]).

Orthogonality of eigenvectors

The series of eigenvectors u_α, $\alpha \geq 1$ satisfies the orthogonality relations "with respect to stiffness" (1.30a) and "with respect to mass" (1.30b)

$$\boxed{\begin{aligned} \int_\Omega \sigma_{ij}(u_\alpha)\epsilon_{ij}(u_\beta)\,dx &= \delta_{\alpha\beta}\lambda_\alpha\mu_\alpha && (a) \\ \int_\Omega \rho u_\alpha\cdot u_\beta\,dx &= \delta_{\alpha\beta}\mu_\alpha && (b) \end{aligned}} \tag{1.30}$$

where $\delta_{\alpha\beta}$ is Kronecker's symbol.

For $\alpha = \beta$, these relationships simply express the definition of μ_α and the property (1.28).
For $\alpha \neq \beta$, two cases have to be considered:

1. **Distinct eigenvalues $\lambda_\alpha \neq \lambda_\beta$**
   We have to show that $\int_\Omega \sigma_{ij}(u_\alpha)\epsilon_{ij}(u_\beta)\,dx = \int_\Omega \rho u_\alpha \cdot u_\beta\,dx = 0$. By applying the respective variational properties ((1.26) of u_α (with $v = u_\beta$) and of u_β (with $v = u_\alpha$), we obtain:

$$\begin{aligned} \int_\Omega \sigma_{ij}(u_\alpha)\epsilon_{ij}(u_\beta)\,dx &= \lambda_\alpha \int_\Omega \rho u_\alpha \cdot u_\beta\,dx \qquad (a) \\ \int_\Omega \sigma_{ij}(u_\beta)\epsilon_{ij}(u_\alpha)\,dx &= \lambda_\beta \int_\Omega \rho u_\beta \cdot u_\alpha\,dx \qquad (b) \end{aligned} \tag{1.31}$$

   On subtracting term by term and using the symmetry of the bilinear forms $\int_\Omega \sigma_{ij}(u)\epsilon_{ij}(v)\,dx$ and $\int_\Omega \rho u \cdot v\,dx$, we obtain:

   $0 = (\lambda_\alpha - \lambda_\beta)\int_\Omega \rho u_\alpha \cdot u_\beta\,dx$, and consequently, $\int_\Omega \rho u_\alpha \cdot u_\beta\,dx = 0$. Referring to (1.31a), we have $\int_\Omega \sigma_{ij}(u_\alpha)\epsilon_{ij}(u_\beta)\,dx = 0$, which is the proof required.

2. **Multiple eigenvalue** — We take the special case of an eigenvalue λ_α of (finite) multiplicity $N_\alpha > 1$ We denote the corresponding subspace $\mathcal{C}_\alpha \subset \mathcal{C}$. If $u_\alpha^1, u_\alpha^2, \ldots, u_\alpha^{N_\alpha}$ is a basis of $\mathcal{C}_\alpha$, the Gram-Schmidt orthogonalization procedure enables a new basis $u_{\alpha'}^1, u_{\alpha'}^2, \ldots, u_{\alpha'}^{N_\alpha}$ to be constructed, which satisfies the orthogonality conditions $\int_\Omega \rho u_{\alpha'}^m \cdot u_{\alpha'}^n\,dx = 0$ for $m \neq n$. Setting $u_\alpha = u_{\alpha'}^m$ and $v = u_{\alpha'}^n$ in (1.27) satisfied by the elements of $\mathcal{C}_\alpha$, we find $\int_\Omega \sigma_{ij}(u_{\alpha'}^m)\epsilon_{ij}(u_{\alpha'}^n)\,dx = 0$.

Remark — Let us consider the superposition of two natural oscillations of frequencies $\omega_\alpha \neq \omega_\beta$, $au_\alpha \cos\omega_\alpha t + bu_\beta \cos\omega_\beta t$. We can then verify that the conservation of total mechanical energy during this motion is equivalent to the orthogonality conditions (1.30).

Rigid body modes

For $\lambda = 0$, (1.25) has solutions $u^{\mathcal{R}}$ satisfying:

$$\int_\Omega \sigma_{ij}(u^{\mathcal{R}})\epsilon_{ij}(v)\,dx = 0 \quad \forall v \in \mathcal{C} \tag{1.32}$$

Satisfaction of this equation is equivalent to $\epsilon_{ij}(u^{\mathcal{R}}) = 0$, whose solution is:

$$u^{\mathcal{R}} = \vec{T} + \overset{\curvearrowleft}{\theta} \times \overrightarrow{OM} \tag{1.33}$$

where the vectors $\vec{T}$ and $\overset{\curvearrowleft}{\theta}$ represent respectively a translation and a rotation (infinitesimal).

- The existence of solutions $u^{\mathcal{R}}$ results from the *degenerate* nature of the operator $\mathcal{K}(u,v) = \int_\Omega \sigma_{ij}(u)\epsilon_{ij}(v)\,dx$ on $\mathcal{C} \times \mathcal{C}$. In the present case, these solutions will be considered as eigenmodes at zero frequency, also referred as "rigid body modes".

- Note that these solutions constitute a 6-dimensional subspace $\mathcal{C}^{\mathcal{R}}$ of $\mathcal{C}$. Restricting $\int_\Omega \rho u \cdot v \, dx$ to $\mathcal{C}^{\mathcal{R}} \times \mathcal{C}^{\mathcal{R}}$ defines the 6×6 inertia matrix of the structure considered as a rigid body.

Conditions for the orthogonality of rigid body modes — The following relationships (1.30) are applicable:

1. Orthogonality "with respect to the stiffness" (1.30a)1.30 is trivially verified (we substitute $v = u^{\mathcal{R}}$ in (1.32).

2. Orthogonality "with respect to the mass" (1.30b) between a free structure eigenmode u_α and a rigid body mode $u_\beta = u^{\mathcal{R}}$, results from the invariance of the center of gravity together with the nullity of the angular momentum during the natural oscillations $u_\alpha \cos \omega_\alpha t$ ($\omega_\alpha > 0$) of a free elastic body (Cabannes [28], Lanczos [113]). We have indeed, according to (1.30b),

$$\int_\Omega \rho(\vec{T} + \overset{\curvearrowleft}{\theta} \times \overrightarrow{OM}) \cdot u_\alpha \, dx = 0 \qquad \forall \vec{T} \;, \; \forall \overset{\curvearrowleft}{\theta}$$
$$\Updownarrow \tag{1.34}$$
$$\int_\Omega \rho \, \vec{u_\alpha} \; dx = \vec{0} \quad \text{ and } \int_\Omega \rho \, \overrightarrow{OM} \times \vec{u_\alpha} \; dx = \vec{0}$$

Eigenvectors basis

In what follows, we shall be numbering the eigenvalues in increasing order, including the zero eigenvalue, repeating N_α times each eigenvalue λ_α of multiplicity N_α.
It can be shown that the sequence $\{u_\alpha\}$ of respectively, the free structure and of the structure fixed on Σ, normalized with respect to the mass ($\mu_\alpha = 1$) constitutes a hilbertian basis of the space of the "kinematically admissible" values of u, i.e. respectively of $\mathcal{C}$ and of $\mathcal{C}^0$.
In the case of a free structure, the completeness of the basis $\{u_\alpha\}$ — which includes the "rigid body modes" — can be established by considering the problem of "shifted" eigenvalues, defined for $s > 0$, by $\int_\Omega \sigma_{ij}(u)\epsilon_{ij}(v) \, dx + s \int_\Omega \rho u \cdot v \, dx = \lambda' \int_\Omega \rho u \cdot v \, dx$, whose eigenvectors are identical to those of (1.25), and whose positive eigenvalues are $\lambda' = \lambda + s$.

Rayleigh quotient

The Rayleigh quotient is defined for all $u \in \mathcal{C}, u \neq 0$ by:

$$R(u) = \frac{\displaystyle\int_\Omega \sigma_{ij}(u)\epsilon_{ij}(u) \, dx}{\displaystyle\int_\Omega \rho |u|^2 \, dx} \tag{1.35}$$

- Note that, by setting $v = u_\alpha$ in the variational property (1.25), $R(u_\alpha) = \lambda_\alpha$. It can also be shown that $R(u)$ is stationary for $u = u_\alpha$: we do indeed find on differentiating (1.35), and using (1.25), that $R(u_\alpha + \epsilon v) - R(u_\alpha)$ is second order in ϵ.

Comparison of eigenfrequencies

Comparison of the eigenfrequencies of various structures is principally based on two comparison theorems involving extremal properties of eigenvalues.

First comparison theorem — Given two eigenvalue problems with admissible spaces respectively $\mathcal{C}$ and $\mathcal{C}^*$, and with Rayleigh quotients respectively R and R^*, satisfying:

$$\mathcal{C}^* \subseteq \mathcal{C} \quad , R^*(u^*) \geq R(u^*) \quad \forall u^* \in \mathcal{C}^* \tag{1.36}$$

the n-th respective eigenvalues satisfy the inequality:

$$\lambda_n^* \geq \lambda_n \tag{1.37}$$

Demonstration of this property is based on the *min-max property* which enables the nth eigenvalue to be characterized independently of the other eigenvalues (and eigenvectors).
The stated principle covers two distinct situations:

1. firstly, the case of vibration problems $\mathcal{P}$ and $\mathcal{P}^*$ from the same admissible class $\mathcal{C}$, and of respective Rayleigh coefficients R and R^* verifying $R^*(u) \geq R(u)$, $\forall u \in \mathcal{C}$,

2. secondly, the case of a problem $\mathcal{P}^*$ obtained by restricting the admissible class of a problem $\mathcal{P}$ (with the same Rayleigh quotient with $\mathcal{C}^* \subseteq \mathcal{C}$).

   From inequality (1.37) we see that the eigenfrequencies of the discretized problem are higher than those frequencies *of the same rank* from the continuum problem — to within a domain approximation.

Second comparison theorem: "principle" of separation. We compare the eigenvalues of two problems differing only by a certain number of constraints defined by the linear relations $L_1(u) = 0, L_2(u) = 0, \ldots, L_p(u) = 0$.
The principle of separation stipulates that the eigenvalues λ_n^* of the problem submitted to p constraints "separate" the eigenvalues λ_n from the problem without constraints. What we have precisely is:

$$\lambda_n \leq \lambda_n^* \leq \lambda_{n+p} \tag{1.38}$$

As an example, let us compare the natural frequencies of a free structure a part Σ of which is *rigid*, with the frequencies of this same structure fixed on Σ.

The displacements of Σ are described by 6 degrees of freedom (3 translations T_X, T_Y, T_Z and 3 rotations $\theta_X, \theta_Y, \theta_Z$). The clamping condition is equivalent to the nullity of these six degrees of freedom.
For $p = 6$, (1.38) leads to:

$$\lambda_n \leq \lambda_n^* \leq \lambda_{n+6} \tag{1.39}$$

We note that the first six eigenvalues $\lambda_n, n = 1, \ldots, 6$ of the unconstrained problem are zero (free structure), and that, under these conditions, the nth strictly

positive eigenfrequency of the fixed structure is lower than the nth *strictly positive eigenfrequency* of the free structure.
As an illustration, in launch vehicle-spacecraft coupled analysis where the coupling interface is generally considered to be rigid, this result serves to check the calculations.

Bibliography — The demonstration of the min-max property for matrices can be found in Ciarlet [36], p. 12, th. 1.3-1. For a thorough analysis of spectral problems, see Courant & Hilbert [45], Dautray & Lions [54] (vol. 5), Babuska & Osborn [8]. For general theorems of comparison, details can be found in Gould [88], Weinstein & Stenger [212], and for systems discretized by finite elements, Raviart & Thomas [191], Weinberger [211].

1.7 Axisymmetric structures

An *axisymmetric structure* is one which occupies a geometrical domain having a rotational symmetry axis Δ, and whose mechanical characteristics are invariant for all *rotations* about an axis Δ, and for symmetry in all *planes* containing Δ.

Classification of the eigenmodes — We use cylindrical coordinates: z is the rotational axis, u_r, u_z, u_θ are the components (functions of (r, z, θ)) of u in the local frame of reference (r, z, θ). In this case we use the Fourier series expansion of vector displacement u in the form:

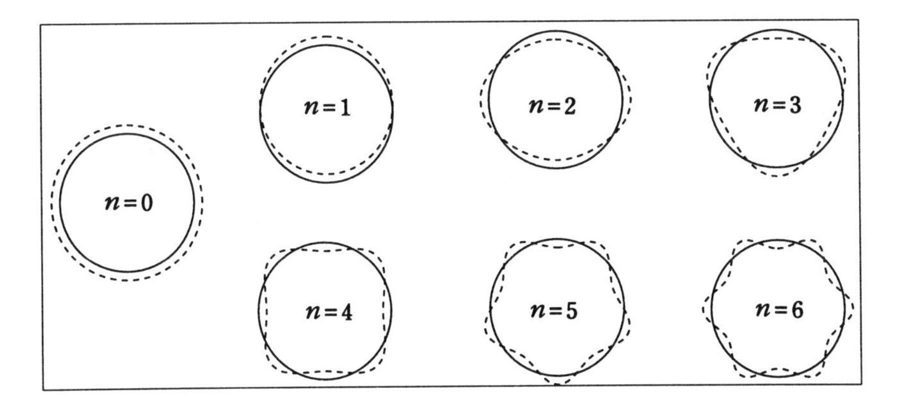

Figure 1.2: Modes of an axisymmetrical structure: deformed states of a section

$$\underbrace{\begin{Bmatrix} u_r \\ u_\theta \\ u_z \end{Bmatrix}}_{\mathcal{C}} = \underbrace{\begin{Bmatrix} U_0^+(r,z) \\ 0 \\ W_0^+(r,z) \end{Bmatrix}}_{\mathcal{C}_0^+} + \underbrace{\begin{Bmatrix} 0 \\ V_0^-(r,z) \\ 0 \end{Bmatrix}}_{\mathcal{C}_0^-} + \underbrace{\begin{Bmatrix} U_n^+(r,z)\cos n\theta \\ V_n^+(r,z)\sin n\theta \\ W_n^+(r,z)\cos n\theta \end{Bmatrix}}_{\mathcal{C}_n^+} + \underbrace{\begin{Bmatrix} U_n^-(r,z)\sin n\theta \\ -V_n^-(r,z)\cos n\theta \\ W_n^-(r,z)\sin n\theta \end{Bmatrix}}_{\mathcal{C}_n^-} \quad (1.40)$$

The eigenvalue problem $\mathcal{P}$ then splits into a series of *decoupled* problems $\mathcal{P}_n^{\pm}$. [1] These problems are obtained by restricting the three-dimensional variational formulation of $\mathcal{P}$, to the admissible subspaces $\mathcal{C}_n^{\pm}$ $(n \geq 0)$ defined below. Their solution involves discretization in a meridian plane.

1. The eigenvalues of $\mathcal{P}_0^+$ are simple and correspond to modal *deformed* shapes which are *invariant with respect to rotation and plane symmetry* (called longitudinal modes in the case of launchers). The eigenvalues $\mathcal{P}_0^-$ are simple and correspond to axisymmetric torsional modes, *rotation-invariant, and antisymmetric with respect to planes containing the axis.*

2. For $n \geq 1$, the eigenvalues $\mathcal{P}_n^+$ and $\mathcal{P}_n^-$ are identical (they correspond to a multiplicity of 2 of the eigenvalues of the three-dimensional problem), and the eigenvectors of $\mathcal{P}_n^-$, are deduced from those of $\mathcal{P}_n^+$ by rotation through $\pi/2n$. As a result, only one resolution is carried out.

   In the case of a slender structure, the modes $n = 1$ are generally called "bending modes". The modes $n \geq 2$ are called "breathing modes" (Fig. 1.2.)

For the case of cylindrical shells, see Leissa [116].

1.8 Systems dependent on one parameter

Infinitesimal perturbations

We consider a problem $\mathcal{P}(\tau)$ defined by a given admissible class $\mathcal{C}$ (*independent of* τ), and two symmetrical bilinear forms $K^\tau(u,v)$, $M^\tau(u,v)$ continuously dependent on a real parameter τ. We denote by $\lambda_\alpha(\tau)$ and by $u_\alpha(\tau)$, the eigenvalues and eigenvectors of $\mathcal{P}$. Under these conditions, the differential of an eigenvalue (assumed to be simple) $\lambda_\alpha(\tau)$ can be written:

$$\mu_\alpha d\lambda_\alpha = dK(u_\alpha, u_\alpha) - \lambda_\alpha dM(u_\alpha, u_\alpha) \tag{1.41}$$

with $dK = \dfrac{\partial K^\tau}{\partial \tau} d\tau$, $dM = \dfrac{\partial M^\tau}{\partial \tau} d\tau$ and $\mu_\alpha = M^\tau(u_\alpha, u_\alpha)$. This formula is obtained on differentiating (1.35) and taking account of the stationarity of $R^\tau(u)$ for $u = u_\alpha(\tau)$.

Remark — If we denote the "unperturbed" problem $\mathcal{P}(0)$, (1.41) enables the first order variation of each eigenvalue corresponding to an infinitesimal value of τ to be evaluated, by calculating the Rayleigh quotient R^τ for $u = u_\alpha(0)$, i.e. for the unperturbed eigenvector of $\mathcal{P}(0)$.

[1] Each problem corresponds to an irreducible representation of the symmetry group of the system here made up of rotations about an axis of revolution and meridian planes of symmetry.

Curve veering phenomenon

Here we study the behaviour of the eigenvalues and eigenvectors of a problem $\mathcal{P}(\tau)$ depending continuously on a parameter τ, whose variational formulation involves an *invariable* admissible class $\mathcal{C}$.
To simplify matters, we make the assumption, among others, that M is invariable [2], and we write $K^\tau(u,v) = K(u,v) + V^\tau(u,v)$, where V is a *symmetric* bilinear form, null for $\tau = 0$. As a result, V^τ is interpreted as a perturbation in the "stiffness" of the system with respect to an "initial" configuration $\mathcal{P}(0)$. The variational formulation of the problem is written:

$$K(u,v) + V^\tau(u,v) = \lambda M(u,v) \qquad u \in \mathcal{C},\ \forall v \in \mathcal{C} \tag{1.42}$$

We envisage a finite variation of the parameter τ, from an unperturbed initial configuration ($\tau = 0$).
We assume that the eigenvalues of $\mathcal{P}(0)$ are simple.
The problem of the intersection of modes consists of studying the occurrence of multiple eigenvalues in terms of τ.
We assume the eigenvectors $u_\alpha \in \mathcal{C}$ of $\mathcal{P}(0)$, known and normalized to 1 in generalized mass ($\mu_\alpha = 1$). As the eigenvectors u_α of $\mathcal{P}(0)$ form a basis of $\mathcal{C}$, we can obtain an approximation of the eigenvectors of $\mathcal{P}(\tau)$ by considering a finite linear combination $u = \sum_{\alpha=1}^{n} q_\alpha u_\alpha$. Applying the method of Ritz-Galerkin to (1.42), we find that the *n-modes approximation* (projection onto the subspace $\mathcal{C}^n \subset \mathcal{C}$ spanned by n eigenvectors of $\mathcal{P}(0)$) verifies the eigenvalue problem (putting $V_{\alpha\beta} = V(u_\alpha, u_\beta)$):

$$\boxed{[\lambda_\alpha \delta_{\alpha\beta} + V_{\alpha\beta}]\, q_\beta = \lambda \delta_{\alpha\beta} q_\beta \qquad 1 \le \alpha \le n} \tag{1.43}$$

One-mode approximation — This approximation consists of using (1.43) with a single vector $n = 1$, and leads to an "independent" estimation of each eigenvalue according to:

$$\lambda_\alpha^*(\tau) = \lambda_\alpha + V_{\alpha\alpha}(\tau) \tag{1.44}$$

Two-modes approximation — We now deal with two consecutive eigenvalues $\lambda_1 < \lambda_2$, for which the previous estimation yields an intersection of eigenvalues $\lambda_1^* = \lambda_2^*$, for a certain value of τ (Fig. 1.3). We examine the behaviour of these eigenvalues in an improved approximation consisting of seeking solutions in the form of a linear combination $u = q_1 u_1 + q_2 u_2$ of the two modes u_1 and u_2. (1.43) leads in this case to:

$$\begin{bmatrix} \lambda_1 + V_{11} & V_{12} \\ V_{12} & \lambda_2 + V_{22} \end{bmatrix} \begin{bmatrix} q_1 \\ q_2 \end{bmatrix} = \lambda \begin{bmatrix} 1 & 0 \\ 0 & 1 \end{bmatrix} \begin{bmatrix} q_1 \\ q_2 \end{bmatrix} \tag{1.45}$$

The characteristic equation is written:

$$(\underbrace{\lambda_1 + V_{11}}_{\lambda_1^*} - \lambda)(\underbrace{\lambda_2 + V_{22}}_{\lambda_2^*} - \lambda) - V_{12}^2 = 0 \tag{1.46}$$

[2]Extension to the case of a mass and stiffness perturbation immediately follows.

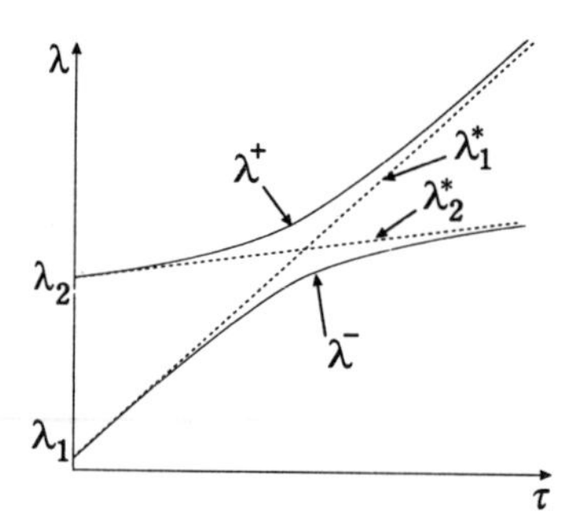

Figure 1.3: Intersection of eigenvalues

where λ_1^* and λ_2^* are the "one-mode" estimations (1.44).
(1.46) has the following solutions for the eigenvalues $\lambda^\pm$:

$$\lambda^\pm = \frac{\lambda_1^* + \lambda_2^* \pm \sqrt{\Delta}}{2} \quad \text{with } \Delta = (\lambda_1^* - \lambda_2^*)^2 + 4V_{12}^2 \tag{1.47}$$

and the eigenvectors (normalized to 1), $q_1^\pm = -V_{12}/\sqrt{(\lambda_1^* - \lambda^\pm)^2 + V_{12}^2}$ and $q_2^\pm = (\lambda_1^* - \lambda^\pm)/\sqrt{(\lambda_1^* - \lambda^\pm)^2 + V_{12}^2}$.
Taking into account (1.46) to eliminate V_{12}, the eigenvectors $u^\pm$ are written:

$$u^\pm = \Delta^{-\frac{1}{4}}(\pm sgn(V_{12})|\lambda^\pm - \lambda_2^*|^{\frac{1}{2}}u_1 + |\lambda^\pm - \lambda_1^*|^{\frac{1}{2}}u_2) \tag{1.48}$$

Discussion — Unlike the one-mode approximation, it can be seen that in general, no double solution exists, since $\Delta = 0 \Longrightarrow \lambda_1^*(\tau) - \lambda_2^*(\tau) = 0$ and $V_{12}(\tau) = 0$ (two equations and a single unknown).

- "Weak interaction": $2|V_{12}| << |\lambda_1^* - \lambda_2^*|$

  1. $\lambda_1^* < \lambda_2^*$ — In this case, from (1.47), $\lambda^- \simeq \lambda_1^*$ and $\lambda^+ \simeq \lambda_2^*$, and from (1.48), $u^- \simeq -u_1$ and $u^+ \simeq +u_2$.
  2. $\lambda_1^* > \lambda_2^*$ — In this case, from (1.47), $\lambda^- \simeq \lambda_2^*$ and $\lambda^+ \simeq \lambda_1^*$, and from (1.48), $u^- \simeq +u_2$ and $u^+ \simeq +u_1$.

- "Strong interaction": $|\lambda_1^* - \lambda_2^*| \sim 2|V_{12}|$.

  This case corresponds to values of τ for which $\lambda_1^* \simeq \lambda_2^*$.

  In this case, from (1.48), the eigenvectors u^+ and u^- result from a "strong hybridization" of u_1 and u_2 ($|q_1^\pm| \sim |q_2^\pm|$).

  In particular, for $\lambda_1^* = \lambda_2^* (= \lambda^*)$, we note that $\lambda^- < \lambda_1^* = \lambda_2^* < \lambda^+$. As a result, the effect of the coupling is to "separate the eigenvalues". We also have $\lambda^+ - \lambda^- = 2|V_{12}|$ and $u^- = (-u_1 + u_2)/\sqrt{2}$, $u^+ = (u_1 + u_2)/\sqrt{2}$.

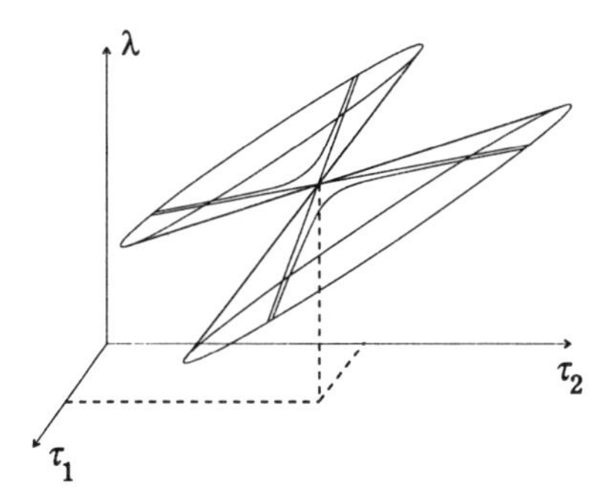

Figure 1.4: Behaviour of the eigenvalues as a function of two parameters

To summarize, by following $\lambda^-(\tau)$ branch, we find that the corresponding eigenvector varies from $u^- = -u_1$ to $u^- = u_2$, passing through $u^- = (-u_1 + u_2)/\sqrt{2}$. In the same way, the eigenvector corresponding to the eigenvalue λ^+, varies from $u^+ = u_2$ to $u^+ = u_1$, passing through $u^+ = (u_1 + u_2)/\sqrt{2}$.
We can interpret this variation of the eigenvectors as a rotation through 90° in the plane u_1, u_2.

The case of systems depending on two parameters (τ_1, τ_2) — In this case, the above analysis yields double eigenvalues. We can then show that, in the neighbourhood of this point, the surface $\lambda(\tau_1, \tau_2)$ is (in general) an elliptical cone (Fig. 1.4). The curves representing the behaviour of the eigenvalues in terms of a single parameter, e.g. for τ_2 = constant, therefore correspond to the intersection of the surface $\lambda(\tau_1, \tau_2)$ with the plane τ_2 = constant. The occurrence of a double eigenvalue corresponds to the case where the plane passes through the vertex of this cone – hence the "accidental" nature of such a degeneracy. For a plane passing near to the vertex, the hyberbolic section presents the same features as those previously encountered in the study of systems depending on one parameter.

The case of systems exhibiting symmetries — We consider the behaviour, in terms of one parameter, of the eigenvalues of a system exhibiting symmetries. Two cases appear, according to whether the perturbation $V(\tau)$ does or does not respect the symmetry of $\mathcal{P}(0)$.

1. **Perturbation respecting the symmetry** — Let us consider, for example, the case of an axisymmetrical structure depending on one parameter (as previously, we are dealing with the case of a perturbation of operators, the admissible class being assumed invariable).

   We have seen that the eigenvalues of $\mathcal{P}_n^+$ are identical to those of $\mathcal{P}_n^-$. Study of the behaviour of an eigenvalue of multiplicity 2 therefore involves studying the behaviour of a simple eigenvalue, which shows that the eigenvalue behaviour curves $\lambda_n(\tau)$ do not intersect.

   On the other hand, crossing is allowed for eigenvalues corresponding to different n (e.g. the eigenfrequencies of bending modes can intersect the eigenfrequencies of axisymmetric modes).

In the general case of symmetrical systems, it can similarly be shown that the eigenvalue behaviour curves corresponding to eigenmodes of a given "type"do not intersect. [3]

2. **Perturbation violating the symmetry** — In this case, the effect of the perturbation is to remove (partly or wholly) the degeneracy of the eigenvalues – which is equivalent to saying that the eigenvalue behaviour curves intersect for $\tau = 0$. The eigenvectors of the perturbed problem are linear combinations of eigenvectors of the unperturbed symmetrical problem. As an example, the modal deformed shapes of a perturbed axisymmetric structure are combinations of the eigenmodes of different circumferential indices.

Additional details can be found in Morand [136], Perkins & Mote [181], Chen & Ginsberg [32], Pierre [185], Min, Igusa & Achenbach [101], and in a general way in Arnold [6] (appendice 10) and Landau & Lifchitz [114] (chap. 11). The problem of eigenvalue multiplicity is studied by asymptotic methods in Sanchez-Hubert & Sanchez-Palencia [195], Haug *et al* [94].

1.9 Harmonic response to prescribed forces and displacements

This problem is analyzed starting from the equations describing the response to forces of the free structure.

Matrix resolution

We start with matrix equations (1.23), which are partitioned, distinguishing the degrees of freedom $\boldsymbol{U}_1$ relating to the nodes located on Σ, from the others denoted $\boldsymbol{U}_2$:

$$\left\{\begin{bmatrix} \boldsymbol{K}_{11} & \boldsymbol{K}_{12} \\ \boldsymbol{K}_{12}{}^T & \boldsymbol{K}_{22} \end{bmatrix} - \omega^2 \begin{bmatrix} \boldsymbol{M}_{11} & \boldsymbol{M}_{12} \\ \boldsymbol{M}_{12}{}^T & \boldsymbol{M}_{22} \end{bmatrix}\right\} \begin{bmatrix} \boldsymbol{U}_1 \\ \boldsymbol{U}_2 \end{bmatrix} = \begin{bmatrix} \boldsymbol{F}_1 \\ \boldsymbol{F}_2 \end{bmatrix} \tag{1.49}$$

Clamped modes and modal reaction forces

For $\boldsymbol{F}_2 = 0$ and $\boldsymbol{U}_1 = 0$, equations (1.49) split into two equations:

$$(\boldsymbol{K}_{22} - \omega^2 \boldsymbol{M}_{22})\boldsymbol{U}_2 = 0 \tag{1.50}$$

$$\boldsymbol{F}_1 = (\boldsymbol{K}_{12} - \omega^2 \boldsymbol{M}_{12})\boldsymbol{U}_2 \tag{1.51}$$

[3] i.e. in the case of an axisymmetric structure, a "type" of modes corresponds to a mode of given circumferential index n.

1. The solutions of the first equation are the eigenmodes of the fixed structure ($\boldsymbol{U}_1 = 0$).

2. The second equation gives the expression for the force of reaction $\boldsymbol{F}_1$, corresponding to an eigenmode $(\omega_\alpha^2, (\boldsymbol{U}_2)_\alpha)$ solution of (1.50).

Prescribed forces and displacements

For a given $\boldsymbol{U}_1$ and $\boldsymbol{F}_2$, (1.49) enables $\boldsymbol{U}_2$ and $\boldsymbol{F}_1$ to be calculated. Since (1.49) then splits into the following two equations:

$$\begin{aligned}(\boldsymbol{K}_{22} - \omega^2 \boldsymbol{M}_{22})\boldsymbol{U}_2 &= \boldsymbol{F}_2 - (\boldsymbol{K}_{12}{}^T - \omega^2 \boldsymbol{M}_{12}{}^T)\boldsymbol{U}_1 && (a)\\ \boldsymbol{F}_1 &= (\boldsymbol{K}_{11} - \omega^2 \boldsymbol{M}_{11})\boldsymbol{U}_1 + (\boldsymbol{K}_{12} - \omega^2 \boldsymbol{M}_{12})\boldsymbol{U}_2 && (b)\end{aligned} \qquad (1.52)$$

Fredholm's alternative

1. either ω^2 does not coincide with an eigenvalue λ_α of $\boldsymbol{K}_{22}\boldsymbol{U}_2 = \lambda \boldsymbol{M}_{22}\boldsymbol{U}_2$ (modes constrained on Σ), and then (1.52a) has a unique solution $\boldsymbol{U}_2$,

2. or ω^2 coincides with an eigenvalue λ_α of $\boldsymbol{K}_{22}\boldsymbol{U}_2 = \lambda \boldsymbol{M}_{22}\boldsymbol{U}_2$, and then (1.52a) has a solution (not unique) if $(\boldsymbol{U}_2)_\alpha$ is orthogonal to the second member of (1.52a), i.e. if $(\boldsymbol{U}_2)_\alpha{}^T(\boldsymbol{F}_2 - (\boldsymbol{K}_{12}{}^T - \omega^2 \boldsymbol{M}_{12}{}^T)\boldsymbol{U}_1) = 0$.

Continuum interpretation

From (1.4), (1.5) and (1.6), for a given ω, F and u_Σ, $u(M)$ verifies the equations:

$$\boxed{\begin{aligned}\sigma_{ij,j}(u) + \rho\omega^2 u_i &= 0 && \text{in } \Omega && (a)\\ \sigma_{ij}(u)n_j &= F_i && \text{on } \partial\Omega \setminus \Sigma && (b)\\ u &= u_\Sigma && \text{on } \Sigma && (c)\end{aligned}} \qquad (1.53)$$

Reaction forces

By definition, the force of reaction R on Σ corresponding to the solution u of (1.53) is:

$$\boxed{R_i = \sigma_{ij}(u)n_j \quad \text{on } \Sigma} \qquad (1.54)$$

Variational interpretation

We begin by establishing a variational property of (1.53ab) and (1.54), satisfied by (u, R). We proceed by the test-functions method. $\mathcal{C}$ denotes the space of smooth functions u in Ω. Multiplying (1.53) by $v \in \mathcal{C}$, integrating over Ω, then applying Green's formula (1.8), and finally taking into account (1.53b) and (1.54), one obtains $\forall v \in \mathcal{C}$:

$$\boxed{\int_\Omega \sigma_{ij}(u)\epsilon_{ij}(v)\,dx - \omega^2 \int_\Omega \rho u\cdot v\,dx - \int_{\partial\Omega\setminus\Sigma} F\cdot v\,d\sigma = \int_\Sigma R\cdot v\,d\sigma} \tag{1.55}$$

1. **Variational property of** u — We eliminate the unknown R of (1.55) by considering functions v vanishing on Σ. We thus introduce the space $\mathcal{C}^0 \subset \mathcal{C}$ defined by:

$$\mathcal{C}^0 = \{u \in \mathcal{C} \mid u = 0 \text{ on } \Sigma\} \tag{1.56}$$

On additionally imposing on u the Dirichlet boundary condition (1.53c), which leads us to consider the space $\mathcal{C}^d$ of the u with a given value u_Σ over Σ, defined by:

$$\mathcal{C}^d = \{u \in \mathcal{C} \mid u = u_\Sigma \text{ on } \Sigma\} \tag{1.57}$$

(1.55) then leads to the following variational property of (1.53):

for a given ω, u_Σ and F, find $u \in \mathcal{C}^d$, such that $\forall v \in \mathcal{C}^0$:

$$\boxed{\int_\Omega \sigma_{ij}(u)\epsilon_{ij}(v)\,dx - \omega^2 \int_\Omega \rho u\cdot v\,dx = \int_{\partial\Omega\setminus\Sigma} F\cdot v\,d\sigma} \tag{1.58}$$

2. **Variational property of reaction forces** — The matrix equation (1.52b) enabling the reaction forces $\boldsymbol{F}_1$ to be calculated — for the solution $\boldsymbol{U}_2$ of (1.52a) — has the following continuum interpretation: the reaction force R is characterized — for a solution u of (1.58) with $v \in \mathcal{C}$ ($v \neq 0$ over Σ) — by the variational property (1.55). [4]

Modal reaction forces

Let us consider an eigenmode of a structure fixed on Σ, i.e. a solution $(\lambda_\alpha, u_\alpha)$ of (1.26).
By putting $\omega^2 = \omega_\alpha^2 = \lambda_\alpha$ and $F = 0$ in (1.55), we find the variational property of the *modal reaction force* $R_i^\alpha = \sigma_{ij}(u_\alpha)n_j$:

$$\boxed{\int_\Sigma R^\alpha\cdot v\,d\sigma = \int_\Omega \sigma_{ij}(u_\alpha)\epsilon_{ij}(v)\,dx - \omega_\alpha^2 \int_\Omega \rho u_\alpha\cdot v\,dx \qquad \forall v \in \mathcal{C}} \tag{1.59}$$

[4] Consider the elastodynamic problem of a "vehicle" Ω_V coupled to a "passenger" Ω_P along an interaction surface Σ, and undergoing external forces F_{Ext} applied to Ω_V — e.g. a launcher undergoing thrust transients and coupled to a satellite.
It can be shown that, from the standpoint of the vibratory response of Ω_P (u, γ at every point of Ω_P), the forces F_{Ext} can be replaced by "equivalent forces" F_{Eq}, which are the dynamic reaction forces on Σ for Ω_V *clamped* on Σ. This method is widely used for the dynamic analysis of launchers coupled to satellites (*cf.* Morand, Le Goarant, Bodagala, Chemoul *et al* [144], [145], [147]).

Introduction of the operator $\boldsymbol{u}^s(\boldsymbol{u}_\Sigma)$

Consider the solution u^s to the elastostatic problem of the structure Ω subject to a prescribed displacement u_Σ on Σ.
The boundary value problem verified by u^s and the corresponding variational formulation — deduced respectively from (1.53) and (1.58) by putting $\omega = 0$ and $F = 0$ on $\partial\Omega \setminus \Sigma$) — are written:

$$\boxed{\begin{array}{rcllr} \sigma_{ij,j}(u^s) &=& 0 & \text{in } \Omega & (a) \\ \sigma_{ij}(u^s)n_j &=& 0 & \text{on } \partial\Omega \setminus \Sigma & (b) \\ u^s &=& u_\Sigma & \text{on } \Sigma & (c) \end{array}} \tag{1.60}$$

$$\boxed{\int_\Omega \sigma_{ij}(u^s)\epsilon_{ij}(v)\, dx = 0 \qquad u^s \in \mathcal{C}^d,\ \forall v \in \mathcal{C}^0} \tag{1.61}$$

where $\mathcal{C}^0$ and $\mathcal{C}^d$ are defined by (1.56) and (1.57) — $\mathcal{C}$ denoting the space of smooth functions in Ω.
We introduce $\mathcal{C}(\Sigma) = \{u(M) \mid M \in \Sigma\}$ "the space of deformed shapes of Σ".
We note that the solution u^s of (1.61) depends linearly on u_Σ.
The mapping $u_\Sigma \in \mathcal{C}(\Sigma) \longrightarrow u^s(u_\Sigma) \in \mathcal{C}$ is called, in mathematics, the *static lifting operator* of u_Σ, and in engineering the *static boundary functions.*
The image of $\mathcal{C}(\Sigma)$ in this mapping constitutes a vector subspace of $\mathcal{C}$ denoted $\mathcal{C}^S_\Sigma$, called the space of "static boundary functions".
The discretization of (1.61) is deduced from (1.49) or (1.52ab), by putting $\omega = 0$ and $\boldsymbol{F}_2 = 0$. By further introducing the notations $\boldsymbol{U}_\Sigma$ in place of $\boldsymbol{U}_1$ and $\boldsymbol{F}_\Sigma$ for the reaction force $\boldsymbol{F}_1$, we obtain: $\begin{bmatrix} \boldsymbol{K}_{11} & \boldsymbol{K}_{12} \\ \boldsymbol{K}_{12}{}^T & \boldsymbol{K}_{22} \end{bmatrix} \begin{bmatrix} \boldsymbol{U}_\Sigma \\ \boldsymbol{U}_2 \end{bmatrix} = \begin{bmatrix} \boldsymbol{F}_\Sigma \\ 0 \end{bmatrix}$, from which:

$$\boxed{\boldsymbol{U}_\Sigma \Longrightarrow \boldsymbol{U}_2 = -\boldsymbol{K}_{22}^{-1}\boldsymbol{K}_{12}{}^T\boldsymbol{U}_\Sigma} \tag{1.62}$$

and

$$\boldsymbol{F}_\Sigma = \underbrace{(\boldsymbol{K}_{11} - \boldsymbol{K}_{12}\boldsymbol{K}_{22}^{-1}\boldsymbol{K}_{12}{}^T)}_{\boldsymbol{K}^s}\boldsymbol{U}_\Sigma \tag{1.63}$$

where $\boldsymbol{K}^s$ is known in structural analysis as the "condensed stiffness matrix".
Static condensation — The calculation

$$\begin{bmatrix} \boldsymbol{A}_{11} & \boldsymbol{A}_{12} \\ \boldsymbol{A}_{12}{}^T & \boldsymbol{A}_{22} \end{bmatrix} \Longrightarrow \boldsymbol{A}_{11} - \boldsymbol{A}_{12}\boldsymbol{A}_{22}^{-1}\boldsymbol{A}_{12}{}^T \tag{1.64}$$

defined for a nonsingular $\boldsymbol{A}_{22}$, is widely used in structural analysis under the name of *static condensation* (a special case of the Gauss elimination algorithm, *cf.* Ciarlet [36]).

Special case — With $\boldsymbol{A}_{11} = 0$, and $\boldsymbol{A}_{22}$ reduced to a single term $-\alpha$, and introducing the notation $\boldsymbol{C} = \boldsymbol{A}_{12}$, (1.64) can be written:

$$\begin{bmatrix} 0 & \boldsymbol{C} \\ \boldsymbol{C}^T & -\alpha \end{bmatrix} \Longrightarrow \frac{1}{\alpha}\boldsymbol{C}\boldsymbol{C}^T \tag{1.65}$$

1.10 Modal analysis of the response to forces

We consider the problem of the harmonic response of a free structure to forces F applied on its boundary $\partial\Omega$ (*cf.* the boundary problem (1.21), the variational formulation (1.22) and its discretized form (1.23)).
We use the so-called "modal superposition" method, which consists of applying the Ritz-Galerkin method, choosing as basis vectors the n first eigenmodes u_α of the *free structure*. We thus put:

$$\boxed{u(M) = \sum_{\alpha=1}^{n} q_\alpha u_\alpha(M)} \tag{1.66}$$

On substituting (1.66) into (1.22) and then putting successively $v = u_1, u_2, \ldots, u_n$ and using the orthogonality relationships (1.30ab), we obtain:

$$\boxed{(-\omega^2 + \omega_\alpha^2)\mu_\alpha q_\alpha = F_\alpha \quad \text{with } F_\alpha = \int_{\partial\Omega} F \cdot u_\alpha \, d\sigma} \tag{1.67}$$

where F_α is the *generalized modal force*. Replacing q_α by this last expression in (1.66), we obtain:

$$\boxed{u = \sum_{\alpha=1}^{n} \frac{1}{-\omega^2 + \omega_\alpha^2} \frac{\int_{\partial\Omega} F \cdot u_\alpha \, d\sigma}{\mu_\alpha} u_\alpha} \tag{1.68}$$

Remarks

- In practice, we begin with a matrix model $\boldsymbol{K}, \boldsymbol{M}, \boldsymbol{F}$. The technique of modal projection consists of applying rule (1.20), choosing as Ritz vectors the n first eigenvectors $\boldsymbol{U}_\alpha$ of the free structure. Under these conditions, according to (1.20), and using the orthogonality conditions (1.30) in the discretized form ${\boldsymbol{U}_\alpha}^T \boldsymbol{K} \boldsymbol{U}_\beta = \delta_{\alpha\beta}\omega_\alpha^2 \mu_\alpha$ and ${\boldsymbol{U}_\alpha}^T \boldsymbol{M} \boldsymbol{U}_\beta = \delta_{\alpha\beta}\mu_\alpha$, we find (1.67) and (1.68) with $\boldsymbol{F}_\alpha = {\boldsymbol{U}_\alpha}^T \boldsymbol{F}$.

- In the case of a free structure, $\omega_\alpha^2 = 0$ for $1 \leq \alpha \leq 6$. The first six terms of (1.68) then represent the response of the structure assimilated to an indeformable solid, i.e. the solution to the following problem set on the subspace $\mathcal{C}^\mathcal{R} \subset \mathcal{C}$ spanned by the rigid body modes: $-\omega^2 \int_\Omega \rho u^\mathcal{R} \cdot v \, dx = \int_\Sigma F \cdot v \, d\sigma$ with $u^\mathcal{R} \in \mathcal{C}^\mathcal{R}$ and for all $v \in \mathcal{C}^\mathcal{R}$, where $u^\mathcal{R} = \vec{T} + \vec{\theta} \times \overrightarrow{OM}$. We note that $\int_\Sigma F \cdot v \, d\sigma$ represents the linear and angular momentums of the applied forces.

- (1.67) and (1.68) apply to the case of a structure fixed on Σ (provided u_α are the eigenmodes of the structure fixed on Σ).

- (1.67) becomes in the transient case (which is equivalent to seeking a solution $u(M,t)$ of (1.15) in the form of $u(M,t) = \sum_{\alpha=1}^{n} q_\alpha(t) u_\alpha(M)$):

$$\boxed{(\ddot{q}_\alpha + \omega_\alpha^2 q_\alpha)\mu_\alpha = F_\alpha(t) \quad \text{with } F_\alpha(t) = \int_{\partial\Omega} F \cdot u_\alpha \, d\sigma} \tag{1.69}$$

  This means solving n independent differential equations (completed by suitable initial conditions).

- The convergence of (1.68) can be improved by introducing appropriate static boundary functions (*cf.* the detailed study by Fraeijs de Veubeke, Géradin, Huck & Hogge [74], particularly in the case of free systems whose stiffness matrix is singular, in the presence of rigid body motions or mechanisms).

- The technique of modal superposition requires prior calculation of the eigenmodes. It is well suited to the so-called "low frequency" domain, i.e. corresponding to the natural frequencies of modes of "not too high" a rank. [5]

1.11 $(u_\Sigma, \{q_\alpha\})$ reduced matrix model of a substructure

We consider a *substructure* Ω, comprising a coupling surface Σ, onto which are applied forces F_Σ (we assume that the forces are null on $\partial\Omega \setminus \Sigma$).
We propose to construct a condensed model of the substructure Ω by projection of (1.22) onto the space spanned by the modes u_α of the structure Ω clamped on Σ and by the static boundary functions $u^s(u_\Sigma)$ (*cf.* §1.9).

Preliminary remarks — As before, $\mathcal{C}$ denotes the space of the smooth u in Ω and $\mathcal{C}^0$ the subspace of the u verifying the condition $u = 0$ on Σ.
We show that any element $u \in \mathcal{C}$ can be written in the form:

$$\boxed{u = u^s(u_\Sigma) + \sum_{\alpha \geq 1} q_\alpha u_\alpha} \tag{1.70}$$

[5] At *high frequencies*, simplified modal bases can be worked out (*cf.* Morand [151]). Other techniques for solving matrix equations (1.23) use double time scale methods (*cf.* Soize, Hutin, Desanti, David & Chabas [202]), methods with multiple scales in time and space (*cf.* Liu, Zhang & Ramirez [123]). See also methods for the statistical analysis of structural vibrations (*cf.* Bourgine [26]; for statistical energy methods *cf.* e.g. Lesueur [117]; for use of a diffusion equation *cf.* Nefske & Sung [204]). In the special case of slender or heterogeneous structures, see the application of asymptotic methods (*cf.* Bensoussan, Lions & Papanicolaou [16] chap. 4, Sanchez-Hubert & Sanchez-Palencia [195]).

where u_Σ is any smooth displacement field defined on Σ, and where $q_\alpha \in \mathbb{R}, \alpha \geq 1$. We have:

- If u^s denotes the solution of the problem of the static response of the structure (1.61) to a prescribed displacement $u = u_\Sigma$ on Σ, $u - u^s$ vanishes on Σ and, as a result, constitutes an element of $\mathcal{C}^0$, which can thus be decomposed on the basis of the u_α.

- Conversely, let v be any element of $\mathcal{C}$ and $v_\Sigma = v|_\Sigma$ (trace of v on Σ), and let $u^s(v_\Sigma)$ be the solution of (1.61) for $u_\Sigma = v_\Sigma$. Under these conditions, $v - u^s(v_\Sigma) \in \mathcal{C}^0$.

Since $\mathcal{C}_\Sigma^S$ denotes the space of static boundary functions defined in §1.9, the above result is equivalent to considering $\mathcal{C}$ as the direct sum of $\mathcal{C}_\Sigma^S \oplus \mathcal{C}^0$.
(1.70) therefore defines a representation of $u \in \mathcal{C}$ by means of $u_\Sigma \in \mathcal{C}(\Sigma)$ and a numerical sequence $\{q_\alpha\}$ (Fig. 1.5).

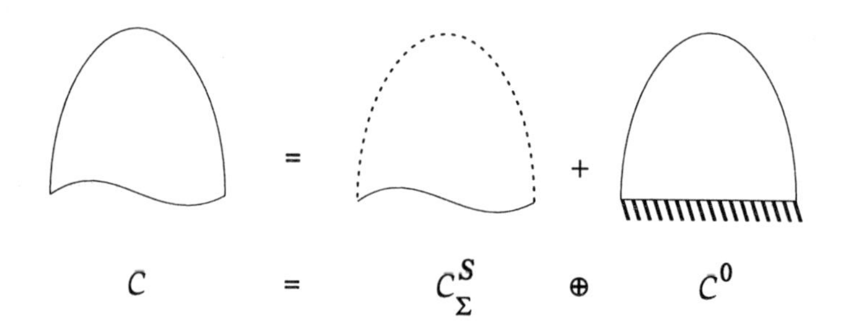

Figure 1.5: Decomposition of $\mathcal{C}$

Conjugate relationships between u_α and u^s — We show that u_α and u^s verify:

$$\boxed{\begin{aligned} \int_\Omega \sigma_{ij}(u^s)\epsilon_{ij}(u_\alpha)\,dx &= 0 & (a)\\ \int_\Omega \rho u^s \cdot u_\alpha\,dx &= -\frac{1}{\omega_\alpha^2}\int_\Sigma R_\alpha \cdot u_\Sigma\,d\sigma & (b)\end{aligned}} \qquad (1.71)$$

1. (1.71a) is obtained by putting $v = u_\alpha$ in the variational formulation (1.61) of u^s, which is allowed since $u_\alpha \in \mathcal{C}^0$.

2. (1.71b) is given by (1.59) on putting $v = u^s$, which is allowed since $u^s \in \mathcal{C}^d \subset \mathcal{C}$, and given that $\int_\Omega \sigma_{ij}(u_\alpha)\epsilon_{ij}(u^s)\,dx$ is null according to (1.71a) using the symmetry of the stiffness operator.

(u_Σ, q_α) reduced formulation

We apply the projection method of Ritz-Galerkin to (1.58), replacing u by its expression (1.70) truncated to n modes and proceeding as follows:

1. We consider the test-functions $u^s(v_\Sigma)$ where v_Σ is an arbitrary displacement defined on Σ.

2. We consider successively the test-functions $v = u_1, v = u_2, \ldots, v = u_n$.

Using the shortened notations $u^s = u^s(u_\Sigma)$ and $v^s = u^s(v_\Sigma)$ we then obtain: $\forall v_\Sigma \in \mathcal{C}_\Sigma^S$ and for $1 \leq \alpha \leq n$:

$$
\begin{aligned}
\int_\Omega \sigma_{ij}(u^s + \sum q_\alpha u_\alpha)\epsilon_{ij}(v^s) - \omega^2 \int_\Omega \rho(u^s + \sum q_\alpha u_\alpha)\cdot v^s &= \int_\Sigma F_\Sigma \cdot v^s \quad (a) \\
\int_\Omega \sigma_{ij}(u^s + \sum q_\alpha u_\alpha)\epsilon_{ij}(u_\alpha) - \omega^2 \int_\Omega \rho(u^s + \sum q_\alpha u_\alpha)\cdot u_\alpha &= \int_\Sigma F_\Sigma \cdot u_\alpha \quad (b)
\end{aligned}
\tag{1.72}
$$

Finally, using the orthogonality relationships (1.30) and the conjugate relationships (1.71) between u^s [6] and u_α, we obtain, $\forall v_\Sigma \in \mathcal{C}_\Sigma^S$ and for $1 \leq \alpha \leq n$:

$$
\begin{aligned}
\mathcal{K}^s(u_\Sigma, v_\Sigma) - \omega^2 \mathcal{M}^s(u_\Sigma, v_\Sigma) + \omega^2 \sum q_\alpha \int_\Sigma \frac{R_\alpha}{\omega_\alpha^2}\cdot v_\Sigma &= \int_\Sigma F_\Sigma \cdot v_\Sigma, \quad (a) \\
(-\omega^2 + \omega_\alpha^2)\mu_\alpha q_\alpha &= -\omega^2 \int_\Sigma \frac{R_\alpha}{\omega_\alpha^2}\cdot u_\Sigma \quad (b)
\end{aligned}
\tag{1.73}
$$

where we have put:

$$
\mathcal{K}^s(u_\Sigma, v_\Sigma) = \int_\Omega \sigma_{ij}(u^s(u_\Sigma))\epsilon_{ij}(u^s(v_\Sigma))\, dx \tag{1.74}
$$

$$
\mathcal{M}^s(u_\Sigma, v_\Sigma) = \int_\Omega \rho u^s(u_\Sigma)\cdot u^s(v_\Sigma)\, dx \tag{1.75}
$$

- $\mathcal{K}^s : \mathcal{C}(\Sigma) \times \mathcal{C}(\Sigma) \rightarrow \mathbb{R}$ is known as the "condensed" stiffness operator over Σ.

- $\mathcal{M}^s : \mathcal{C}(\Sigma) \times \mathcal{C}(\Sigma) \rightarrow \mathbb{R}$ is known as the coherent quasi-static mass operator "condensed on Σ" or Guyan's operator [7] (*cf.* Guyan [91], Irons [103], O' Callahan [29]).

Reduced mass and stiffness matrices

We introduce the matrices which correspond to the bilinear forms found in (1.73) as follows:

$$
\begin{aligned}
\mathcal{K}^s(u_\Sigma, v_\Sigma) &\Longrightarrow \boldsymbol{V_\Sigma}^T \boldsymbol{K}^s \boldsymbol{U}_\Sigma \quad (a) \\
\mathcal{M}^s(u_\Sigma, v_\Sigma) &\Longrightarrow \boldsymbol{V_\Sigma}^T \boldsymbol{M}^s \boldsymbol{U}_\Sigma \quad (b) \\
-\int_\Sigma \frac{R_\alpha \cdot u_\Sigma}{\omega_\alpha^2}\, d\sigma &\Longrightarrow \boldsymbol{C_\alpha}^T \boldsymbol{U}_\Sigma \quad (c)
\end{aligned}
\tag{1.76}
$$

[6] applied successively with $u^s(v_\Sigma)$, then $u^s(u_\Sigma)$.
[7] in mathematics, the "Schur complement".

In matrix form, equations (1.73) can then be written:

$$\underbrace{\left[\begin{array}{c|ccc} \boldsymbol{K}^s & \cdots & 0 & \cdots \\ \hline \vdots & \ddots & & \\ 0 & & \omega_\alpha^2\mu_\alpha & \\ \vdots & & & \ddots \end{array}\right]}_{\widehat{\boldsymbol{K}}} \begin{bmatrix} \boldsymbol{U}_\Sigma \\ \vdots \\ q_\alpha \\ \vdots \end{bmatrix} - \omega^2 \underbrace{\left[\begin{array}{c|ccc} \boldsymbol{M}^s & \cdots & \boldsymbol{C}_\alpha & \cdots \\ \hline \vdots & \ddots & & \\ \boldsymbol{C}_\alpha{}^T & & \mu_\alpha & \\ \vdots & & & \ddots \end{array}\right]}_{\widehat{\boldsymbol{M}}} \begin{bmatrix} \boldsymbol{U}_\Sigma \\ \vdots \\ q_\alpha \\ \vdots \end{bmatrix} = \begin{bmatrix} \boldsymbol{F}_\Sigma \\ \vdots \\ 0 \\ \vdots \end{bmatrix} \tag{1.77}$$

(1.77) can be found by the projection rule (1.20) applied to the matrix model $\boldsymbol{K}, \boldsymbol{M}$ of the structure.
To do this, we consider the partitioning (1.49) of $\boldsymbol{U}$. In discretized form, (1.70) is interpreted as a projection $(\boldsymbol{U}_1, \boldsymbol{U}_2) \longrightarrow (\boldsymbol{U}_\Sigma, q_\alpha)$ which consists of putting:

$$\left\{\begin{array}{lcll} \boldsymbol{U}_1 & = & \boldsymbol{U}_\Sigma & (a) \\ \boldsymbol{U}_2 & = & -\boldsymbol{K}_{22}^{-1}\boldsymbol{K}_{12}{}^T\boldsymbol{U}_\Sigma + \sum_{\alpha=1}^{n} q_\alpha \boldsymbol{U}_\alpha & (b) \end{array}\right. \tag{1.78}$$

Indeed, from (1.62), $-\boldsymbol{K}_{22}^{-1}\boldsymbol{K}_{21}\boldsymbol{U}_\Sigma$ represents the static displacement $\boldsymbol{U}_2^s$ (at the internal nodes of the structure) for a given displacement $\boldsymbol{U}_\Sigma$.
Equations (1.78) can be written in matrix form:

$$\begin{bmatrix} \boldsymbol{U}_1 \\ \boldsymbol{U}_2 \end{bmatrix} = \underbrace{\left[\begin{array}{c|c} \boldsymbol{I} & 0 \\ \hline -\boldsymbol{K}_{22}^{-1}\boldsymbol{K}_{12}{}^T & \cdots \boldsymbol{U}_\alpha \cdots \end{array}\right]}_{\boldsymbol{H}} \begin{bmatrix} \boldsymbol{U}_\Sigma \\ \boldsymbol{q} \end{bmatrix} \tag{1.79}$$

On applying the projection rule (1.20) to the matrix model of the structure, and using the property of "stiffness decoupling" of the constrained modes and static boundary functions (1.71a), together with the orthogonality relationships (1.30), we then arrive at (1.77).

Dynamic substructuring

Consider the vibrations of a free structure Ω decomposed into two substructures Ω_1 and Ω_2, in contact along a surface Σ, subject to harmonic forces F_2 of circular frequency ω, applied on $\Sigma_2 = \partial\Omega_2 \setminus \Sigma$.
The above modal reduction scheme leads to a dynamic substructuring procedure which consists of constructing a reduced model of the type (1.77) for each substructure.
It then remains to assemble the mass and stiffness matrices obtained in this way according to the degrees of freedom relative to Σ. [8]

[8] In the case of a launch vehicle consisting of an axisymmetric central body and axisymmetric boosters, this technique can be applied to the coupling of superelements associated with each body, for each value of the Fourier index, *cf.* Chemoul & Morand [33].

We can equally well assemble a model condensed in u_Σ, q_α with a non-condensed matrix model having the same degrees of freedom over Σ.
We note that the condensed matrices of each substructure are constructed independently of each other — the only requirement being the compatibility of the meshes along Σ.

In what follows, we shall establish a dynamic substructuring scheme involving only the generalized modal coordinates (Fig. 1.6). We denote by $\boldsymbol{K}_2$ and $\boldsymbol{M}_2$

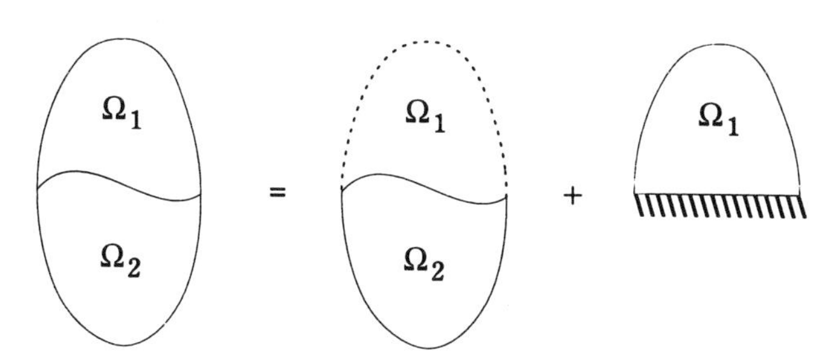

Figure 1.6: Dynamic substructuring scheme

the stiffness and relative mass matrices with $\boldsymbol{U}^{(2)}$ degrees of freedom, of the substructure Ω_2 discretized by finite elements (including the nodal values of u on Σ).
Referring back to (1.77), we consider the "superelement" $(\widehat{\boldsymbol{K}_1}, \widehat{\boldsymbol{M}_1})$ describing the substructure Ω_1, denoting by $\boldsymbol{q}_1 = \{q_{\alpha_1}\}$, the vector of the n_1 generalized coordinates corresponding to the modes of Ω_1 constrained on Σ. The matrix equations of the coupled system are then:

$$\begin{aligned}
(\boldsymbol{K}_2 + \boldsymbol{K}_1^s)\,\boldsymbol{U}^{(2)} - \omega^2\,(\boldsymbol{M}_2 + \boldsymbol{M}_1^s)\,\boldsymbol{U}^{(2)} &= \omega^2 \sum_{\alpha_1=1}^{n_1} \boldsymbol{C}_{\alpha_1}^{(1)} q_{\alpha_1} + \boldsymbol{F}_2 \quad (a) \\
(-\omega^2 + \omega_{\alpha_1}^2)\mu_{\alpha_1} q_{\alpha_1} &= \omega^2 \boldsymbol{C}_{\alpha_1}^{(1)^T} \boldsymbol{U}^{(2)} \quad (b)
\end{aligned} \tag{1.80}$$

In the dynamic substructuring technique, we keep the vector $\boldsymbol{q}_1 = \{q_{\alpha_1}\}$, and substitute for the variables $\boldsymbol{U}^{(2)}$, the vector $\boldsymbol{q}_2$ of the n_2 generalized coordinates $\{q_{\alpha_2}\}$ as follows:

$$\boxed{\boldsymbol{U}^{(2)} = \sum_{\alpha_2=1}^{n_2} q_{\alpha_2} \boldsymbol{U}_{\alpha_2}^{(2)}} \tag{1.81}$$

where $\{\boldsymbol{U}_{\alpha_2}^{(2)}\}$ is a projection basis chosen so as to diagonalize the matrix blocks in relation to the degrees of freedom $\boldsymbol{U}^{(2)}$ in equations (1.80). The technique of projection consists of applying rule (1.20) to matrix equations (1.80), the projection matrix $\boldsymbol{H}$ being defined by:

$$\boldsymbol{H} = \left[\begin{array}{c|c} \boldsymbol{U}_1^{(2)} \cdots \boldsymbol{U}_{n_2}^{(2)} & 0 \\ \hline 0 & \boldsymbol{I}_{n_1} \end{array}\right] \tag{1.82}$$

where $\boldsymbol{I}_{n_1}$ is the unit matrix of rank n_1, and where $\boldsymbol{U}^{(2)}_{\alpha_2}$ are the eigenvector solutions of:

$$(\boldsymbol{K}_2+\boldsymbol{K}^s_1)\,\boldsymbol{U}^{(2)} = \lambda(\boldsymbol{M}_2+\boldsymbol{M}^s_1)\boldsymbol{U}^{(2)} \tag{1.83}$$

with the following orthogonality relationships:

$$\boldsymbol{U}^{(2)^T}_{\alpha_2}(\boldsymbol{M}_2+\boldsymbol{M}^s_1)\boldsymbol{U}^{(2)}_{\alpha'_2} = \delta_{\alpha_2\alpha'_2}\ ,\ \ \boldsymbol{U}^{(2)^T}_{\alpha_2}(\boldsymbol{K}_2+\boldsymbol{K}^s_1)\boldsymbol{U}^{(2)}_{\alpha'_2} = \omega^2_{\alpha_2}\delta_{\alpha_2\alpha'_2} \tag{1.84}$$

Equations (1.80) then give:

$$\boxed{\begin{bmatrix} \textbf{Diag}\ \omega^2_{\alpha_2} & 0 \\ 0 & \textbf{Diag}\ \omega^2_{\alpha_1} \end{bmatrix}\begin{bmatrix} \boldsymbol{q}_2 \\ \boldsymbol{q}_1 \end{bmatrix} - \omega^2\begin{bmatrix} \boldsymbol{I}_{n_2} & [C_{\alpha_2\alpha_1}] \\ [C_{\alpha_1\alpha_2}] & \boldsymbol{I}_{n_1} \end{bmatrix}\begin{bmatrix} \boldsymbol{q}_2 \\ \boldsymbol{q}_1 \end{bmatrix} = \begin{bmatrix} \boldsymbol{f} \\ 0 \end{bmatrix}} \tag{1.85}$$

where $\boldsymbol{f}$ denotes the vector of components $f_{\alpha_2} = \boldsymbol{U}^{(2)^T}_{\alpha_2}\boldsymbol{F}$, and where we write:

$$C_{\alpha_1\alpha_2} = \boldsymbol{C}^{(2)^T}_{\alpha_2}\boldsymbol{U}^{(1)}_{\alpha_1} \tag{1.86}$$

- We stress that the generalized modal coordinates $\boldsymbol{q}_1$ are relative to the modes of the "physical substructure" Ω_1 constrained on Σ.
- We stress that the generalized modal coordinates $\boldsymbol{q}_2$ are relative to the coupled system $\Omega_1 \cup \Omega_2$, in the "quasi-static" approximation which consists of describing the physical substructure Ω_1 by the mass and stiffness matrices $\boldsymbol{K}^s_1$ and $\boldsymbol{M}^s_1$. This quasi-static behaviour of Ω_1, which introduces no degrees of freedom belonging to Ω_1, *can be termed non-resonant*.
- $C_{\alpha_1\alpha_2}$ characterizes the "mass matrix" coupling of the two sub-structures. From (1.76c) and (1.81), we can check that this coupling term has the following continuum interpretation:

$$C_{\alpha_1\alpha_2} = -\int_\Sigma \frac{R_{\alpha_1}\cdot u_{\alpha_2}|_\Sigma}{\omega^2_{\alpha_1}}\,d\sigma \tag{1.87}$$

  where R_{α_1} is the reaction force of the mode α_1 (normalized to 1 in generalized mass). As a result, $C_{\alpha_1\alpha_2}$ is interpreted as the work of the modal reaction force of a constrained mode of $C_{\alpha_1\alpha_2}$ Ω_1, in the displacement $u_{\alpha_2}|_\Sigma$ of the interaction surface Σ corresponding to an eigenmode of the coupled system $\Omega_1 \cup \Omega_2$ (where Ω_1 is taken into consideration in the quasistatic approximation).

"Strong" interaction and "weak" interaction — The "strong" or "weak" interaction between two modes, due to siffness coupling, has been discussed above. Here we extend this analysis to the case of coupling by the mass matrix.

We state that the coupling between two modes of each substructure of values $\lambda_{\alpha_1} = \omega^2_{\alpha_1}$ and $\lambda_{\alpha_2} = \omega^2_{\alpha_2}$ is weak if $2|C_{\alpha_1\alpha_2}| \ll \frac{|\lambda_{\alpha_1}-\lambda_{\alpha_2}|}{\lambda_{\alpha_1}}$. If this condition is satisfied within a sufficiently large band of frequencies, the modes of the coupled

system, whose frequency lies within this band, are close — in frequency and deformed shape — to the modes of each substructure.
If this is not the case, the modes of the coupled system have to be sought as linear combinations of the modes of each substructure. In particular, if $2|C_{\alpha_1\alpha_2}| \sim \frac{|\lambda_{\alpha_1}-\lambda_{\alpha_2}|}{\lambda_{\alpha_1}}$, the eigenvectors of the coupled system arise from a "strong hybridization" of the modes of each substructure.

Bibliography — We generally distinguish the methods involving the modes of constrained substructures (*cf.* Hurty [100], Craig & Bampton [48], Benfield & Hruda [14], Leung [118], Bourquin & d'Hennezel [27]), or modes of free substructures (*cf.* Mac Neal [157], Rubin [193], Destuynder [57]). A synthesis of the methods of dynamic substructures can be found in Martinez & Miller [130], Craig [49].

1.12 Modal analysis of the response to prescribed displacements

We consider the problem of the harmonic response of a structure Ω subject to *given* displacements u_Σ on Σ ($F = 0$ on $\partial\Omega\backslash\Sigma$). Equations (1.73) apply provided F_Σ is considered as a reaction force.
Calculating q_α in terms of the given value of u_Σ by means of (1.73b) and substituting the expression obtained in (1.70), we have:

$$\boxed{u = u^s - \sum_{\alpha=1}^{n} \frac{\omega^2}{-\omega^2 + \omega_\alpha^2} \left(\int_\Sigma \frac{R_\alpha \cdot u_\Sigma}{\omega_\alpha^2 \mu_\alpha} \, d\sigma \right) u_\alpha} \tag{1.88}$$

Furthermore, (1.73a) constitutes a variational characterization of the reaction force F_Σ. Substituting (1.88) in (1.73) we have:

$$\boxed{\int_\Sigma F \cdot v_\Sigma \, d\sigma = \mathcal{K}^s(u_\Sigma, v_\Sigma) - \omega^2 \left[\mathcal{M}^s(u_\Sigma, v_\Sigma) + \sum_\alpha \frac{\omega^2}{-\omega^2 + \omega_\alpha^2} \mathcal{M}_\alpha(u_\Sigma, v_\Sigma) \right]} \tag{1.89}$$

in which we have introduced the *modal mass* M_α defined by:

$$\boxed{\mathcal{M}_\alpha(u_\Sigma, v_\Sigma) = \left(\int_\Sigma \frac{R_\alpha \cdot u_\Sigma}{\omega_\alpha^2 \mu_\alpha} \, d\sigma \right) \left(\int_\Sigma \frac{R_\alpha \cdot v_\Sigma}{\omega_\alpha^2 \mu_\alpha} \, d\sigma \right)} \tag{1.90}$$

Dynamic impedance operators — The (linear) mapping which associates the reaction force F_Σ to u_Σ is an operator of the "dynamic impedance" type known as the *dynamic stiffness* of the structure [9] along Σ. From (1.89), $\int_\Sigma F \cdot v_\Sigma \, d\sigma$ defines a bilinear *symmetric* form $\mathcal{M}_\Sigma(u_\Sigma, v_\Sigma)$ known as the "dynamic impedance" of which (1.74) is the modal expansion (*cf.* Morand [143], Hughes [98], Poelaert [188], and Lancaster [112]) for the connection between this problem and the spectral expansion of matrices).

[9] In the literature, "impedance" can also denote the relation between force and velocity.

Resonance and damping

The relation (1.68) enables the analysis of "vibration tests" consisting of subjecting the structure to harmonic *forces* continuously varying in frequency within a given range (vibration tests of aircraft, launch vehicles, etc.). We note the following results:

- The expansion (1.68) is singular for $\omega = \omega_\alpha$ (resonance). Furthermore, we note that the contribution of a given mode u_α is proportional to a *shape factor*; the more "appropriate" the spatial variations of F to that of the modal deformed shape, the higher the shape factor.

- Actually the response is limited by damping phenomena in the structure. These phenomena are generally taken into account *in the case of a weakly damped elastic structures*, by considering u and F to be complex amplitudes, and carrying out the substitution $\omega_\alpha^2 \to \omega_\alpha^2 + 2j\xi_\alpha\omega\omega_\alpha, \quad (j^2 = -1)$.

  This substitution is equivalent to modifying (1.69) according to $(\ddot{q}_\alpha + 2\xi_\alpha\omega_\alpha\dot{q}_\alpha + \omega_\alpha^2 q_\alpha)\mu_\alpha = F_\alpha(t)$, where $\xi_\alpha \ll 1$ is a dimensionless quantity determined experimentally, and known as the *critical damping* (*cf.* Ewins [66]).

Similarly, (1.88) enables analysis of a "vibration test" with applied displacements (e.g. a satellite on a vibrating table).

- In this case, the resonant frequencies correspond to the eigenfrequencies of the structure fixed on Σ.

  Moreover, the contribution of a mode u_α is proportional to a *shape factor*, and the better the spatial variations u_Σ "fit" that of the modal reaction force R_α, the higher the shape factor.

- The damping is taken into account by introducing a modal dissipation term through the substitution $\omega_\alpha^2 + 2j\xi_\alpha\omega\omega_\alpha$ à ω_α^2 in (1.73b).

For a deeper treatment of the theory of vibrations, the following list of works is by no means exhaustive: Argyris & Mlejnek [5], Courant & Hilbert [45], Clough & Penzien [40], Géradin & Rixen [80], Lalanne & Ferraris [111], Meirovich [131], [132] Petyt [183], Roseau [192].

CHAPTER 2

Linearized equations of small movements of inviscid fluids

2.1 Introduction

This chapter is devoted to a review of the linearized dynamic equations describing the small movements of an inviscid fluid. For a more detailed study, the reader is referred to the basic monographs of Germain [83], Landau & Lifchitz [115], Whitham [213], Lighthill [119].
After introducing the eulerian and lagrangian fluctuations p and $p_{\mathcal{L}}$ respectively used in the derivation of the conditions for a free surface, we establish the general equations in terms of p and the potential of displacements φ in the following two cases:

- incompressible fluid in the presence of gravity,
- compressible fluid in the absence of gravity.

2.2 Linearized dynamic equations

Notations

Configuration	equilibrium	instant t
Position of the particle	M	M'
Fluid domain	Ω_F	Ω'_F
Free Surface	Γ	Γ'
Wall contact surface	$\Sigma = \partial\Omega_F \setminus \Gamma$	$\Sigma' = \partial\Omega'_F \setminus \Gamma'$
Density	ρ_F	ρ'_F
Pressure	$P_0(M)$	$P(M', t)$

Euler Equations

The study of an inviscid fluid involves a description of the stresses by a pressure field within the fluid. Under these conditions, the dynamics of an inviscid fluid are governed by Euler's equation:

$$\nabla P = \rho'_F \, (\vec{g} - \gamma^F) \quad \text{in } \Omega'_F \tag{2.1}$$

where $\gamma^F(M', t)$ is the instantaneous acceleration in $M' \in \Omega'_F$, and $\vec{g} = -g i_z$ is the gravity vector (with i_z the vertical unit vector and g assumed to be constant). At equilibrium, the hydrostatic pressure P_0 is the solution of the usual equation:

$$\boxed{\nabla P_0 = -\rho_F \, g \, i_z \quad \text{in } \Omega_F} \tag{2.2}$$

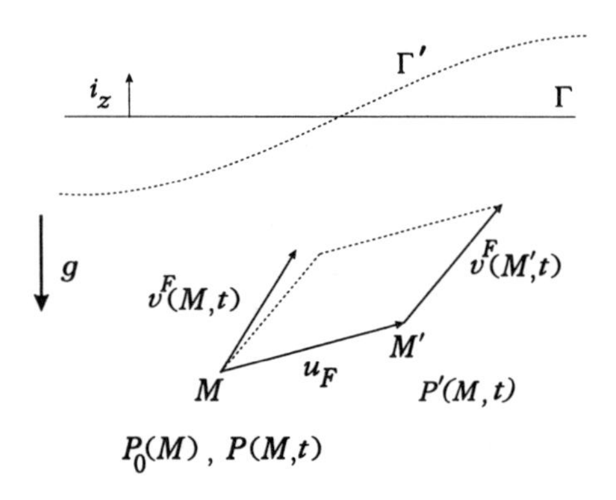

Figure 2.1: Geometrical configuration

Linearization

Linearization consists of deducing from equations (2.1) and (2.2), the equations which are satisfied, *in the equilibrium domain* Ω_F, by the variations of the various quantities relative to their equilibrium values. In the present case, we are concerned with the *field of displacement* and *variations of pressure*.

Linearization hypotheses — We consider small amplitude movements about an equilibrium state, in galilean frame. The small movement hypothesis consists of considering the displacement and pressure variations — together with their space and time derivatives — as equivalent infinitesimal quantities.

Fluid displacement field u^F

We denote by $u^F(M, t)$ the displacement of a particle located at M at rest. If $M' \in \Omega'_F$ denotes its instantaneous position, we then have (Fig. 2.1):

$$\boxed{M' = M + u^F(M,t), \quad \forall M \in \Omega_F} \tag{2.3}$$

Eulerian p and lagrangian $p_{\mathcal{L}}$ pressure fluctuations

We let $P(M',t)$ denote the instantaneous pressure in M', and we shall use the terms *euleriam pressure fluctuation* $p\,(M,t)$ and *lagrangian pressure fluctuation* $p_{\mathcal{L}}\,(M,t)$, to denote the quantities defined in *all points M of the equilibrium domain Ω_F* by:

$$p(M,t) = P(M,t) - P_0(M) \tag{2.4}$$

$$p_{\mathcal{L}}(M,t) = P(M',t) - P_0(M) \quad \text{where } M' = M + u^F(M,t) \tag{2.5}$$

Remarks

- $p\,(M,t)$, whose definition involves the instantaneous value of the pressure at a point in the equilibrium domain Ω_F, is only rigorously defined for $M \in \Omega_F \cap \Omega'_F$. However, since u^F is assumed to be infinitesimal, we may consider that (2.4) defines p at all points inside Ω_F.
- On the other hand, $p_{\mathcal{L}}\,(M,t)$ is defined at all points M *of the equilibrium domain* (for movements of any amplitude).

Relation between $p_{\mathcal{L}}$ and p

We can compare $p_{\mathcal{L}}$ and p which are defined in the same domain Ω_F. Carrying out the limited expansion of the difference between (2.5) and (2.4), we obtain, to within the second order in u^F:

$$p_{\mathcal{L}} - p = \nabla P(M,t)\cdot u^F(M,t) \tag{2.6}$$

Futhermore, using (2.4), and assuming that p and ∇p are infinitely small of the same order as u^F, and using also (2.2) and (2.6), we obtain the following relation, verified at all points $M \in \Omega_F$:

$$\boxed{p_{\mathcal{L}} - p = -\rho_F\, g\, u_z^F \quad \forall M \in \Omega_F} \tag{2.7}$$

Remarks

- It is important to note that *in the presence of a gravitational field*, and despite the small-movement approximation, the pressure fluctuations p — measured at a geometric point — and $p_{\mathcal{L}}$ — following a particle in the course of its movement — do not coincide (*cf.* Lighthill [119], Wilcox [214]).
- On the other hand, *when the effects of gravity are neglected*, we have:

$$\boxed{p_{\mathcal{L}} = p} \tag{2.8}$$

- Finally, we note that (2.7) enables the definition of p to be extended to points M on the boundary of Ω_F .

Linearized Euler equations

We shall show that, in the following two situations:

1. **Incompressible homogeneous fluid, in the presence of gravity,**
2. **Compressible homogeneous fluid, in the absence of gravity.**

p and u^F verify the *linearized Euler equations*:

$$\boxed{\nabla p = -\rho_F \frac{\partial^2 u^F}{\partial t^2} \quad \text{in } \Omega_F} \tag{2.9}$$

Derivation — A "lagrangian" approach would consist of "pulling back" (2.1) by *reciprocal mapping* into the domain Ω_F according to $M \leftarrow M' = M + u^F(M,t)$, then, after subtraction of the equilibrium equation (2.2), retaining the first order terms, to deduce from this the required equations.

Here, we use a simplified "eulerian" derivation, which consists of comparing (2.1) and (2.2) at a geometric point $M \in \Omega_F \cap \Omega'_F$.

The first step in the derivation relies on the fact that, if $M' = M + u^F$ denotes as above the instantaneous position of a particle located in M at rest, the instantaneous velocity $v^F(M',t)$ and acceleration $\gamma^F(M',t))$ respectively can be assimilated to the velocity $v^F(M,t)$ and acceleration $\gamma^F(M,t)$:

$$\begin{aligned} v^F(M,t) &\simeq v^F(M',t) = \frac{\partial u^F(M,t)}{\partial t} && (a) \\ \gamma^F(M,t) &\simeq \gamma^F(M',t) = \frac{\partial^2 u^F(M,t)}{\partial t^2} && (b) \end{aligned} \tag{2.10}$$

Indeed, since $M' = M + u^F(M,t)$, we have $v^F(M',t) = \frac{\partial u^F(M,t)}{\partial t}$. Furthermore, $v_i^F(M + u^F,t) - v_i^F(M,t) \simeq \nabla v_i^F(M,t) \cdot u^F(M,t)$ which show that $(v_i^F(M',t) - v_i^F(M,t)) \cdot n$ is second order in u^F (according to the linearization hypotheses), which thus establishes (2.10a) (Fig. 2.1). We proceed similarly to establish (2.10b).

In the second step of the derivation, we have to distinguish the following two cases:

1. **Incompressible fluid with weight** — We assume the fluid to be *incompressible* and *homogeneous*, i.e. $\rho'_F = \rho_F =$ constant.

   We consider a fixed geometric point M inside the equilibrium domain (belonging to $\Omega_F \cap \Omega'_F$), then we take the difference between (2.1) and (2.2) which, from (2.4) and (2.10b), leads to (2.9).

2. **Compressible weightless fluid** — We here assume that the effects of gravity are negligible and the fluid is homogeneous.

   In that case, (2.1) reduces to $\nabla P(M',t) = -\rho'_F(M',t)\gamma^F(M',t)$ in Ω'_F. At a fixed geometric point M inside the equilibrium domain (belonging to

$\Omega_F \cap \Omega'_F$), and taking into account (2.10b), equation (2.1) can be written $\nabla P(M,t) = -\rho'_F(M,t)\frac{\partial^2 u^F(M,t)}{\partial t^2}$. Subtracting from the latter expression the equilibrium equation (2.2) — which here reduces to $\nabla P_0 = 0$ — and replacing ρ'_F by its equilibrium value ρ_F in the term $\rho'_F\gamma^F$ (since γ^F is infinitesimal), we obtain the linearized Euler equation (2.9).

- In the case of a *weightless* compressible fluid, equation (2.9) is verified by $p_{\mathcal{L}}$, which assimilates in this case to p.

  On the other hand, in the case of an incompressible fluid with weight (heavy fluid), (2.9) is verified by p and not by $p_{\mathcal{L}}$ (*cf.* (2.7)).

- In the general case of a *compressible fluid with weight* (generally inhomogeneous because of stratification effects due to the gravitational field), equation (2.9) does not apply to (*cf.* Wilcox [214]).

Condition for linearized incompressibility

We recall that incompressibility classically results from the condition $div\, v^F(M',t) = 0$ at every point of Ω'_F. Applying this condition to every point $M \in \Omega_F \cap \Omega'_F$, and using (2.3), we obtain the linearized condition:

$$\boxed{div\, u^F = 0 \quad \text{in } \Omega_F} \tag{2.11}$$

Constitutive law for a weightless incompressible fluid

We begin by defining the lagrangian fluctuation $\rho_{\mathcal{L}}$ of the density, in a similar way to (2.5) as follows:

$$\rho_{\mathcal{L}}(M,t) = \rho'_F(M',t) - \rho_F, \quad \text{with } M' = M + u^F(M,t) \tag{2.12}$$

The fluid is assumed to be barotropic. The constitutive law then relates $p_{\mathcal{L}}$ and $\rho_{\mathcal{L}}$ with:

$$p_{\mathcal{L}} = c^2\, \rho_{\mathcal{L}} \tag{2.13}$$

Furthermore, the equation for the conservation of mass can be written $\rho'_F J = \rho_F$ where J is the Jacobian of the transformation $M \to M'$. For an u^F infinitesimal, we recall that $J \simeq 1 + div\, u^F$, which enables us to write from (2.12):

$$\rho_{\mathcal{L}} = -\rho_F div\, u^F \quad \text{in } \Omega_F \tag{2.14}$$

By substituting (2.14) into (2.13) we obtain the behaviour law relating $p_{\mathcal{L}}$ and u^F. Recalling that, when the effects of gravity are neglected, $p_{\mathcal{L}} = p$, we thus have:

$$\boxed{p = -\rho_F c^2 div\, u^F} \tag{2.15}$$

Wall contact condition

We assume that the particles in contact with the wall are at equilibrium, and remain so during movement (Fig. 2.2).
For an inviscid fluid, the wall condition results from the slipping of the fluid, which is equivalent to the normal velocities on the instantaneous configuration Σ' of the fluid-structure contact surface being equal. If we denote by $v^S(M', t)$ the velocity of the solid particle located at $M' \in \Sigma'$, we thus have:

$$v^F(M', t) \cdot n' = v^S(M', t) \cdot n' \quad \forall M' \in \Sigma' \tag{2.16}$$

where n' denotes the normal to Σ'. Within the hypothesis of small movements

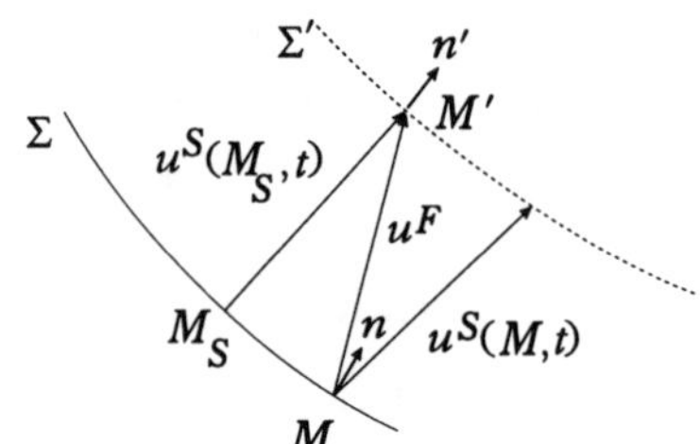

Figure 2.2: Wall contact condition

that we are considering here, we assume that the displacements of particles of liquids and solids are infinitesimal and of the same order of magnitude.
We shall show that the linearization of (2.16) results in the equality of the normal components of the displacements of the structure and the fluid.
We know that $v^F(M', t) = \frac{\partial u^F(M,t)}{\partial t}$ where $M \in \Sigma$ denotes the equilibrium position of the fluid particle.
If we denote by M_S the equilibrium position of the solid particle located at M' at a time t, we similarly have $v^S(M', t) = \frac{\partial u^S(M_S,t)}{\partial t}$.
Moreover, since $\overrightarrow{M_S M} = u^S(M_S, t) - u^F(M, t)$ is infinitesimal, we have $u^S(M_S, t) \simeq u^S(M, t)$.
Finally, if n denotes the normal to M at Σ, $n' - n$ is infinitely small and of the first order. Equation (2.16) then gives:

$$\frac{\partial u^F(M,t)}{\partial t} \cdot n = \frac{\partial u^S(M,t)}{\partial t} \cdot n \quad \forall M \in \Sigma \tag{2.17}$$

which on integrating gives the contact condition:

$$\boxed{u^F \cdot n = u^S \cdot n \quad \text{on } \Sigma} \tag{2.18}$$

From (2.17), $(u^S - u^F) \cdot n$ is independent of time and consequently null if we take the rest configuration as the initial configuration.
Special case of a fixed wall — In this case, (2.18) becomes:

$$\boxed{u^F \cdot n = 0 \quad \text{on } \Sigma} \tag{2.19}$$

Free surface condition

We consider the case of a fluid under gravity having a free surface in contact with a medium at constant (atmospheric) pressure P_{atm} (Fig. 2.3). We neglect the effects of surface tension (which will be examined in chapter 4).
We can therefore state $P(M',t) = P_{atm}$ on the instantaneous free surface Γ', which from (2.5) is written:

$$\boxed{p_{\mathcal{L}}(M,t) = 0 \quad \forall M \in \Gamma} \tag{2.20}$$

Taking account of relation (2.7), (2.20) then gives:

$$\boxed{p = \rho_F g\, u_z^F \quad \text{on } \Gamma} \tag{2.21}$$

Given the hypothesis of small movements, the inclination of the instantaneous tangential plane is a first order infinitesimal quantity; the vertical component u_z^F of the displacement u^F of a point M on the free surface Γ thus coincides, to within the second order, with the vertical elevation η of the free surface in M.

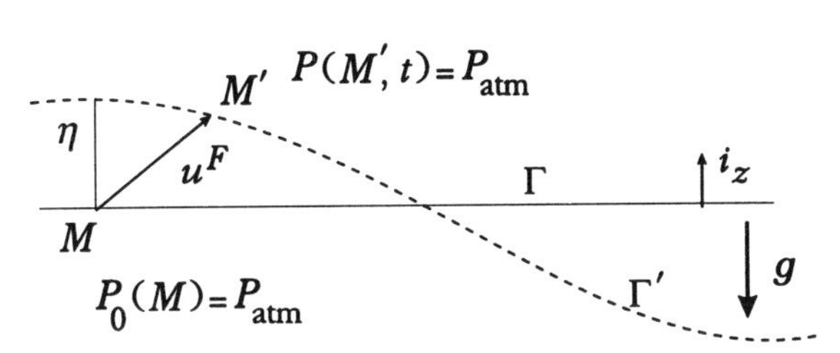

Figure 2.3: Free surface condition

2.3 Potential of displacements φ

Here we consider small harmonic movements, for which the various quantities studied vary sinusoidally about their equilibrium value. We then have:

$$u^F(M,t)=u^F(M)\cos\omega t \;, \quad p(M,t)=p(M)\cos\omega t \tag{2.22}$$

The linearized Euler equation (2.9) becomes:

$$\boxed{\nabla p - \rho_F \omega^2 u^F = 0} \tag{2.23}$$

First definition of φ

For $\omega \neq 0$, we see from (2.23) that $u^F = \nabla(\frac{p}{\rho_F \omega^2})$, in other words that $\varphi = \frac{p}{\rho_F \omega^2}$ is a displacement potential *uniquely* defined by:

$$\boxed{p = \rho_F \omega^2 \varphi} \tag{2.24}$$

equation (2.23) then reducing to:

$$\boxed{u^F = \nabla \varphi} \tag{2.25}$$

Remark: limiting case $\omega = 0$

- From (2.24) we observe that the displacement potential φ is not generally defined for $\omega = 0$: for $\omega = 0$, (2.23) leads to $p = p^0 =$ constant, which leads in general to $\varphi = \infty$ (unless p^0 is null).

Second definition of φ

An alternative definition of the potential of displacements φ consists of putting:

$$\boxed{\begin{array}{ll} p = \rho_F \omega^2 \varphi + \pi & (a) \\ l(\varphi) = 0 \quad \text{with } l(1) \neq 0 & (b) \end{array}} \tag{2.26}$$

where π is a constant and $l(\varphi)$ an arbitrary linear relation.

1. For $\omega \neq 0$, we note that relations (2.26ab) enable φ and π to be uniquely defined from p, and that from (2.23) it is obvious that $u^F = \nabla \varphi$.

   This potential therefore differs from the potential defined by (2.24) of constant value which is a function of ω. This constant depends on the choice of unicity condition $l(\varphi)$.

2. In the limiting case $\omega = 0$, we have $\pi = p^0 =$ constant, which may lead us to suppose that $\varphi = (p - \pi)/\rho_F \omega^2$, which for $\omega = 0$ appears in indeterminate form $0/0$, with a finite value, unlike the first definition of φ.

3. Relations (2.26) can be written in the following equivalent form:

$$\boxed{\begin{array}{ll} \nabla p - \rho_F \omega^2 \nabla \varphi = 0 & (a) \\ l(\varphi) = 0 \quad \text{with } l(1) \neq 0 & (b) \end{array}} \tag{2.27}$$

4. Comparing (2.27) with the linearized Euler equation (2.23), we note that the present definition of φ is equivalent to putting in (2.23):

$$\boxed{\begin{array}{lll} u^F = \nabla\varphi & & (a) \\ l(\varphi) = 0 & \text{with } l(1) \neq 0 & (b) \end{array}} \tag{2.28}$$

The introduction of φ is thus equivalent to restricting the study to small movements deriving from a potential, which is further equivalent to eliminating unwanted solutions reminiscent of existing non-rotational stationary solutions for domains which are not simply connected, or permanent rotational movements (which are outside the scope of the present vibratory analysis).

2.4 (p, φ) general equations

Incompressibility condition

Replacing u^F by $\nabla\varphi$ in (2.11) we find that φ verifies the Laplace equation:

$$\boxed{\Delta\varphi = 0 \quad \text{in } \Omega_F} \tag{2.29}$$

Constitutive law for a weightless compressible fluid

Replacing u^F by $\nabla\varphi$ in (2.15) we obtain:

$$\boxed{p = -\rho_F c^2 \, \Delta\varphi \quad \text{in } \Omega_F} \tag{2.30}$$

Wall contact condition

Replacing u^F by $\nabla\varphi$ in (2.18) we obtain:

$$\boxed{\frac{\partial\varphi}{\partial n} = u^S \cdot n \quad \text{on } \Sigma} \tag{2.31}$$

Special case of a fixed wall

$$\boxed{\frac{\partial\varphi}{\partial n} = 0 \quad \text{on } \Sigma} \tag{2.32}$$

Free surface condition

Replacing u^F by $\nabla\varphi$ in (2.21) we obtain:

$$\boxed{p = \rho_F g \frac{\partial \varphi}{\partial z} \quad \text{on } \Gamma} \tag{2.33}$$

In chapter 4, we examine in detail the generalization of this equation to the case of a fluid subject to surface tension forces.

Case of a surface of separation of two fluids

We consider two inviscid fluids 1 and 2, assumed to be non miscible, in contact, at equilibrium, along a horizontal surface Γ (in the absence of surface tension). We denote by $p_{\mathcal{L}}^1$ and u^{F_1} (*resp.* $p_{\mathcal{L}}^2$ and u^{F_2}), the lagrangian fluctuation of pressure and the displacement in medium 1 (*resp.* 2). The conditions for coupling along Γ are then written:

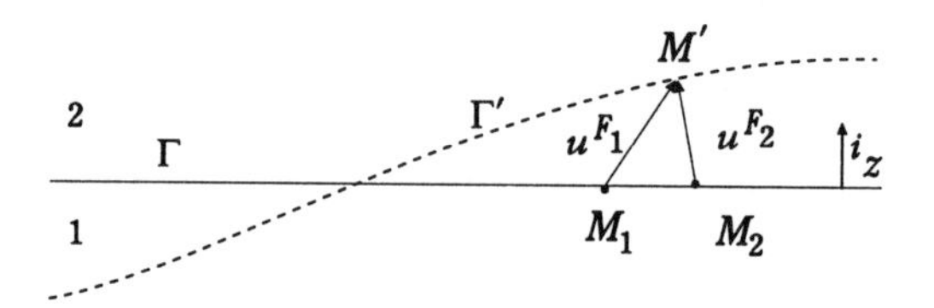

Figure 2.4: Surface of separation of two fluids

$$\begin{aligned} u^{F_1} \cdot i_z &= u^{F_2} \cdot i_z & (a) \\ p_{\mathcal{L}}^1 &= p_{\mathcal{L}}^2 & (b) \end{aligned} \tag{2.34}$$

where i_z is the unit normal to Γ. The first relation results from the slipping condition for the two fluids. The second relation is established as follows (Fig. 2.4). We denote by M_1 and M_2 the respective equilibrium positions — on Γ — of the particle of fluid 1 and the particle of fluid 2 located at M' on the instantaneous surface of separation Γ'. We thus have $p_{\mathcal{L}}^1(M_1,t) = P^1(M',t) - P_0(M_1)$ and $p_{\mathcal{L}}^2(M_2,t) = P^2(M',t) - P_0(M_2)$.
The condition for continuity of pressure in $M' \in \Gamma'$ is then written $P^1(M',t) = P^2(M',t)$, from which $p_{\mathcal{L}}^2(M_2,t) - p_{\mathcal{L}}^1(M_1,t) = P_0(M_2) - P_0(M_1) \sim \nabla P_0 \cdot \overrightarrow{M_1M_2}$, which is null according to (2.2) and (2.34a), which gives (2.34b).
Relation (2.7) enables interpretation of (2.34b) in terms of eulerian pressure fluctuations p^1 and p^2 and of vertical displacement u_z^F of Γ; when gravity is neglected ($g = 0$), we obtain $p_1 = p_2$.

CHAPTER 3

Sloshing modes

3.1 Introduction

We are concerned with harmonic vibrations of an inviscid imcompressible liquid contained in a reservoir, having a free surface Γ, in the presence of gravity and in the absence of surface tension.

As examples, we can cite the sloshing of liquids in satellite launchers subjected to winds and gusts, in the tanks of nuclear reactors, and in cryogenic liquid transport tanks. For general sloshing problems encountered in the aerospace field, see Abramson [1] and Moiseev & Rumyantsev [135].

First of all we pose the boundary value problem in terms of a potential of displacements φ for the response of a liquid to an arbitrary harmonic deformation of the wall.

Next, we establish the variational formulation in terms of φ of the spectral boundary value problem of the sloshing modes of a liquid in a rigid motionless tank, together with the matrix equations resulting from the finite element discretization.

Finally, the formulation of this problem in terms of the trace of φ on Γ is used to compare the eigenfrequencies of tanks of different shapes, and for the harmonic response of the liquid to a prescribed wall motion which introduces impedance, modal sloshing masses and stiffnesses operators of the liquid.

3.2 Harmonic response to a wall displacement u_N

We consider the harmonic response of an inviscid incompressible homogeneous liquid in a cavity occupying a bounded domain Ω_F, having a free surface Γ, and subject, on Σ, to a prescribed displacement u^S of the wall, of normal component $u^S \cdot n = u_N$ and circular frequency ω (Fig. 3.1(a)).

Equations in terms of p, φ

The equations in pressure p and displacement potential φ are written:

$$\boxed{\begin{array}{rcllr} \nabla p &=& \rho_F \omega^2 \nabla\varphi & \text{in } \Omega_F & (a) \\ \Delta\varphi &=& 0 & \text{in } \Omega_F & (b) \\ \dfrac{\partial\varphi}{\partial n} &=& u_N & \text{on } \Sigma = \partial\Omega_F & (c) \\ p &=& \rho_F g \dfrac{\partial\varphi}{\partial z} & \text{on } \Gamma & (d) \\ l(\varphi) &=& 0 & \text{with } l(1) \neq 0 & (e) \end{array}} \tag{3.1}$$

where $l(\varphi)$ is an arbitrary linear relation to ensure the unicity of φ.

- (3.1a) results from Euler's equation linearized for displacements deriving from a potential (equation (2.27).
- (3.1b) results from the incompressibility of the fluid (equation (2.29)).
- (3.1c) is the wall contact condition (equation (2.31)) where we have put $u_N = u^S \cdot n$.
- (3.1d) is the free surface condition (2.33).

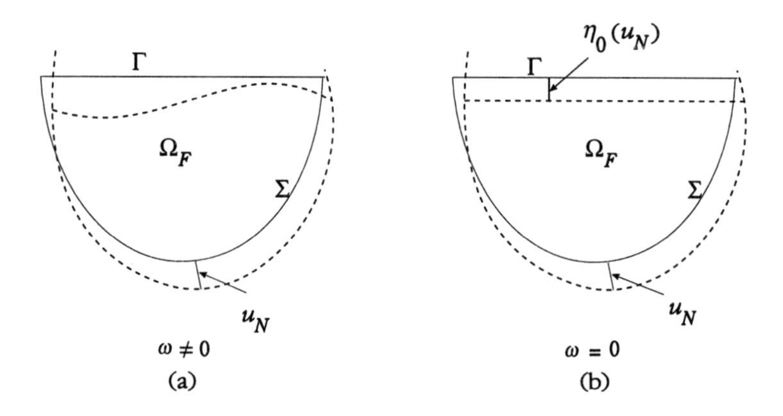

Figure 3.1: Liquid subject to a wall displacement

Solution for $\omega = 0$

For $\omega = 0$, and a given u_N, (3.1a) is written $\nabla p = 0$, which gives a constant value p_0 of p which can be calculated in terms of u_N. Let us first show that there exists a relation between p and u_N (independent of ω). Integrating (3.1b) in Ω_F and taking account of (3.1c) and (3.1d), we indeed obtain:

$$\int_\Gamma p\, d\sigma = -\rho_F g \int_\Sigma u_N\, d\sigma \tag{3.2}$$

This relation enables the value of p_0 to be determined:

$$p^0 = -\frac{\rho_F g}{\text{Area}(\Gamma)} \int_\Sigma u_N\, d\sigma \tag{3.3}$$

Boundary value problem in terms of φ

We aim first to eliminate p, and next to arrive at a formulation yielding a well-posed boundary value problem for $\omega = 0$. We know that equations (3.1) are equivalent — in conformity with the second definition (2.26) of φ — to:

$$\boxed{\begin{array}{ll} p = \rho_F \omega^2 \varphi + \pi & (a) \\ l(\varphi) = 0 \quad \text{with } l(1) \neq 0 & (b) \end{array}} \tag{3.4}$$

Relation between p, φ and u_N — Substituting (3.4a) into (3.2), we find:

$$\pi = -\frac{\rho_F \omega^2}{\text{Area}(\Gamma)} \int_\Gamma \varphi\, d\sigma - \frac{\rho_F g}{\text{Area}(\Gamma)} \int_\Sigma u_N\, d\sigma \tag{3.5}$$

from which we have the following expression for p in terms of φ and u_N:

$$p = \rho_F \omega^2 (\varphi - \frac{1}{\text{Area}(\Gamma)} \int_\Gamma \varphi\, d\sigma) - \frac{\rho_F g}{\text{Area}(\Gamma)} \int_\Sigma u_N d\sigma \tag{3.6}$$

We note that φ and $\varphi +$ constant both give the same value for p.
In what follows we choose:

$$l(\varphi) = \int_\Gamma \varphi d\sigma \tag{3.7}$$

Substituting (3.6) into (3.1a),we obtain the following boundary value problem in terms of φ:

$$\boxed{\begin{array}{rcllr} \Delta\varphi &=& 0 & \text{in } \Omega_F & (a) \\ \dfrac{\partial \varphi}{\partial n} &=& u_N & \text{on } \Sigma & (b) \\ \dfrac{\partial \varphi}{\partial z} &=& \dfrac{\omega^2}{g}\varphi - \dfrac{1}{\text{Area}(\Gamma)} \displaystyle\int_\Sigma u_N\, d\sigma & \text{on } \Gamma & (c) \\ \displaystyle\int_\Gamma \varphi\, d\sigma &=& 0 & & (d) \end{array}} \tag{3.8}$$

knowing that from (3.6),(3.8d),(3.3), p is given by:

$$\boxed{p = \rho_F \omega^2 \varphi - \frac{\rho_F g}{\text{Area}(\Gamma)} \int_\Sigma u_N d\sigma \quad (= \rho_F \omega^2 \varphi + p^0)} \tag{3.9}$$

Solution φ^0 for $\omega = 0$

We shall now show that the boundary value problem in φ for $\omega = 0$ is well-posed. Let us consider the solution φ^0 of (3.8) for $\omega = 0$:

$$\begin{array}{|rcllr|}\hline \Delta\varphi^0 & = & 0 & \text{in } \Omega_F & (a) \\ \dfrac{\partial\varphi^0}{\partial n} & = & u_N & \text{on } \Sigma & (b) \\ \dfrac{\partial\varphi^0}{\partial z} & = & -\dfrac{1}{\text{Area}(\Gamma)}\displaystyle\int_\Sigma u_N\, d\sigma & \text{on } \Gamma & (c) \\ \displaystyle\int_\Gamma \varphi^0\, d\sigma & = & 0 & & (d) \\ \hline \end{array} \qquad (3.10)$$

This is a well-posed Neumann problem, the condition for whose existence is trivially verified, and whose unicity is guaranteed by (3.10d).

- To φ^0 corresponds the displacement $u^0 = \nabla\varphi^0$.
- From (3.10c), the elevation of the free surface $\partial\varphi^0/\partial z$ is a constant, denoted η^0 (Fig. 3.1(b)):

$$\boxed{\eta^0 = -\frac{1}{\text{Area}(\Gamma)}\int_\Sigma u_N\, d\sigma} \qquad (3.11)$$

- p^0 given by (3.6) (for $\omega = 0$) is interpreted as the hydrostatic variation of pressure corresponding to η^0.

Remarks on the definition of φ — If instead of the "second definition" (2.26) of φ, we use the "first definition" (2.24) $\varphi = p/(\rho_F\omega^2)$, which is equivalent to replacing in (3.1), equation (3.1a) by $p = \rho_F\omega^2\varphi$, and if we ignore equation (3.1e), the corresponding problem in φ is written:

$$\begin{array}{|rcllr|}\hline \Delta\varphi & = & 0 & \text{in } \Omega_F & (a) \\ \dfrac{\partial\varphi}{\partial n} & = & u_N & \text{on } \Sigma & (b) \\ \dfrac{\partial\varphi}{\partial z} & = & \dfrac{\omega^2}{g}\varphi & \text{on } \Gamma & (c) \\ \hline \end{array} \qquad (3.12)$$

- (3.12c) coincides with the classic hydrodynamic *surface waves* equation.
- For $\omega \neq 0$, equations (3.12) are equivalent to those of (3.8) (they yield the same value of p and of $u^F = \nabla\varphi$).
- For $\omega = 0$, the problem (3.12) is ill-posed: it is a Neumann problem the condition for whose existence is not verified, since in general $\int_\Sigma u_N\, d\sigma \neq 0$.

  We note on the other hand that use of the potential defined in (2.26) leads to equation (3.8c), which differs from the classic surface waves equation

(3.12c) by the presence of the "non local" term $-(1/Area(\Gamma)) \int_\Sigma u_N \, d\sigma$. It is this term which ensures the existence condition of the Neumann problem obtained for $\omega = 0$.

Boundary value problem in p — Taking the divergence (3.1a) and using (3.1b), we obtain:

$$\boxed{\begin{array}{rcllr} \Delta p & = & 0 & \text{in } \Omega_F & (a) \\ \dfrac{\partial p}{\partial n} & = & \rho_F \omega^2 u_N & \text{on } \Sigma & (b) \\ \dfrac{\partial p}{\partial z} & = & \dfrac{\omega^2}{g} p & \text{on } \Gamma & (c) \end{array}} \tag{3.13}$$

- (3.13c) coincides with the classic equation for surface waves in p. For $\omega \neq 0$, we can verify that equations (3.13) are equivalent to (3.8).

- For $\omega = 0$, problem (3.13) is ill-posed as u_N does not appear in the equations. Indeed, we find an indeterminate constant value of p (the physical value being given by (3.3)).

3.3 Variational formulation in terms of φ for sloshing modes

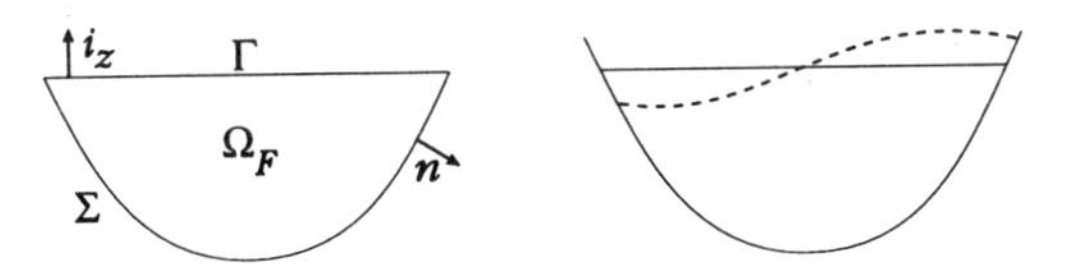

Figure 3.2: Sloshing mode

Boundary value spectral problem

The solutions of the spectral problem deduced from (3.8) for $u_N = 0$ (fixed wall condition) are known as sloshing modes. Putting:

$$\boxed{\lambda = \frac{\omega^2}{g}} \tag{3.14}$$

the boundary value spectral problem is written (Fig. 3.2):

find λ and $\varphi \neq 0$ such that:

$$\boxed{\begin{array}{llll} \Delta\varphi &= 0 & \text{in } \Omega_F & (a) \\ \dfrac{\partial\varphi}{\partial n} &= 0 & \text{on } \Sigma & (b) \\ \dfrac{\partial\varphi}{\partial z} &= \lambda\varphi & \text{on } \Gamma & (c) \end{array}} \quad (3.15)$$

$$\boxed{\int_\Gamma \varphi\, d\sigma = 0 \qquad (d)}$$

Remarks

- To begin with, we verify that $\lambda = 0$ is not a solution of (3.15): for $\lambda = 0$, equations (3.15abc) have the solution $\varphi =$ constant — which is zero according to (3.15b).
- We now consider the spectral problem reduced to equations (3.15abc) (i.e. ignoring (3.15d).
  1. This problem has the solution $\lambda = 0$, $\varphi =$ constant.
  2. On the other hand, let us show that the solutions to the problem for $\lambda \neq 0$, naturally satisfy condition (3.15d), and consequently coincide with the solutions of (3.15abcd).

     On integrating (3.15a) and apply Stokes's formula, and taking (3.15b) into account, we find $\int_\Gamma \frac{\partial\varphi}{\partial z}\, d\sigma = 0$, which, from (3.15c) and for $\lambda \neq 0$, leads to $\int_\Gamma \varphi\, d\sigma = 0$.

Variational formulation

Following the above remark, and for the sake of simplification, we shall deal with solutions to the boundary value spectral problem (3.15), ignoring condition (3.15d), which only differs from the problem posed, in the presence of the *non physical unwanted solution* $\lambda = 0, \varphi =$ constant.
We proceed formally by the test-functions method.

1. First of all, we introduce the space $\mathcal{C}$ of the functions $\varphi(M)$, $M \in \Omega_F$, "sufficiently" smooth. Multiplying (3.15a) by an arbitrary $\delta\varphi \in \mathcal{C}$, then integrating over the domain Ω_F, and finally applying the following Green formula:

$$\int_{\Omega_F} \Delta\varphi\delta\varphi\, dx = \int_{\partial\Omega_F} \frac{\partial\varphi}{\partial n}\delta\varphi\, d\sigma - \int_{\Omega_F} \nabla\varphi\cdot\nabla\delta\varphi\, dx \qquad (3.16)$$

we obtain:

$$\int_{\Omega_F} \nabla\varphi\cdot\nabla\delta\varphi\, dx - \int_{\partial\Omega_F} \frac{\partial\varphi}{\partial n}\delta\varphi\, d\sigma = 0 \qquad (3.17)$$

2. Secondly, distinguishing the contributions relative to Σ and Γ in the integral on $\partial\Omega_F$, and taking into account equation (3.15b) and (3.15c), we then obtain

$$\boxed{\int_{\Omega_F} \nabla\varphi\cdot\nabla\delta\varphi\, dx - \lambda \int_{\Gamma} \varphi\delta\varphi\, d\sigma = 0 \ \forall\delta\varphi \in \mathcal{C}} \tag{3.18}$$

The variational formulation of (3.15abc) can then be stated: find $\lambda \in \mathbb{R}, \varphi \in \mathcal{C}$ such that property (3.18) is verified $\forall\delta\varphi \in \mathcal{C}$.

- We note that this formulation involves the two *symmetric bilinear* forms $\int_{\Omega_F} \nabla\varphi\cdot\nabla\delta\varphi\, dx$ and $\int_\Gamma \varphi\delta\varphi\, d\sigma$ on $\mathcal{C} \times \mathcal{C}$.

- The eigenvalues λ are positive: putting $\delta\varphi = \varphi$ in (3.18), we see that $\lambda = \int_{\Omega_F} |\nabla\varphi|^2\, dx / \int_\Gamma \varphi^2\, d\sigma$, confirming that λ is the quotient of two positive numbers.

Converse — Let us show that if λ and φ verify (3.18), then λ and φ satisfy the equations (3.15abc). Going through the calculations formally by means of Green's formula (3.16), and separating the relative contributions Σ and Γ in the integral on $\partial\Omega_F$, we find:

$$-\int_{\Omega_F} \Delta\varphi\, \delta\varphi\, dx + \int_\Sigma \delta\varphi \frac{\partial\varphi}{\partial n}\, d\sigma + \int_\Gamma \delta\varphi\, (\frac{\partial\varphi}{\partial z} - \lambda\varphi)\, d\sigma = 0 \quad \forall\delta\varphi \in \mathcal{C} \tag{3.19}$$

To retrieve the local equations, we proceed in two steps:

1. On choosing test-functions $\delta\varphi$ which vanish on $\partial\Omega_F = \Sigma \cup \Gamma$, (3.19) reduces to its first term, whose nullity leads to (3.15a).

2. We refer to (3.19) and consider the test-functions $\delta\varphi \in \mathcal{C}$, $\delta\varphi \neq 0$ on $\partial\Omega_F$. The first term is identically null since we have just established that φ satisfies (3.15a). The nullity of the remaining integrals then leads to equations (3.15b) and (3.15c).

Remark — The space $\mathcal{C}$ required in variational formulation (3.18) is the Sobolev space $H^1(\Omega_F)$.

Orthogonality properties

We consider two solutions to (3.18) $(\lambda_\alpha, \varphi_\alpha)$ and $(\lambda_\beta, \varphi_\beta)$ with $\lambda_\alpha \neq \lambda_\beta$. Let us show that:

$$\boxed{\begin{array}{rcll} \int_\Gamma \varphi_\alpha\varphi_\beta\, d\sigma & = & 0 & (a) \\ \int_{\Omega_F} \nabla\varphi_\alpha\cdot\nabla\varphi_\beta\, dx & = & 0 & (b) \end{array}} \tag{3.20}$$

Applying variational property (3.18) on φ_α (with $\delta\varphi = \varphi_\beta$) and φ_β (with $\delta\varphi = \varphi_\alpha$), we have:

$$\boxed{\begin{array}{ll} \displaystyle\int_{\Omega_F} \nabla\varphi_\alpha \cdot \nabla\varphi_\beta \, dx = \lambda_\alpha \int_\Gamma \varphi_\alpha\varphi_\beta \, d\sigma & (a) \\ \displaystyle\int_{\Omega_F} \nabla\varphi_\beta \cdot \nabla\varphi_\alpha \, dx = \lambda_\beta \int_\Gamma \varphi_\beta\varphi_\alpha \, d\sigma & (b) \end{array}} \tag{3.21}$$

then on subtracting term by term, and using the symmetry of the bilinear forms encountered, we obtain $(\lambda_\alpha - \lambda_\beta) \int_\Gamma \varphi_\alpha\varphi_\beta \, d\sigma = 0$.
As a result, $\int_\Gamma \varphi_\alpha\varphi_\beta \, d\sigma = 0$, from which, on referring to (3.21), $\int_{\Omega_F} \nabla\varphi_\alpha \cdot \nabla\varphi_\beta \, dx = 0$, which completes the demonstration.

- (3.20ab) constitutes the orthogonality relations of the eigensolutions.

- We have seen that all solutions $\lambda_\alpha \neq 0, \varphi_\alpha$ satisfy $\int_\Gamma \varphi_\alpha \, d\sigma = 0$.

  We recall that $\varphi =$ constant is a solution of (3.15abc) (and consequently of (3.18)) for $\lambda = 0$ (non physical solution mentioned previously).

  As a result, $\int_\Gamma \varphi_\alpha \, d\sigma = 0$ is here interpreted as a special case of the orthogonality condition (3.20a) of the solutions $\lambda_\alpha > 0, \varphi_\alpha$ and of the non physical "unwanted mode" $\varphi =$ constant.

- We can eliminate this unwanted solution by adding condition (3.15d) to the equations (3.15abc), which is equivalent to restricting the variational formulation (3.18) to the admissible class $\mathcal{C}$ *of zero mean value functions on* Γ.

  We emphasize that this type of constraint consisting of an orthogonality relation on an eigensolution — corresponding to a zero eigenvalue in the present case — results in the elimination of this solution in the modified spectral problem, *without modifying the other solutions*.

  In numerical applications, we can therefore use formulation (3.18).

Energy interpretation

We consider an eigensolution $\lambda_\alpha = \frac{\omega_\alpha^2}{g} > 0, \varphi_\alpha$. We denote by $u_\alpha = \nabla\varphi_\alpha$ the displacement of the liquid, and by $\eta_\alpha = \frac{\partial\varphi_\alpha}{\partial z} = \frac{\omega_\alpha^2}{g}\varphi_\alpha$, the displacement coresponding to a free surface.
From (3.18), and for $\delta\varphi = \varphi_\alpha$, we have:

$$\int_{\Omega_F} |\nabla\varphi_\alpha|^2 \, dx = \frac{\omega_\alpha^2}{g} \int_\Gamma \varphi_\alpha^2 \, d\sigma \tag{3.22}$$

Let us show that this relation results from the conservation of total mechanical energy of the fluid during a harmonic oscillation, i.e. the sum of the kinetic and potential energy.

We therefore have to express the kinetic and potential energy in terms of φ_α.

Kinetic energy — Knowing that the instantaneous displacement is written $u(M,t) = u_\alpha(M)\cos\omega_\alpha t$, the kinetic energy $E_C = \frac{1}{2}\int_{\Omega_F} \rho_F |\frac{\partial u}{\partial t}|^2 \, dx$ is expressed as:

$$E_C = \left(\frac{\omega_\alpha^2}{2}\int_{\Omega_F} \rho_F |u_\alpha|^2 \, dx\right)\sin^2\omega_\alpha t = \left(\frac{\omega_\alpha^2}{2}\int_{\Omega_F} \rho_F |\nabla\varphi_\alpha|^2 \, dx\right)\sin^2\omega_\alpha t \quad (3.23)$$

Potential energy — First of all we shall directly express the variation of potential energy of the liquid as a quadratic functional of the vertical elevation from the free surface. Since the liquid is incompressible, the only potential energy

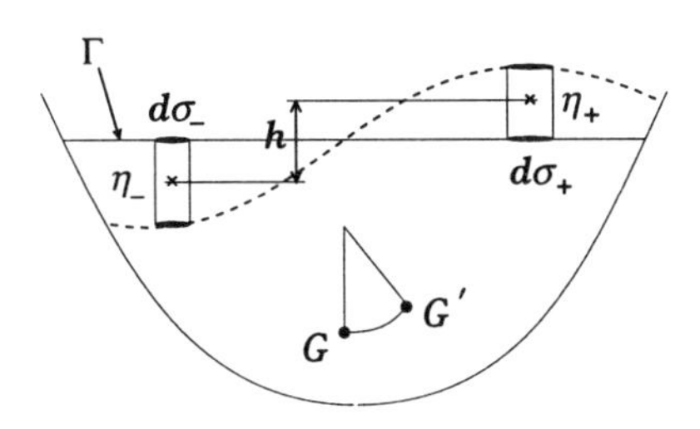

Figure 3.3: Elevation of the center of gravity

involved E_P corresponds to the work of the weight forces, which is written in terms of the instantaneous vertical elevation Z of the center of gravity of the liquid at rest as follows:

$$E_P = MgZ \quad (3.24)$$

where M denotes the total mass of liquid contained in Ω_F. Let us show that E_P can be expressed in terms of the vertical elevation η from the free surface as follows:

$$E_P = \frac{1}{2}\int_\Gamma \rho_F g \eta^2 \, d\sigma \quad (3.25)$$

The elevation of the center of gravity results from the "decanting" of the liquid occupying the domain Ω_{F-}, located under Γ, in the domain Ω_{F+}, located above Γ (Fig. 3.3). As these volumes are equal (incompressible liquid), we can subdivide them into infinitesimal cylindrical domains, of pairs of volumes equal to $d\Omega_{F-} = d\Omega_{F+}$, of respective heights η_- and η_+, and of respective cross sections $d\sigma_-$ and $d\sigma_+$. The mass dm of these volume elements can then be written:

$$dm = \rho_F |\eta_-| d\sigma_- = \rho_F |\eta_+| d\sigma_+ \quad (3.26)$$

The elementary work dE_P of the weight forces corresponding to the decanting $d\Omega_{F-} \Longrightarrow d\Omega_{F+}$ (associated with the elevation $h = 1/2(|\eta_-| + \eta_+)$), is:

$$dE_P = dm\, g\, \frac{|\eta_-| + \eta_+}{2} \quad (3.27)$$

Using (3.26), dE_P is written:

$$dE_P = \rho_F\, g\frac{\eta_-^2}{2}d\sigma_- + \rho_F\, g\frac{\eta_+^2}{2}d\sigma_+ \tag{3.28}$$

Summing these elementary contributions, we obtain expression (3.25) which can be written:

$$E_P = \frac{1}{2}\int_\Gamma \rho_F g \eta^2\, d\sigma = \frac{1}{2}\int_\Gamma \rho_F g u_z^2\, d\sigma \tag{3.29}$$

where we have used the fact that for movements of low amplitude, we can assimilate the vertical elevation η from the free surface, to the vertical component $u_z = u{\cdot}i_z$ of u.
For the oscillatory movement under consideration, we therefore have:

$$E_P = \left(\frac{\rho_F \omega^4}{2g}\int_\Gamma \varphi_\alpha^2\, d\sigma\right)\cos^2\omega_\alpha t \tag{3.30}$$

Conservation of total energy — From (3.22), the factors of $\sin^2\omega_\alpha t$ in E_C and $\cos^2\omega_\alpha t$ in E_P are *identical*, leading to a constant value for $E_C + E_P$ during the oscillation:

- The property of conservation of total mechanical energy extends to any sloshing movement, which can always be considered as a superposition of movements of the type $\sum_{\alpha=1}^{N}\kappa_\alpha\varphi_\alpha\cos\omega_\alpha t$ of different frequencies.

  We can show precisely that the total energy coincides with the sum of the energies associated with each particular oscillation (the coupling terms being null from the orthogonality relations (3.20)).

- From (3.25), we see that the vertical elevation of the center of gravity of the liquid is a quadratic functional of η (this property results from the fact that the equilibrium condition — where the free surface Γ is horizontal — minimizes the potential energy of the liquid in the gravitational field).

  On the other hand, we can demonstrate that a sloshing oscillation is accompanied by a horizontal displacement from the center of gravity, and thus by a sinusoidal variation in the momentum of the liquid, which can be expressed as a linear functional of η.

Generalized rigidity and mass

1. We call the following quantity the *generalized modal mass* μ_α: [1]

$$\boxed{\mu_\alpha = \int_{\Omega_F}\rho_F|u_\alpha|^2\, dx = \int_{\Omega_F}\rho_F|\nabla\varphi_\alpha|^2\, dx} \tag{3.31}$$

[1] By analogy with the definition of the generalized mass of a structural mode of vibration $\mu = \int_{\Omega_S}\rho|u^S|^2\, dx$.

2. We call the following quantity the *generalized modal rigidity* γ_α:

$$\boxed{\gamma_\alpha = \int_\Gamma \rho_F g \eta_\alpha^2 \, d\sigma = \frac{\rho_F \omega_\alpha^4}{g} \int_\Gamma \varphi_\alpha^2 \, d\sigma} \tag{3.32}$$

(γ_α = twice the maximum potential energy)

From (3.22), μ_α and γ_α are thus related by:

$$\boxed{\gamma_\alpha = \mu_\alpha \omega_\alpha^2} \tag{3.33}$$

With these definitions, the orthogonality relations (3.20) can be written indifferently:

$$\begin{array}{c}
\boxed{\begin{array}{rcll}
\int_\Gamma \rho_F g \, \eta_\alpha \eta_\beta \, d\sigma & = & \delta_{\alpha\beta} \omega_\alpha^2 \mu_\alpha & (a) \\
\int_{\Omega_F} \rho_F u_\alpha \cdot u_\beta \, dx & = & \delta_{\alpha\beta} \mu_\alpha & (b)
\end{array}} \\
\Updownarrow \\
\boxed{\begin{array}{rcll}
\int_\Gamma \varphi_\alpha \varphi_\beta \, d\sigma & = & \delta_{\alpha\beta} \dfrac{g}{\rho_F} \dfrac{\mu_\alpha}{\omega_\alpha^2} & (c) \\
\int_{\Omega_F} \nabla\varphi_\alpha \cdot \nabla\varphi_\beta \, dx & = & \delta_{\alpha\beta} \dfrac{\mu_\alpha}{\rho_F} & (d)
\end{array}}
\end{array} \tag{3.34}$$

Normalization — The eigenvectors φ_α are defined to within one multiplying factor C. Taking $C = 1/\sqrt{\mu_\alpha}$, we define an eigenvector (defined to within a factor of ± 1) of generalized mass equal to 1.

3.4 Spectral problem discretized by finite elements

The discretization by finite elements of the symmetric bilinear forms of (3.18) introduces the following matrices:

$$\boxed{\begin{array}{rcll}
\int_{\Omega_F} \nabla\varphi \cdot \nabla\delta\varphi \, dx & \Longrightarrow & \delta\boldsymbol{\Phi}^T \boldsymbol{F} \boldsymbol{\Phi} & (a) \\
\int_\Gamma \varphi \, \delta\varphi \, d\sigma & \Longrightarrow & \delta\boldsymbol{\Phi}^T \boldsymbol{S} \boldsymbol{\Phi} & (b)
\end{array}} \tag{3.35}$$

The matrix form of (3.18) is then written:

$$\boxed{\boldsymbol{F}\boldsymbol{\Phi} = \lambda \boldsymbol{S} \boldsymbol{\Phi}} \tag{3.36}$$

Kernel of F — We note that $\int_{\Omega_F} \nabla\varphi \cdot \nabla\delta\varphi \, dx = 0$ if and only if φ = constant. Consequently, the kernel of $\boldsymbol{F}$ is of dimension 1 (the matrix $\boldsymbol{F}(N \times N)$ is of rank

$N-1$). If we denote by $\mathbf{1}_N$ the vector of $\mathbb{R}^N$ with components equal to 1, we have precisely:

$$\boldsymbol{F}\boldsymbol{\Phi} = 0 \Longleftrightarrow \boldsymbol{\Phi} = \text{ constant} \times \mathbf{1}_N \tag{3.37}$$

As a result, all diagonal blocks of $\boldsymbol{F}$ of dimension less than or equal to $N-1$ are non-singular (submatrix obtained by eliminating at least one row and its corresponding column).

Structure of S — The matrix $\boldsymbol{S}$ results from the discretization of $\int_\Gamma \varphi\delta\varphi \, d\sigma$, which only involves the nodal values of φ on Γ.
Consequently, the rows and columns of $\boldsymbol{S}$ belonging to nodes located on the free surface Γ are zeros.

Condensation of the matrix problem — We define a partitioning of $\boldsymbol{\Phi} = (\boldsymbol{\Phi}_1, \boldsymbol{\Phi}_2)$, where $\boldsymbol{\Phi}_1$ denotes the N_1 nodal values of φ corresponding to the nodes located on the free surface Γ. (3.36) is then written:

$$\boxed{\begin{bmatrix} \boldsymbol{F}_{11} & \boldsymbol{F}_{12} \\ \boldsymbol{F}_{12}{}^T & \boldsymbol{F}_{22} \end{bmatrix} \begin{bmatrix} \boldsymbol{\Phi}_1 \\ \boldsymbol{\Phi}_2 \end{bmatrix} = \lambda \begin{bmatrix} \boldsymbol{S}_{11} & 0 \\ 0 & 0 \end{bmatrix} \begin{bmatrix} \boldsymbol{\Phi}_1 \\ \boldsymbol{\Phi}_2 \end{bmatrix}} \tag{3.38}$$

We note from the discussion on the kernel of $\boldsymbol{F}$ that $\boldsymbol{F}_{22}$ is non-singular, and from (3.38) we have:

$$\boxed{\boldsymbol{\Phi}_2 = -\boldsymbol{F}_{22}^{-1}\boldsymbol{F}_{12}{}^T\boldsymbol{\Phi}_1} \tag{3.39}$$

The eigenvalue problem (3.38) is then equivalent to the "condensed" problem:

$$\boxed{\left(\boldsymbol{F}_{11} - \boldsymbol{F}_{12}\boldsymbol{F}_{22}^{-1}\boldsymbol{F}_{12}{}^T\right)\boldsymbol{\Phi}_1 = \lambda \boldsymbol{S}_{11}\boldsymbol{\Phi}_1} \tag{3.40}$$

We see that (3.40) involves only the values of φ on Γ, the values of φ in Ω_F being retrieved for each eigenvalue by means of (3.39).

Practical application

- We begin by noting that the presence of the solution $\lambda = 0$ does not perturb the spectrum of positive eigenvalues (*cf.* the discussion § 3.3).
- Number of eigenvalues: (3.40) possesses N_1 eigenvalues — including the zero eigenvalue. Consequently, we can say that equation (3.38) possesses $N - N_1$ "infinite" eigenvalues related to the presence of $N - N_1$ rows of zeros in the "mass" matrix.
- In practice, however, it is simpler to solve (3.38) directly rather than first to construct the condensed form (3.40); the zero terms of the "mass" matrix do not interfere with the solution of the equations (*cf.* e.g. Berger, Boujot & Ohayon [17], Fu [77]).

3.5 Axisymmetric reservoirs

We examine here the case of a reservoir of revolution. We operate in cylindrical coordinates: z is the axis of revolution, r, θ, z denote the local frame of reference. We can expand φ into a Fourier series as follows:

$$\underbrace{\varphi}_{\mathcal{C}} = \underbrace{\varphi_0(r,z)}_{\mathcal{C}_0} + \underbrace{\varphi_n^+(r,z)\cos n\theta}_{\mathcal{C}_n^+} + \underbrace{\varphi_n^-(r,z)\sin n\theta}_{\mathcal{C}_n^-} \tag{3.41}$$

where $n \geq 0$ is an integer. We show that the boundary value problem $\mathcal{P}$ splits up into a series of two-dimensional problems $\mathcal{P}_n^\pm$ (for $n \geq 1$) and $\mathcal{P}_0$, set in the meridian plane.
The variational formulation corresponding to $\mathcal{P}_0$ and $\mathcal{P}_n^\pm$ is obtained by restricting the variational formulation (3.18) to the admissible subspaces $\mathcal{C}_0$, $\mathcal{C}_n^\pm$ of $\mathcal{C}$ (respectively).

1. **Axisymmetric modes** — The axisymmetric modes correspond to non-degenerate eigenvalues of the three-dimensional problem.

2. **Non-axisymmetric modes** — For $n \geq 1$, the eigenvalues of $\mathcal{P}_n^+$ and $\mathcal{P}_n^-$ are identical (they correspond to a multiplicity of 2 of the eigenvalues of the three-dimensional problem), and the eigenvectors of $\mathcal{P}_n^-$ are deduced from those of $\mathcal{P}_n^+$ by rotation through $\pi/2n$.

Consequently, the numerical application is carried out, for example, for modes of the type $\mathcal{C}_n^+$.

3.6 Formulation in terms of $\varphi|_\Gamma = f$

We have already noted that (3.40) only involves the nodal values $\mathbf{\Phi}_1$ of φ on Γ. In what follows we shall establish a variational formulation in terms of the value φ_Γ of φ on Γ, and we shall show that (3.40) is a discretized form of this.

Definition of the mapping $A : \varphi|_\Gamma \longrightarrow \frac{\partial\varphi}{\partial z}|_\Gamma$

We consider here the non homogeneous Neumann-Dirichlet problem, which consists of seeking a harmonic function φ in Ω_F with a given value f on Γ, and whose normal derivative is null on Σ:

$$\begin{array}{|lcllr|} \Delta\varphi & = & 0 & \text{in } \Omega_F & (a) \\ \dfrac{\partial\varphi}{\partial n} & = & 0 & \text{on } \Sigma & (b) \\ \varphi|_\Gamma & = & f & & (c) \end{array} \tag{3.42}$$

1. **Definition of the operator φ_f** — The preceding equations enable the following linear mapping to be defined:

$$\boxed{f \longrightarrow \varphi_f} \tag{3.43}$$

which, for any f on Γ, defines the corresponding φ in Ω_F.

We shall denote by φ_f the solution of (3.42) for a given f (in mathematics, φ_f is a *lifting* of f, where f is the *trace* of φ on Γ).

2. We then introduce the linear mapping A which associates to f the value of the normal derivative on Γ of the operator φ_f:

$$\boxed{f \xrightarrow{A} \left.\frac{\partial \varphi_f}{\partial z}\right|_{\Gamma}} \tag{3.44}$$

With the preceding notation, the boundary value problem (3.15abc) is then written:

$$\boxed{\left.\frac{\partial \varphi_f}{\partial z}\right|_{\Gamma} = \lambda f} \tag{3.45}$$

or, using the operator A defined in (3.44):

$$\boxed{Af = \lambda f} \tag{3.46}$$

Variational formulation — We introduce the space $\mathcal{C}_\Gamma$ of the smooth functions defined on Γ.
Multiplying (3.45) by a test-function $\delta f \in \mathcal{C}_\Gamma$ and integrating on Γ, we obtain:

$$\int_\Gamma \frac{\partial \varphi_f}{\partial z} \delta f \, d\sigma = \lambda \int_\Gamma f \delta f \, d\sigma \quad \forall \delta f \in \mathcal{C}_\Gamma \tag{3.47}$$

Symmetry of $\int_\Gamma \frac{\partial \varphi_f}{\partial z} \delta f \, d\sigma$ — We show that $\int_\Gamma \frac{\partial \varphi_f}{\partial z} \delta f \, d\sigma$ is a symmetric bilinear form on $\mathcal{C}_\Gamma \times \mathcal{C}_\Gamma$.
We denote by $\varphi_{\delta f}$ the lifting of $\delta f \in \mathcal{C}_\Gamma$ defined by (3.42), (3.43). Applying Green's formula, we obtain:

$$\int_\Gamma \frac{\partial \varphi_f}{\partial z} \delta f \, d\sigma = \int_{\Omega_F} \nabla \varphi_f \cdot \nabla \varphi_{\delta f} \, dx - \int_{\Omega_F} \varphi_{\delta f} \Delta \varphi_f \, dx - \int_\Sigma \frac{\partial \varphi_f}{\partial n} \varphi_{\delta f} \, d\sigma \tag{3.48}$$

The last two terms are null by definition of φ_f, from which we have:

$$\boxed{\int_\Gamma \frac{\partial \varphi_f}{\partial z} \delta f \, d\sigma = \int_{\Omega_F} \nabla \varphi_f \cdot \nabla \varphi_{\delta f} \, dx} \tag{3.49}$$

which leads the stated symmetry property.

(3.47) can thus be written:

$$\boxed{\int_{\Omega_F} \nabla\varphi_f \cdot \nabla\varphi_{\delta f}\, dx = \lambda \int_\Gamma f\delta f\, d\sigma \quad f \in \mathcal{C}_\Gamma \;,\; \forall \delta f \in \mathcal{C}_\Gamma} \tag{3.50}$$

Remarks

- The variational formulation (3.50) in terms of surface variables $f \in \mathcal{C}_\Gamma$, can be interpreted as the restriction of the variational formulation (3.18) to the space $\hat{\mathcal{C}}_\Gamma$ of the functions which satisfy $\Delta\varphi = 0$ in Ω_F and $\partial\varphi/\partial n = 0$ on Σ. $\hat{\mathcal{C}}_\Gamma$ is identifiable with $\mathcal{C}_\Gamma$.

- We verify that $f =$ constant is a solution of (3.50) for $\lambda = 0$ ("unwanted" solution mentioned above), and that for an eigenvalue $\lambda_\alpha \neq 0$, $\int_\Gamma f_\alpha\, d\sigma = 0$, (by taking a constant δf in (3.50)).

- **Sloshing modes basis** — We can show that the eigenvectors $\{f_\alpha\}$ of (3.50) — including the constant solution corresponding to the zero eigenvalue — form a basis of $\mathcal{C}_\Gamma$.

  For sake of brevity, we shall designate by *sloshing modes basis*, the set of solutions corresponding to the strictly positive eigenvalues. Since these solutions or eigenmodes satisfy $\int_\Gamma f\, d\sigma = 0$, it could be easily established that they form a basis of $\mathcal{C}_\Gamma^*$, defined by:

$$\boxed{\mathcal{C}_\Gamma^* = \left\{ f \in \mathcal{C}_\Gamma \mid \int_\Gamma f\, d\sigma = 0 \right\}} \tag{3.51}$$

  To each f_α, there coresponds a lifting φ_{f_α} in Ω_F, which is none other than φ_α (*cf.* (3.42) and (3.43)).

  We shall use the term *sloshing mode* loosely to mean both the function $f_\alpha = \varphi_\alpha|_\Gamma$ and the mapping $\varphi_\alpha = \varphi_{f_\alpha}$, and we shall say that the φ_α constitute a basis of the subspace $\hat{\mathcal{C}}_\Gamma$ of the harmonic functions in Ω_F whose normal derivative is null on Σ.

  The formulation (3.50) has been used for the mathematical study of the spectrum (*cf.* Boujot [25], references included).

Discretization

The discretization of $\int_\Gamma f\delta f\, d\sigma$ coincides with (3.35b), and leads to the matrix $\boldsymbol{S}_{11}$ of (3.38).
The direct discretization of $\int_{\Omega_F} \nabla\varphi_f \cdot \nabla\varphi_{\delta f}\, dx$ requires an explicit construction of A.
One way consists in using a boundary integral representation of φ_f in terms of f by means of Green's functions. This approach leads to various boundary integral methods involving a discretization of the boundary of the fluid domain (Σ and

Γ) and different possible choice of unknown field (see, for example, Siekmann & Schilling [200]).
In what follows, we show that (3.39) can be interpreted as performing a particular discretization of the operator (3.43).

Discretization of the φ_f operator

We begin with the variational formulation of the Neumann-Dirichlet problem (3.42) which is classic and is written:

$$\int_{\Omega_F} \nabla\varphi \cdot \nabla\delta\varphi \, dx = 0 \quad \varphi|_\Gamma = f, \;\; \forall\delta\varphi \text{ null on } \Gamma \tag{3.52}$$

whose discretization involves the matrix $\boldsymbol{F}$ defined by (3.35a).
Explicitly, denoting by $\boldsymbol{\Phi}_1 = \boldsymbol{f}$ the nodal values of $\varphi = f$ on Γ, and by $\boldsymbol{\Phi}_2$ (and respectively $\delta\boldsymbol{\Phi}_2$) the other nodal unknowns, the discretized form of (3.52) is written:

$$\begin{bmatrix} 0 & \delta\boldsymbol{\Phi}_2{}^T \end{bmatrix} \begin{bmatrix} \boldsymbol{F}_{11} & \boldsymbol{F}_{12} \\ \boldsymbol{F}_{12}{}^T & \boldsymbol{F}_{22} \end{bmatrix} \begin{bmatrix} \boldsymbol{f} \\ \boldsymbol{\Phi}_2 \end{bmatrix} = 0 \quad \forall\delta\boldsymbol{\Phi}_2 \tag{3.53}$$

leading to:

$$\boldsymbol{\Phi}_2 = -\boldsymbol{F}_{22}^{-1}\boldsymbol{F}_{12}{}^T\boldsymbol{f} \tag{3.54}$$

This last relation corresponds exactly to (3.39).
subsubsectionDiscretization of $\int_{\Omega_F} \nabla\varphi_f \cdot \nabla\varphi_{\delta f}\, dx$ We partition the nodal values $\boldsymbol{\Phi}$ of φ, distinguishing the nodal values $\boldsymbol{\Phi}_1$ relative to Γ from the other nodal values $\boldsymbol{\Phi}_2$. (3.35a) is then written:

$$\int_{\Omega_F} \nabla\varphi \cdot \nabla\delta\varphi \, d\sigma \Longrightarrow \begin{bmatrix} \delta\boldsymbol{\Phi}_1^T & \delta\boldsymbol{\Phi}_2^T \end{bmatrix} \begin{bmatrix} \boldsymbol{F}_{11} & \boldsymbol{F}_{12} \\ \boldsymbol{F}_{12}{}^T & \boldsymbol{F}_{22} \end{bmatrix} \begin{bmatrix} \boldsymbol{\Phi}_1 \\ \boldsymbol{\Phi}_2 \end{bmatrix} \tag{3.55}$$

When, according to (3.54), we replace $\boldsymbol{\Phi}_2$ in terms of $\boldsymbol{\Phi_1}$ denoted $\boldsymbol{f}$ (and respectively $\delta\boldsymbol{\Phi}_2$ in terms of $\delta\boldsymbol{\Phi}_1$ denoted $\delta\boldsymbol{f}$) in the above expression, we find:

$$\int_{\Omega_F} \nabla\varphi_f \cdot \nabla\varphi_{\delta f} \, d\sigma \Longrightarrow \delta\boldsymbol{f}^T \left(\boldsymbol{F}_{11} - \boldsymbol{F}_{12}\boldsymbol{F}_{22}^{-1}\boldsymbol{F}_{12}{}^T\right)\boldsymbol{f} \tag{3.56}$$

Consequently, the condensed matrix equation (3.40) can be interpreted as a particular discretization of (3.50).

3.7 Comparison of eigenfrequencies

We are interested in the following two problems:

1. Overestimation of the exact eigenvalues by the eigenvalues of the discretized problem.

2. Comparison of eigenfrequencies of reservoirs of various shapes.

This comparison is based on the use of the Rayleigh quotient associated with the variational formulation (3.50) which is written:

$$R(f) = \frac{\displaystyle\int_{\Omega_F} |\nabla \varphi_f|^2 \, dx}{\displaystyle\int_\Gamma f^2 \, d\sigma} = \frac{\displaystyle\min_{\varphi|\Gamma = f} \int_{\Omega_F} |\nabla \varphi|^2 \, dx}{\displaystyle\int_\Gamma f^2 \, d\sigma} \quad , f \in \mathcal{C}_\Gamma \tag{3.57}$$

The stationarity of the Rayleigh quotient (defined by the first equality) results directly from (3.50).
The second equality arises from the classic extremal property $\int_{\Omega_F} |\nabla \varphi_f|^2 \, dx = \min_{\varphi|\Gamma=f} \int_{\Omega_F} |\nabla \varphi|^2 \, dx$ of the solution φ_f of the non homogeneous Neumann-Dirichlet problem (3.42).
Overestimation of eigenvalues — We assume that the discretized domain Ω_F^h coincides with Ω_F. Under these conditions, the Rayleigh quotient of the discretized problem is obtained by restricting the Rayleigh quotient to $\mathcal{C}_\Gamma^h \subset \mathcal{C}_\Gamma$. Consequently, from the theorem of comparison (1.37), the *nth* approximate eigenvalue λ_n^h is higher than the exact eigenvalue of the same rank.

Comparison of eigenvalues for different domains

The physical problem consists of comparing the eigenfrequencies of reservoirs of various shapes (*cf.* Moiseev & Rumyantsev [135]).
This comparison is possible if, after an appropriate translation of the liquid domains in question, the free surfaces coincide ($\Gamma' = \Gamma$), and if moreover one of the liquid domains is included in the other: $\Omega'_F \subset \Omega_F$ (Fig. 3.4). The method

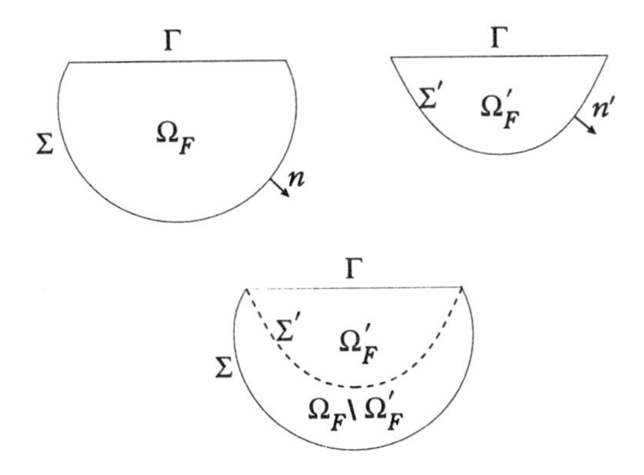

Figure 3.4: Reservoirs of various shapes

that we use here consists of applying the first comparison theorem, which is based on the comparison between Rayleigh quotients, in the present case, with *identical admissible classes* $\mathcal{C}_\Gamma$.
We denote by φ'_f, the lifting of f defined by (3.43), i.e. the solution of $\Delta\varphi'_f = 0$ in Ω'_F, $\partial\varphi'_f/\partial n' = 0$ on Σ', and $\varphi'_f = f$ on Γ.

The respective Rayleigh coefficients are:

$$R(f) = \frac{\int_{\Omega_F} |\nabla \varphi_f|^2 \, dx}{\int_\Gamma f^2 \, d\sigma}, \quad R'(f) = \frac{\int_{\Omega'_F} |\nabla \varphi'_f|^2 \, dx}{\int_\Gamma f^2 \, d\sigma}, \quad , \; f \in \mathcal{C}_\Gamma \tag{3.58}$$

Let us show that

$$R(f) \geq R'(f), \quad \forall f \in \mathcal{C}_\Gamma \tag{3.59}$$

Since the denominators of R' and R are identical, this comparison is equivalent to showing that:

$$I = \int_{\Omega_F} |\nabla \varphi_f|^2 \, dx - \int_{\Omega'_F} |\nabla \varphi'_f|^2 \, dx \geq 0 \tag{3.60}$$

I can be written:

$$I = \int_{\Omega_F \backslash \Omega'_F} |\nabla \varphi_f|^2 \, dx + \int_{\Omega'_F} (|\nabla \varphi_f|^2 - |\nabla \varphi'_f|^2) \, dx \tag{3.61}$$

By using the identity $a^2 - b^2 = (a-b)^2 + 2b(a-b)$, with $a = \nabla \varphi$ and $b = \nabla \varphi'$, we obtain:

$$I = \int_{\Omega_F \backslash \Omega'_F} |\nabla \varphi_f|^2 \, dx + \int_{\Omega'_F} |\nabla \varphi_f - \nabla \varphi'_f|^2 \, dx + 2 \int_{\Omega'_F} \nabla \varphi'_f \cdot (\nabla \varphi_f - \nabla \varphi'_f) \, dx \tag{3.62}$$

We shall show that the last term is null.
We begin with the variational property of φ'_f: $\int_{\Omega'_F} \nabla \varphi' \nabla \delta \varphi' \, dx = 0$, with $\varphi'|_\Gamma = f$ and $\delta \varphi'$ vanishing on Γ.
Taking $\delta \varphi' = \varphi_f - \varphi'_f$, which verifies the condition $\delta \varphi|_\Gamma = 0$, we then obtain the stated result.
I, the sum of the two positive terms, is thus positive.
According to the first comparison theorem, if λ'_n (and respectively λ_n) denotes the *nth* eigenvalue of the problem $\mathcal{P}'$ (and respectively $\mathcal{P}$), we then have:

$$\boxed{\lambda_n \geq \lambda'_n} \tag{3.63}$$

Remarks

- The *nth* sloshing eigenfrequency of the "large reservoir" is thus higher than the *nth* sloshing eigenfrequency of the "small reservoir" (*cf.* Morand [141]). This result, where the eigenfrequencies increase with mass of liquid, may appear paradoxical at first sight.

- The above result enables the sloshing eigenfrequencies corresponding to a domain Ω_F of complex shape to be bounded by the eigenfrequencies corresponding to domains Ω'_F, Ω''_F of simpler shapes, whose eigenvalues λ'_n and λ''_n are known analytically or numerically. If the domains verify the inclusion condition: $\Omega'_F \subset \Omega_F \subset \Omega''_F$, then the corresponding eigenvalues verify the inequality: $\lambda'_n \leq \lambda_n \leq \lambda''_n$.

3.8 Modal analysis of the vibratory response of the liquid

Variational formulation of the response to a prescribed displacement u_N

We consider the harmonic response of a liquid to a given motion of the wall of normal amplitude u_N on Σ, and of circular frequency ω.
We begin with equations (3.8), and we introduce the admissible space $\mathcal{C}$ of the smooth functions defined in Ω_F, and $\mathcal{C}^*$ the subspace of $\mathcal{C}$ defined as follows:

$$\mathcal{C}^* = \{\varphi \in \mathcal{C} \mid \int_\Gamma \varphi \, d\sigma = 0\} \tag{3.64}$$

Multiplying (3.8a) by a test-function $\delta\varphi \in \mathcal{C}^*$, integrating in Ω_F, then applying Green's formula (3.16), and taking into account (3.8b) and (3.8c), we obtain the variational formulation of the problem:
For a given ω and u_N, find $\varphi \in \mathcal{C}^*$ such that:

$$\boxed{\int_{\Omega_F} \nabla\varphi \cdot \nabla\delta\varphi \, dx - \frac{\omega^2}{g} \int_\Gamma \varphi\delta\varphi \, d\sigma = \int_\Sigma u_N \delta\varphi \, d\sigma \quad \forall \delta\varphi \in \mathcal{C}^*} \tag{3.65}$$

Converse — We have to verify, by going through the calculations formally, that all solutions φ of (3.65) are solutions of the boundary value problem (3.8).
We begin by noting that we can write (3.65) in an equivalent form obtained by taking as a test-function in (3.65) the function $\delta\varphi \in \mathcal{C}^*$ defined from any function $\psi \in \mathcal{C}$ by $\delta\varphi = \psi - \frac{1}{\text{Area}(\Gamma)} \int_\Gamma \psi \, d\sigma$, which gives, for $\varphi \in \mathcal{C}^*$ and for all $\psi \in \mathcal{C}$:

$$\begin{aligned} \int_{\Omega_F} \nabla\varphi \cdot \nabla\psi \, dx - \frac{\omega^2}{g} \int_\Gamma \varphi\psi \, d\sigma = \\ \int_\Sigma u_N \psi \, d\sigma - \int_\Gamma \left(\frac{1}{\text{Area}(\Gamma)} \int_\Sigma u_N \, d\sigma \right) \psi \, d\sigma \end{aligned} \tag{3.66}$$

We find equations (3.8) on applying Green's formula to the first integral, and proceeding in the usual way.

- If $\boldsymbol{U}$ denotes the vector of the nodal values of u, (3.65) is written in discretized form:

$$\boxed{\boldsymbol{F\Phi} - \frac{\omega^2}{g} \boldsymbol{S\Phi} = \boldsymbol{C}^T \boldsymbol{U} \quad \text{with } \boldsymbol{L}^T \boldsymbol{\Phi} = 0} \tag{3.67}$$

  where $\delta\boldsymbol{\Phi}^T \boldsymbol{C}^T \boldsymbol{U}$ discretizes $\int_\Sigma u_N \delta\varphi \, d\sigma$, and where $\boldsymbol{L}^T \boldsymbol{\Phi} = 0$ discretizes the constraint $\int_\Gamma \varphi \, d\sigma = 0$.

- In what follows, the variational property (3.65) is reformulated in terms of surface variables for the modal analysis of the problem.

We shall consider two alternative solutions to the problem.

First modal decomposition

We let:

$$\boxed{\varphi = \varphi^0 + \underline{\varphi}} \tag{3.68}$$

where φ^0 is the solution of problem (3.10).
From (3.8) and (3.10), $\underline{\varphi}$ verifies the boundary value problem:

$$\begin{array}{rcllr}
\Delta\underline{\varphi} & = & 0 & \text{in } \Omega_F & (a) \\
\dfrac{\partial\underline{\varphi}}{\partial n} & = & 0 & \text{on } \Sigma & (b) \\
\dfrac{\partial\underline{\varphi}}{\partial z} & = & \dfrac{\omega^2}{g}\underline{\varphi} + \dfrac{\omega^2}{g}\varphi^0 & \text{on } \Gamma & (c) \\
\displaystyle\int_\Gamma \underline{\varphi}\, d\sigma & = & 0 & & (d)
\end{array} \tag{3.69}$$

We denote by $\underline{f}$ and f^0 the values of $\underline{\varphi}$ and φ^0 on Γ. Using the operator A and φ_f defined by (3.43),(3.44), equations (3.69) are written:

$$A\underline{f} = \frac{\partial\varphi_{\underline{f}}}{\partial z} = \frac{\omega^2}{g}\underline{f} + \frac{\omega^2}{g} f^0\Big|_\Gamma \quad \text{with} \int_\Gamma \underline{f}\, d\sigma = 0 \tag{3.70}$$

The variational formulation of (3.70) is then written, for $\underline{f} \in \mathcal{C}^*_\Gamma$, and $\forall\delta\underline{f} \in \mathcal{C}^*_\Gamma$:

$$\int_\Gamma \frac{\partial\varphi_{\underline{f}}}{\partial z}\delta\underline{f}\, d\sigma - \frac{\omega^2}{g}\int_\Gamma \underline{f}\delta\underline{f}\, d\sigma = \frac{\omega^2}{g}\int_\Gamma f^0\delta\underline{f}\, d\sigma \tag{3.71}$$

which from (3.49), can be written, for $\underline{f} \in \mathcal{C}^*_\Gamma$ and $\forall\delta\underline{f} \in \mathcal{C}^*_\Gamma$:

$$\boxed{\int_{\Omega_F} \nabla\varphi_{\underline{f}}\cdot\nabla\varphi_{\delta\underline{f}}\, dx - \frac{\omega^2}{g}\int_\Gamma \underline{f}\delta\underline{f}\, d\sigma = \frac{\omega^2}{g}\int_\Gamma f^0\delta\underline{f}\, d\sigma} \tag{3.72}$$

- We note that the homogeneous problem associated with (3.72), obtained on putting $u_N = 0$, — which involves $\varphi^0 = 0$ and therefore $f^0 = 0$ — coincides with the spectral problem (3.50) restricted to $\mathcal{C}^*_\Gamma \subset \mathcal{C}_\Gamma$, which has no zero eigenvalue.
- We have already indicated that the solutions $\{f_\alpha\}$ corresponding to $\lambda_\alpha > 0$, constitute a basis of $\mathcal{C}^*_\Gamma$.

We can write:

$$\underline{f} = \sum_{\alpha\geq 1} \kappa_\alpha f_\alpha \Longrightarrow \underline{\varphi} = \sum_{\alpha\geq 1} \kappa_\alpha\varphi_\alpha \tag{3.73}$$

where $\{\kappa_\alpha\}$ is a generalized system of coordinates. Substituting this expansion into (3.72) and successively putting $\delta\underline{f} = \varphi_1|_\Gamma, \varphi_2|_\Gamma, \ldots$, and taking account of the orthogonality relations (3.34cd), we finally obtain:

$$(-\omega^2 + \omega_\alpha^2)\kappa_\alpha\mu_\alpha = \omega^2\int_\Gamma \rho_F\frac{\omega_\alpha^2}{g} f^0\varphi_\alpha\, d\sigma \tag{3.74}$$

The last term can be transformed, if we recall that $f^0 = \varphi^0|_\Gamma$, and then apply the following Green formula:

$$\int_{\Omega_F} (u\Delta v - v\Delta u)\, dx = \int_{\partial\Omega_F} (u\frac{\partial v}{\partial n} - v\frac{\partial u}{\partial n})\, d\sigma \tag{3.75}$$

with $u = \varphi^0$ and $v = \varphi_\alpha$.
If we then use equations (3.10) verified by φ^0 and (3.15) verified by φ_α, we obtain the following *conjugate relation* between φ^0 and φ_α:

$$\int_\Gamma \varphi^0 \varphi_\alpha\, d\sigma = \frac{g}{\omega_\alpha^2} \int_\Sigma u_N \varphi_\alpha\, d\sigma \tag{3.76}$$

(3.74) is then written:

$$\boxed{(-\omega^2 + \omega_\alpha^2)\kappa_\alpha \mu_\alpha = \omega^2 \int_\Sigma \rho_F u_N \varphi_\alpha\, d\sigma} \tag{3.77}$$

From (3.68), we have therefore

$$\boxed{\varphi = \varphi^0 + \sum_{\alpha \geq 1} \frac{\omega^2}{-\omega^2 + \omega_\alpha^2} \frac{\int_\Sigma \rho_F u_N \varphi_\alpha\, d\sigma}{\mu_\alpha} \varphi_\alpha} \tag{3.78}$$

Remark

The steps used to establish (3.78) are formally equivalent to the following heuristic procedure:
We substitute the development:

$$\boxed{\varphi = \varphi^0 + \sum_{\alpha \geq 1} \kappa_\alpha \varphi_\alpha} \tag{3.79}$$

into the variational property (3.65) and we successively put $\delta\varphi = \varphi_1, \varphi_2, \ldots$, then we use the orthogonality relations (3.34cd) and the property (3.76).

Second decomposition

We let:

$$\boxed{\varphi = \varphi^\infty + \overline{\varphi}} \tag{3.80}$$

where φ^∞ is the solution to the following Neumann-Dirichlet problem: [2]

$$\boxed{\begin{array}{llll} \Delta\varphi^\infty = 0 & & \text{in } \Omega_F & (a) \\ \dfrac{\partial \varphi^\infty}{\partial n} = & u_N & \text{on } \Sigma & (b) \\ \varphi^\infty = & 0 & \text{on } \Gamma & (c) \end{array}} \tag{3.81}$$

[2] φ^∞ coincides with the displacement potential introduced in chapter 5, for the study of hydroelastic vibrations without gravity (*cf.* equations (5.1abc)).

From (3.8) and (3.81), $\overline{\varphi}$ verifies:

$$
\begin{aligned}
\Delta\overline{\varphi} &= 0 && \text{in } \Omega_F \quad (a)\\
\frac{\partial\overline{\varphi}}{\partial n} &= 0 && \text{on } \Sigma \quad (b)\\
\frac{\partial\overline{\varphi}}{\partial z} &= \frac{\omega^2}{g}\overline{\varphi} - \left.\frac{\partial\varphi^\infty}{\partial z}\right|_\Gamma - \frac{1}{\text{Area}(\Gamma)}\int_\Sigma u_N\,d\sigma && \text{on } \Gamma \quad (c)\\
\int_\Gamma \overline{\varphi}\,d\sigma &= 0 && (d)
\end{aligned}
\tag{3.82}
$$

We denote by $\overline{f}$ the value of $\overline{\varphi}$ on Γ. Using respectively definitions (3.43) and (3.44) of A and of φ_f, (3.82) is written:

$$
A\overline{f} = \frac{\partial\varphi_{\overline{f}}}{\partial z} = \frac{\omega^2}{g}\overline{f} - \left.\frac{\partial\varphi^\infty}{\partial z}\right|_\Gamma - \frac{1}{\text{Area}(\Gamma)}\int_\Sigma u_N\,d\sigma \quad \text{with } \int_\Gamma \overline{f}\,d\sigma = 0 \tag{3.83}
$$

- The problem (3.81) can be considered as the boundary value problem obtained by letting $\lambda = \infty$ in (3.8) (we divide (3.8c) by λ, then let $\lambda = \infty$, which gives $\varphi = 0$ on Γ), hence the notation φ^∞.

The variational formulation of (3.83) is then written, for $\overline{f} \in \mathcal{C}^*_\Gamma$, and $\forall\delta\overline{f} \in \mathcal{C}^*_\Gamma$: [3]

$$
\int_\Gamma \frac{\partial\varphi_{\overline{f}}}{\partial z}\,\delta\overline{f}\,d\sigma - \frac{\omega^2}{g}\int_\Gamma \overline{f}\delta\overline{f}\,d\sigma = -\int_\Gamma \left.\frac{\partial\varphi^\infty}{\partial z}\right|_\Gamma \delta\overline{f}\,d\sigma \tag{3.84}
$$

which from (3.49), is written, for $\overline{f} \in \mathcal{C}^*_\Gamma$ and $\forall\delta\overline{f} \in \mathcal{C}^*_\Gamma$:

$$
\boxed{\int_{\Omega_F} \nabla\varphi_{\overline{f}}\cdot\nabla\varphi_{\delta\overline{f}}\,dx - \frac{\omega^2}{g}\int_\Gamma \overline{f}\delta\overline{f}\,d\sigma = -\int_\Gamma \left.\frac{\partial\varphi^\infty}{\partial z}\right|_\Gamma \delta\overline{f}\,d\sigma} \tag{3.85}
$$

As before $\overline{f} \in \mathcal{C}^*_\Gamma$ can be expanded onto the basis (of the traces f_α) of the sloshing modes according to:

$$
\overline{f} = \sum_{\alpha\geq 1} \tau_\alpha f_\alpha \tag{3.86}
$$

where $\{\tau_\alpha\}$ constitutes a (second) generalized system of coordinates. Substituting this expansion into (3.85) and putting successively $\delta\overline{f} = \varphi_1|_\Gamma, \varphi_2|_\Gamma, \ldots$ and taking account of the orthogonality relations (3.34cd), we finally obtain:

$$
(-\omega^2 + \omega_\alpha^2)\tau_\alpha\mu_\alpha = -\omega_\alpha^2 \int_\Gamma \rho_F \frac{\partial\varphi^\infty}{\partial z}\varphi_\alpha\,d\sigma \tag{3.87}
$$

Applying Green's formula (3.75) to the above integral with $u = \varphi^\infty$ and $v = \varphi_\alpha$, and using equations (3.81) and (3.15) verified by φ^∞ and φ_α, we obtain the following *conjugate relation* between φ^∞ and φ_α:

$$
\int_\Gamma \rho_F \frac{\partial\varphi^\infty}{\partial z}\varphi_\alpha\,d\sigma = -\int_\Sigma \rho_F u_N \varphi_\alpha\,d\sigma \tag{3.88}
$$

[3] We recall that $\mathcal{C}^*_\Gamma$ is defined by (3.51).

(3.87) is then written:

$$\boxed{(-\omega^2 + \omega_\alpha^2)\tau_\alpha\mu_\alpha = \omega_\alpha^2 \int_\Sigma \rho_F u_N \varphi_\alpha \, d\sigma} \tag{3.89}$$

Referring to (3.80), φ is written:

$$\boxed{\varphi = \varphi^\infty + \sum_{\alpha \geq 1} \frac{\omega_\alpha^2}{-\omega^2 + \omega_\alpha^2} \frac{\int_\Sigma \rho_F u_N \varphi_\alpha \, d\sigma}{\mu_\alpha} \varphi_\alpha} \tag{3.90}$$

Remark

The steps used to establish (3.90) are formally equivalent to the following heuristic procedure:
We substitute the development:

$$\boxed{\varphi = \varphi^\infty + \sum_{\alpha \geq 1} \tau_\alpha \varphi_\alpha} \tag{3.91}$$

into the variational property (3.65) and we successively put $\delta\varphi = \varphi_1, \varphi_2, \ldots,$ then we use the orthogonality relations (3.34cd) and the property (3.88).

Interpretation of φ^∞

From (3.90), φ^∞ can be considered as the limit, for $\omega \longrightarrow \infty$, of the harmonic response of the liquid in the following sense:

$$\varphi^\infty = \lim_{n \to \infty} \lim_{\omega \to \infty} \varphi^{n,\omega} \tag{3.92}$$

where $\varphi^{n,\omega}$ denotes the approximation of the harmonic response φ on the basis truncated to the first n modes.

3.9 Impedance operator of the liquid

We have seen that a displacement u_N of Σ, generates, according to (3.9), a eulerian pressure fluctuation $p = \rho_F \omega^2 \varphi + p^0$.
In what follows, we are concerned with the *dynamic component* $p_{dyn} = \rho_F \omega^2 \varphi$ of p on Σ.
From (3.8), φ is linearly dependent on u_N. This solution will be denoted φ_{u_N}.
We then consider the virtual work of the dynamic pressure forces $\int_\Sigma p_{dyn} \delta u_N \, d\sigma$.
Replacing p_{dyn} by $\rho_F \omega^2 \varphi$, we define the following bilinear form in $(u_N, \delta u_N)$:
$\omega^2 \int_\Sigma \rho_F \varphi_{u_N} \delta u_N \, d\sigma$. We then go on to analyze this operator by using the basis of the normal modes, together with φ^0 and φ^∞. The results obtained will be used in chapter 9 for the modal analysis of the vibrations of an elastic structure containing a liquid, in the presence of gravity.

Symmetry properties

We show that:

$$\int_{\Sigma} \rho_F \varphi_{u_N} \delta u_N \, d\sigma = \int_{\Sigma} \rho_F \varphi_{\delta u_N} u_N \, d\sigma \tag{3.93}$$

Letting $\delta\varphi = \varphi_{\delta u_N}$ in the variational property (3.65) of φ_{u_N}, we obtain, after multiplication by ρ_F:

$$\boxed{\int_{\Omega_F} \rho_F \nabla \varphi_{u_N} \cdot \nabla \varphi_{\delta u_N} \, dx - \frac{\omega^2}{g} \int_{\Gamma} \rho_F \varphi_{u_N} \varphi_{\delta u_N} \, d\sigma = \int_{\Sigma} \rho_F u_N \varphi_{\delta u_N} \, d\sigma} \tag{3.94}$$

which reveals the stated symmetry property and enables the bilinear symmetric form, $\mathcal{M}_B^{\omega}(u_N, \delta u_N)$ called the "impedance operator" or "dynamic mass" of the liquid, to be defined for a given ω as follows:

$$\boxed{\mathcal{M}_B^{\omega}(u_N, \delta u_N) = \int_{\Sigma} \rho_F \varphi_{u_N} \delta u_N \, d\sigma} \tag{3.95}$$

Mass operator $\mathcal{M}_B^0$

The solution (3.10), which is linearly dependent on u_N, will be denoted $\varphi_{u_N}^0$ in what follows (the variational formulation of (3.10) is obtained, for example, by letting $\omega = 0$ in (3.65)).
For $\omega = 0$, (3.94) defines the bilinear form of "hydrostatic" mass $\mathcal{M}_B^0$ (positive definite):

$$\boxed{\mathcal{M}_B^0(u_N, \delta u_N) = \int_{\Sigma} \rho_F \varphi^0 \delta u_N \, d\sigma = \int_{\Omega_F} \rho_F \nabla \varphi_{u_N}^0 \cdot \nabla \varphi_{\delta u_N}^0 \, dx} \tag{3.96}$$

Discretization of $\mathcal{M}_B^0$

We wish to express $\varphi_{u_N}^0$ in terms of u_N, in order to calculate the discretized form of (3.96).
We know that for a given $\boldsymbol{U}$, $\boldsymbol{\Phi}^0$ is obtained by solving the linear equation (3.67) for $\omega = 0$, which is written $\boldsymbol{F\Phi} = \boldsymbol{C}^T\boldsymbol{U}$ with $\boldsymbol{L}^T\boldsymbol{\Phi} = 0$.
— A first method consists of writing $\boldsymbol{L}^T\boldsymbol{\Phi} = L_1\phi_1 + \cdots + L_N\phi_N$, and to *take the constraint explicitly into account* by doing the change of variables which consists of expressing a nodal value — for example ϕ_1, assuming $L_1 \neq 0$ — in terms of other nodal values.
This is equivalent to letting $\boldsymbol{\Phi} = \boldsymbol{H\Phi}'$, where $\boldsymbol{\Phi}'$ is an N component vector, with:

$$\boldsymbol{H} = \left[\begin{array}{c|cccc} -\frac{L_2}{L_1} & -\frac{L_3}{L_1} & \cdots & & -\frac{L_N}{L_1} \\ \hline \vdots & \ddots & & & \\ 0 & & \boldsymbol{I} & & \\ \vdots & & & & \ddots \end{array}\right] \tag{3.97}$$

$\boldsymbol{F\Phi} = \boldsymbol{C}^T\boldsymbol{U}$ is then written in the following equivalent form:

$$\underbrace{\boldsymbol{H}^T\boldsymbol{F}\boldsymbol{H}}_{\boldsymbol{F}'}\,\boldsymbol{\Phi}' = \underbrace{\boldsymbol{H}^T\boldsymbol{C}^T}_{\boldsymbol{C}'^T}\boldsymbol{U} \tag{3.98}$$

From (3.35a), the quadratic form corresponding to $\mathcal{M}_B^0$ is written in the discretized form $\rho_F\boldsymbol{\Phi}^T\boldsymbol{F\Phi}$ where $\boldsymbol{\Phi}$ is a function of $\boldsymbol{U}$, from which we have the expression for the hydrostatic mass matrix: $\boldsymbol{M}_B^0 = \rho_F\boldsymbol{C}'\boldsymbol{F}'^{-1}\boldsymbol{F}\boldsymbol{F}'^{-1}\boldsymbol{C}'^T$.

— Another method consists of replacing condition (3.10d) — which here plays the role of unicity condition — in the boundary value problem (3.10), by any condition of the type $l(\varphi) = 0$ [4]. We introduce the space $\mathcal{C}$ of the smooth functions in Ω_F and the space $\mathcal{C}^l \subset \mathcal{C}$ of the functions verifying the constraint $l(\varphi) = 0$.
Applying the test-functions method to this problem, we obtain the variational property of $\varphi \in \mathcal{C}^l$, verified for all $\delta\varphi \in \mathcal{C}$:

$$\int_{\Omega_F} \nabla\varphi\cdot\nabla\delta\varphi\, dx = \int_\Sigma u_N\delta\varphi\, d\sigma - \frac{1}{\text{Area}(\Gamma)}\left(\int_\Sigma u_N\, d\sigma\right)\left(\int_\Gamma \delta\varphi\, d\sigma\right) \tag{3.99}$$

We note that (3.99) is trivially verified by $\delta\varphi$ = constant. Consequently, the variational formulation can be restricted to $\mathcal{C}^l$, considering $\varphi \in \mathcal{C}^l$ and $\delta\varphi \in \mathcal{C}^l$. [5]
In discretized form, taking as the constraint the cancellation of the first nodal value ϕ_1 of $\boldsymbol{\Phi} = (\phi_1, \boldsymbol{\Phi}_*)$, (3.99) is written:

$$\delta\boldsymbol{\Phi}^T\boldsymbol{F\Phi} = \delta\boldsymbol{\Phi}^T\left(\boldsymbol{C}^T\boldsymbol{U} - \boldsymbol{R}^T\boldsymbol{U}\right) \text{ with } \phi_1 = 0, \text{ and } \delta\phi_1 = 0 \tag{3.100}$$

where $\delta\boldsymbol{\Phi}^T\,\boldsymbol{R}^T\boldsymbol{U}$ discretizes $\frac{1}{\text{Area}(\Gamma)}\left(\int_\Sigma u_N\, d\sigma\right)\left(\int_\Gamma \delta\varphi\, d\sigma\right)$.
We deduce from this $\boldsymbol{\Phi}_* = \boldsymbol{F}_*^{-1}\left(\boldsymbol{C}_*^T - \boldsymbol{R}_*^T\right)\boldsymbol{U}$, where the matrices $\boldsymbol{C}_*$, $\boldsymbol{R}_*$, and $\boldsymbol{F}_*$ are obtained by eliminating on the one hand the row of $\boldsymbol{C}^T$ and of $\boldsymbol{R}^T$, and on the other hand the row and column of $\boldsymbol{F}$, which corresponds to ϕ_1.
From (3.35a), the quadratic form corresponding to $\mathcal{M}_B^0$ can be written in discretized form $\rho_F\boldsymbol{\Phi}^T\boldsymbol{F\Phi} = \rho_F\boldsymbol{\Phi}_*^T\boldsymbol{F}_*\boldsymbol{\Phi}_*$, where $\boldsymbol{\Phi}_*$ is a function of $\boldsymbol{U}$.
By letting $\boldsymbol{E}_* = \boldsymbol{C}_* - \boldsymbol{R}_*$, we finally arrive at the following expression for the *hydrostatic* mass matrix:

$$\boldsymbol{M}_B^0 = \rho_F\boldsymbol{E}_*\boldsymbol{F}_*^{-1}\boldsymbol{E}_*^T \tag{3.101}$$

We note that this matrix can be obtained by applying the Gauss elimination algorithm (1.64) to the matrix $\begin{bmatrix} 0 & \boldsymbol{E}_* \\ \boldsymbol{E}_*^T & -\boldsymbol{F}_* \end{bmatrix}$.

---

[4] We note that in the case of the harmonic boundary value problem (3.8), such a substitution must be accompanied by the following modification of (3.8c) which results from the general relation (3.6) between p, φ and u_N: $\frac{\partial\varphi}{\partial z} = \frac{\omega^2}{g}(\varphi - \frac{1}{\text{Area}(\Gamma)}\int_\Gamma \varphi\, d\sigma) - \frac{1}{\text{Area}(\Gamma)}\int_\Sigma u_N\, d\sigma$.

[5] $\varphi \in \mathcal{C}$, can be uniquely written in the form $\varphi = \varphi^l$ + constant, with $\varphi^l \in \mathcal{C}^l$, as can be seen from $\varphi^l = \varphi - l(\varphi)/l(1)$, avec $l(1) \neq 0$.

Mass operator $\mathcal{M}_B^\infty$

The solution of (3.81), which is linearly dependent on u_N will be denoted $\varphi_{u_N}^\infty$ in what follows. We then define the bilinear form $\mathcal{M}_B^\infty = \int_\Sigma \varphi_{u_N}^\infty u_N \, d\sigma$ whose symmetry results from the following variational property of the solution φ^∞ of (3.81) (after multiplication by ρ_F):

$$\int_{\Omega_F} \rho_F \nabla \varphi_{u_N}^\infty \cdot \nabla \delta\varphi \, dx = \int_\Sigma \rho_F u_N \delta\varphi \, d\sigma \,, \quad \varphi_{u_N}^\infty \in \mathcal{C}^0, \, \forall \delta\varphi \in \mathcal{C}^0 \tag{3.102}$$

where $\mathcal{C}^0 = \{\varphi \mid \varphi = 0 \text{ on } \Gamma\}$.
$\mathcal{M}_B^\infty$ is thus expressed: [6]

$$\boxed{\mathcal{M}_B^\infty(u_N, \delta u_N) = \int_\Sigma \rho_F \varphi_{u_N}^\infty \delta u_N \, d\sigma = \int_{\Omega_F} \rho_F \nabla \varphi_{u_N}^\infty \cdot \nabla \varphi_{\delta u_N}^\infty \, dx} \tag{3.103}$$

Modal decompositions of $\mathcal{M}_B^\omega$

First decomposition

On substituting (3.78) into (3.95), we find:

$$\mathcal{M}_B^\omega(u_N, \delta u_N) = \mathcal{M}_B^0(u_N, \delta u_N) + \sum_{\alpha \geq 1} \frac{\omega^2}{-\omega^2 + \omega_\alpha^2} \mathcal{M}_\alpha(u_N, \delta u_N) \tag{3.104}$$

where $\mathcal{M}_\alpha$ is the *modal sloshing mass* operator defined by:

$$\boxed{\mathcal{M}_\alpha(u_N, \delta u_N) = \frac{1}{\mu_\alpha} \left(\int_\Sigma \rho_F \varphi_\alpha u_N \, d\sigma \right) \left(\int_\Sigma \rho_F \varphi_\alpha \delta u_N \, d\sigma \right)} \tag{3.105}$$

Second decomposition

On substituting (3.90) into (3.95), we find:

$$\boxed{\mathcal{M}_B^\omega(u_N, \delta u_N) = \mathcal{M}_B^\infty(u_N, \delta u_N) + \sum_{\alpha \geq 1} \frac{\omega_\alpha^2}{-\omega^2 + \omega_\alpha^2} \mathcal{M}_\alpha(u_N, \delta u_N)} \tag{3.106}$$

Summation rule for modal sloshing masses

Putting $\omega = 0$ in (3.106) we find:

$$\boxed{\mathcal{M}_B^0 - \mathcal{M}_B^\infty = \sum_{\alpha \geq 1} \mathcal{M}_\alpha} \tag{3.107}$$

which shows that the serie of modal sloshing masses converges.

---

[6] It is sufficient to use the preceding variational property for $\delta\varphi = \varphi_{\delta u_N}^\infty$. We shall see that $\mathcal{M}_B^\infty$ coincides with the added mass operator $\mathcal{M}_A$ introduced in chapter 5, in the study of hydroelastic vibrations without gravity (*cf.* equation (5.25).

Alternative interpretation of the $\mathcal{M}_\alpha$ summation rule

We let $\chi = \varphi^0_{u_N} - \varphi^\infty_{u_N}$. From (3.10) and (3.81), χ verifies the following boundary value problem:

$$\begin{array}{|rcllr|} \Delta\chi & = & 0 & \text{in } \Omega_F & (a) \\ \dfrac{\partial\chi}{\partial n} & = & 0 & \text{on } \Sigma & (b) \\ \dfrac{\partial\chi}{\partial z} & = & \eta^0_{u_N} - \eta^\infty_{u_N} & \text{on } \Gamma & (c) \\ \displaystyle\int_\Gamma \chi\, d\sigma & = & 0 & & (d) \end{array} \tag{3.108}$$

where η^0 is the elevation (which is here constant) of the free surface corresponding to $\varphi^0_{u_N}$ (*cf.* (3.11)), and where $\eta^\infty = \frac{\partial\varphi^\infty_{u_N}}{\partial z}$.
χ is a harmonic function with a zero normal derivative on Σ. It can therefore be expanded on the basis formed by the sloshing modes.
To obtain this expansion, it is sufficient to let $\omega = 0$ in (3.90), giving:

$$\boxed{\chi = \varphi^0_{u_N} - \varphi^\infty_{u_N} = \sum_{\alpha\geq 1} \frac{1}{\mu_\alpha}\left(\int_\Sigma \rho_F \varphi_\alpha u_N\, d\sigma\right)\varphi_\alpha} \tag{3.109}$$

By calculating $\int_{\Omega_F} \rho_F |\nabla\chi|^2\, dx = \int_{\Omega_F} \rho_F |\nabla(\varphi^0_{u_N} - \varphi^\infty_{u_N})|^2\, dx$, we obtain:

$$\int_{\Omega_F} \rho_F |\nabla\chi|^2\, dx = \mathcal{M}^0_B(u_N, u_N) - \mathcal{M}^\infty_B(u_N, u_N) \tag{3.110}$$

which involves the quadratic forms associated with the bilinear forms $\mathcal{M}^0_B$ and $\mathcal{M}^\infty_B$ introduced above.
Using (3.110) and the definition (3.105) of the modal mass operator, we then find the summation rule (3.107) for the corresponding quadratic forms.

Summation rule for sloshing stiffnesses

We calculate the modal development of the potential energy $\int_\Gamma \rho_F g |\frac{\partial\chi}{\partial z}|^2\, d\sigma$ (*cf.* (3.29)).
We first note the relation:

$$\int_\Gamma \rho_F g |\eta^\infty - \eta^0|^2\, d\sigma = \int_\Gamma \rho_F g |\eta^\infty|^2\, d\sigma - \int_\Gamma \rho_F g |\eta^0|^2\, d\sigma \tag{3.111}$$

This relation is established by using the identity $(a-b)^2 = a^2 - b^2 - 2b(a-b)$ with $a = \frac{\partial\varphi^\infty_{u_N}}{\partial z}$ and $b = \frac{\partial\varphi^0_{u_N}}{\partial z}$, then applying Green's formula (3.16), and finally using equations (3.10) verified by $\varphi^0_{u_N}$, and (3.81) verified by $\varphi^\infty_{u_N}$.

Taking the normal derivative of (3.109) on Γ, we obtain the modal development of $\frac{\partial \chi}{\partial z}$: [7]

$$\frac{\partial \chi}{\partial z} = \eta^0 - \eta^\infty = \sum_{\alpha \geq 1} \frac{1}{\mu_\alpha} \left(\int_\Sigma \rho_F \varphi_\alpha u_N \, d\sigma \right) \eta_\alpha \tag{3.112}$$

where $\eta_\alpha = \frac{\partial \varphi_\alpha}{\partial z}$.
Substituting (3.112) into the first member of (3.111), we obtain:

$$\boxed{\mathcal{K}_B^\infty(u_N, u_N) \; - \; \mathcal{K}_B^0(u_N, u_N) \; = \; \sum_{\alpha \geq 1} \mathcal{K}_\alpha} \tag{3.113}$$

where $\mathcal{K}_B^\infty(u_N, u_N) = \int_\Gamma \rho_F g |\eta^\infty|^2 \, d\sigma$ and $\mathcal{K}_B^0(u_N, u_N) = \int_\Gamma \rho_F g |\eta^0|^2 \, d\sigma$, and where the "modal sloshing stiffness" operator of the αth sloshing eigenmode is defined as follows:

$$\mathcal{K}_\alpha = \omega_\alpha^2 \mathcal{M}_\alpha \tag{3.114}$$

Remark

We note that, since ω_α^2 is a positive increasing series tending to infinity, the convergence of the $\mathcal{K}_\alpha$ series implies a *fast* convergence of the serie of sloshing masses $\mathcal{M}_\alpha$ (*cf.* e.g. Morand [143]).

Special case of rigid reservoirs

1. In the case of a vertical translation $u = i_z$ of the reservoir, we can verify that $\int_\Sigma u_N \varphi \, d\sigma = 0$.

   To do this, it is sufficient, in a first step, to apply Green's formula (3.75) with $u = \varphi_\alpha$ and $v = z$. Then, in a second step, we take into account of equations (3.15). As a result, in the linear approximation, the vibratory response of the liquid is null (translational movement of the liquid as a whole).

   In other words, only rotational or horizontal translational movements, induce vibratory deformations of the free surface.

2. In the special case of motions of an axisymmetric reservoir in a meridian plane, the impedance (3.106) can be identified with that of a rigid body coupled to a set of simple pendulums:
   — the inertia matrix of this body is the restriction of $\mathcal{M}_B^\infty(u_N, u_N)$ to rigid body movements,
   — the length of each pendulum is defined by the frequency of each mode. Its point of suspension on the axis of revolution, and the mobile mass, result from identifying the operator $\mathcal{M}_\alpha$, restricted to rigid body motions, with the similar expressions for a pendulum.

---

[7] This serie expansion can be directly established from an expansion of η on the basis formed by the eigenvectors $\varphi_\alpha|_\Gamma$. In the next chapter, with surface tension effects, we shall formulate the sloshing problem in terms of η.

Models of this type are still used in dynamic bending analysis of launchers, for attitude control.

3.10 Open problems

Among the problems under investigation are:

- convergence analysis of the modal methods presented above,
- numerical study of weakly non-linear vibrations: oscillations of finite amplitude, parametric resonance phenomena in forced oscillations of walls, especially vertical or rotational excitation (*cf.* Luke [125], Nakayama & Washizu [156], Fox & Kuttler [73], Holmes [97], Meserole & Fortini [133], Miles & Henderson [134]),
- damping due to the presence of "anti-sloshing" devices creating localized dissipation phenomena within the fluid.

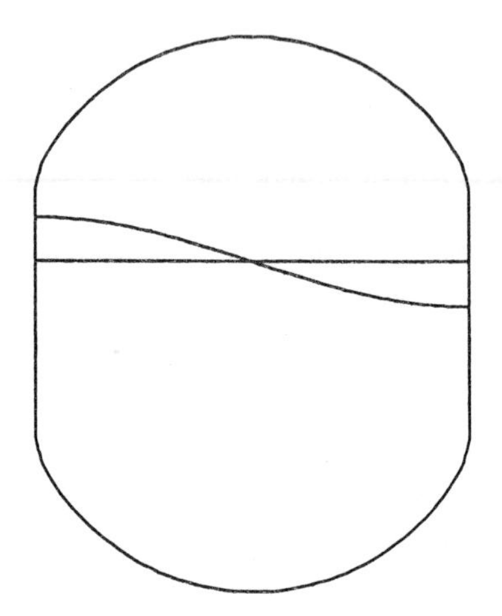

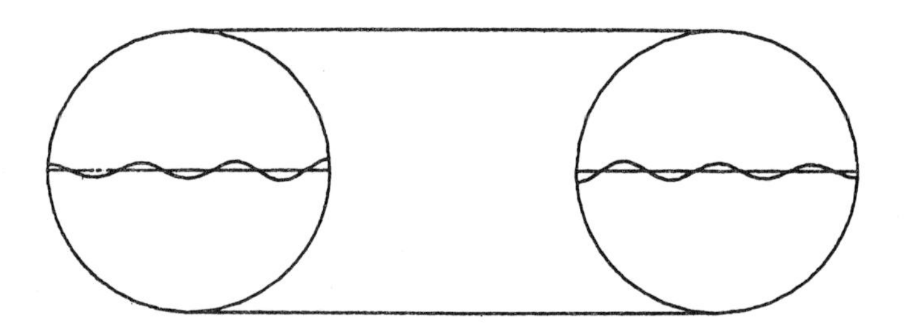

Example of sloshing modes: free surface deformation of the first sloshing mode $n = 1$ of the liquid oxygen tank of the launcher Ariane 5, and the 7th. mode $n = 1$ of a toroidal tank partially filled with liquid.

CHAPTER 4

Sloshing under surface tension

4.1 Introduction

We are interested in the harmonic vibrations of an inviscid incompressible liquid contained in a reservoir, having a free surface Γ, in the presence of both surface tension and gravity.
An example is the vibration modes of liquid ergols in weightless conditions involved in the problems of control of satellites stabilized along the three axes.
First of all, we establish the conditions of linearized free surface and contact angle, using surface theory.
We next pose the boundary value problem governing the response of a liquid to an arbitrary harmonic deformation of the wall, in terms of the potential of the displacements φ and the normal displacement field to the free surface η.
Thirdly, we establish the variational formulation in (φ, η) of the spectral boundary value problem of the sloshing modes of the liquid in a rigid motionless reservoir, together with the matrix equations resulting from a finite element discretization.
Finally, the formulation of this problem in terms of η on Γ, is used to compare the eigenfrequencies of reservoirs of various shapes and for the modal analysis of the reponse to prescribed motions of the wall.

4.2 Review of capillarity theory

Laplace's law and the contact angle condition

Laplace's law — We denote by Γ the surface of separation of a liquid medium "1" (occupying the domain Ω_F) and a gaseous medium "2", and by R_α and R_β the principle radii of curvature of Γ, taken as positive if the centers of curvature are located in the liquid medium "1".
Laplace's law expresses the discontinuity in pressure $[\![P]\!] = P^{(1)} - P^{(2)}$ at the crossover of Γ:

$$\boxed{[\![P]\!] = \sigma\left(\frac{1}{R_\alpha} + \frac{1}{R_\beta}\right)} \tag{4.1}$$

where σ is the surface tension constant [1] characteristic of the interface between the mediums "1" and "2" under consideration (*cf.* Landau & Lifchitz [115]).

Contact angle— We denote by γ the *line of contact* of three mediums, liquid "1", gas "2" and solid "3". We denote by Σ_L (and respectively Σ_G) the liquid–solid and gas–solid contact surface. We assume that $\Sigma = \Sigma_L \cup \Sigma_G$ is a smooth surface.

Figure 4.1 represents a section of the domains by the plane normal to γ. Denoting by σ_{13}, σ, and σ_{23} the coefficients of surface tension associated with the surfaces of separation between the various mediums considered, the condition of constancy of angle of contact — i.e. the dihedral angle formed by the tangent planes to Γ and Σ along γ — are written [2]:

$$\boxed{\cos\delta = \frac{\sigma_{23} - \sigma_{13}}{\sigma}} \tag{4.2}$$

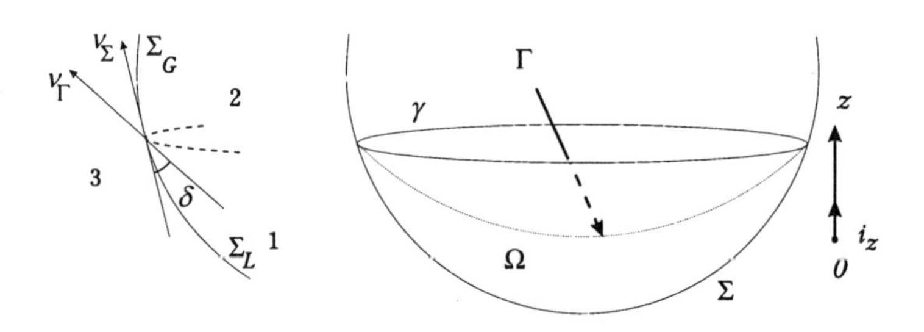

Figure 4.1: Equilibrium configuration

Equilibrium condition of a free surface in the presence of gravity

We consider an incompressible liquid at equilibrium in a rigid reservoir in a quasi-inertial reference frame (we assume the Coriolis terms to be negligible), in contact with a gas at a pressure $P^{(2)}$ assumed constant. We denote by P_0 the equilibrium value of $P^{(1)}$, which verifies the equation $\nabla P_0 = -\rho_F g\, i_z$, then $P_0 = -\rho_F gz + \text{constant}$ with $z = i_z \cdot \overrightarrow{OM}$. Laplace's law (4.1) then leads to the equation:

$$\boxed{\sigma\left(\frac{1}{R_\alpha} + \frac{1}{R_\beta}\right) + \rho_F g\, z = \text{constant} \quad \text{on } \Gamma} \tag{4.3}$$

[1] For a viscous fluid with σ non constant, *cf.* Nezit [161].

[2] The physical data σ, σ_{13} and σ_{23} are assumed to verify $\sigma > |\sigma_{23} - \sigma_{13}|$.

Review of surface theory

We consider a surface $\mathbb{S}$ with orthogonal coordinates α, β (Fig. 4.2(a)). We introduce the unitary vectors i_α, i_β defined by $\frac{\partial M}{\partial \alpha} = A i_\alpha$ and $\frac{\partial M}{\partial \beta} = B i_\beta$. The unitary normal to $\mathbb{S}$ is defined by $n = i_\alpha \times i_\beta$. We thus have $dM = \frac{\partial M}{\partial \alpha} d\alpha + \frac{\partial M}{\partial \beta} d\beta = A i_\alpha d\alpha + B i_\beta d\beta$. Furthermore, if we denote the element of area by $d\sigma$, we can write $n\, d\sigma = (\frac{\partial M}{\partial \alpha} \times \frac{\partial M}{\partial \beta}) d\alpha d\beta$.
Now, if η is a scalar field defined on $\mathbb{S}$, and $u_\| = u_\alpha i_\alpha + u_\beta i_\beta$ is a vector field tangential to $\mathbb{S}$, we can introduce the *surface gradient* vector $\nabla_S \eta$, and the *surface divergence* of $u_\|$ defined respectively by:

$$\nabla_S \eta = \frac{1}{A}\frac{\partial \eta}{\partial \alpha} i_\alpha + \frac{1}{B}\frac{\partial \eta}{\partial \beta} i_\beta \quad div_S u_\| = \frac{1}{AB}\left(\frac{\partial (u_\alpha B)}{\partial \alpha} + \frac{\partial (u_\beta A)}{\partial \beta}\right) \tag{4.4}$$

For a surface $S \in \mathbb{S}$ of boundary ∂S, we have *Stokes' formula*:

$$\int_S div_S u_\| \, d\sigma = \oint_{\partial S = \gamma} u_\| \cdot \nu \, dl \tag{4.5}$$

where dl is the length element, ν the normal external to the curve ∂S situated in the plane tangential to $\mathbb{S}$ (geodesic normal), ∂S being oriented by the tangent vector t such that (t, n, ν_Γ) form a positive trihedral (Fig. 4.2(b)). The tensor of

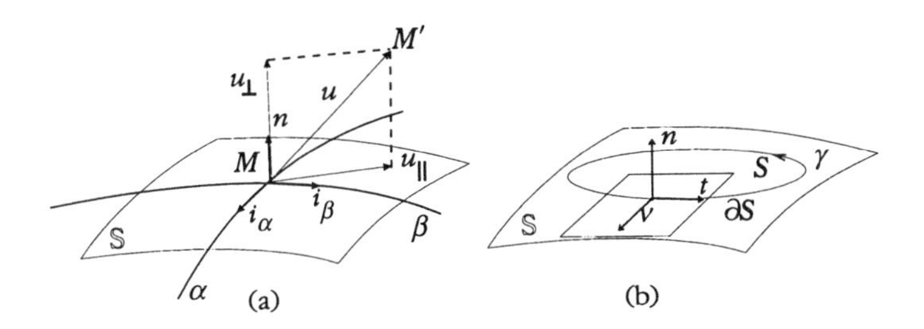

Figure 4.2: Surface coordinates

curvature $\widehat{K} = \frac{\partial n}{\partial M}$ of the tangent plane in M at $\mathbb{S}$ is defined by $dn = \dfrac{\partial n}{\partial M} dM$ $= \frac{\partial n}{\partial \alpha} d\alpha + \frac{\partial n}{\partial \beta} d\beta$.
We assume in what follows that (α, β) is a system of Gauss orthogonal coordinates, which enables us to write $\frac{\partial n}{\partial M} i_\alpha = \frac{1}{R_\alpha} i_\alpha$ and $\frac{\partial n}{\partial M} i_\beta = \frac{1}{R_\beta} i_\beta$, where R_α and R_β are the principle radii of curvature in M. In other words, the tensor of curvature $\frac{\partial n}{\partial M}$ is represented in the coordinates i_α, i_β by the matrix $\begin{bmatrix} \frac{1}{R_\alpha} & 0 \\ 0 & \frac{1}{R_\beta} \end{bmatrix}$.
The invariants $(1/2)\left(\frac{1}{R_\alpha} + \frac{1}{R_\beta}\right)$ and $\frac{1}{R_\alpha R_\beta}$ are known respectively as the mean curvature and the total curvature of the surface at M. We then have the classical relations:

$\frac{\partial n}{\partial \alpha} = \frac{A}{R_\alpha} i_\alpha$	$\frac{\partial i_\alpha}{\partial \alpha} = -\frac{1}{B}\frac{\partial A}{\partial \beta} i_\beta - \frac{A}{R_\alpha} n$	$\frac{\partial i_\beta}{\partial \alpha} = \frac{1}{B}\frac{\partial A}{\partial \beta} i_\alpha$
$\frac{\partial n}{\partial \beta} = \frac{B}{R_\beta} i_\beta$	$\frac{\partial i_\alpha}{\partial \beta} = \frac{1}{A}\frac{\partial B}{\partial \alpha} i_\beta$	$\frac{\partial i_\beta}{\partial \beta} = -\frac{1}{A}\frac{\partial B}{\partial \alpha} i_\alpha - \frac{B}{R_\beta} n$

(4.6)

For a detailed treatment of the surface theory, see for example Berger & Gostiaux [18], Cartan [30].

Potential energy

Consider the quantity W defined by: [3]

$$W = \sigma \mathcal{A}(\Gamma) + \sigma_{13}\mathcal{A}(\Sigma_L) + \sigma_{23}\mathcal{A}(\Sigma_G) + \rho_F g V Z \tag{4.7}$$

where $\mathcal{A}(\Gamma)$, $\mathcal{A}(\Sigma_L)$ and $\mathcal{A}(\Sigma_G)$ are the respective areas of Γ, Σ_L and Σ_G, V the total volume of liquid and Z the elevation of the center of gravity .

The minimum property — We shall verify that if equations (4.2) and (4.3) are satisfied at all points of a surface Γ of contour γ situated on Σ, the surface of Γ is a solution of the problem of minimisation of W, in the class of surfaces Γ', whose line boundary γ' is situated on Σ, and which delimit with Σ a *constant volume* V.

We denote by $W^{(1)}(u)$ and $V^{(1)}(u)$ the first order variations in u of W and V, and we introduce the linearized condition of the bearing of the edge γ of the free surface Γ on the wall Σ which is written $u \cdot n = 0$.

We thus have to verify that $W^{(1)}(u) = 0$ for all u satisfying the constraints $V^{(1)}(u) = 0$ and $u \cdot n = 0$ on Σ, or further, on introducing the Lagrange multiplier μ associated with the constraint $V^{(1)}(u) = 0$, that:

$$W^{(1)}(u) \; - \; \mu V^{(1)}(u) \; = \; 0 \quad \forall u \text{ verifying } u \cdot n = 0 \text{ on } \Sigma \tag{4.8}$$

From (4.7), the calculation of $W^{(1)}(u)$ involves the first order variations in u of $\mathcal{A}(\Gamma)$, $\mathcal{A}(\Sigma_L)$, $\mathcal{A}(\Sigma_G)$, Z and V, denoted respectively: $\mathcal{A}^{(1)}_{\Gamma}(u)$, $\mathcal{A}^{(1)}_{\Sigma_L}(u)$, $\mathcal{A}^{(1)}_{\Sigma_G}(u)$, $Z^{(1)}(u)$ and $V^{(1)}(u)$.

Calculation of $\mathcal{A}^{(1)}_{\Gamma}(u)$, $\mathcal{A}^{(1)}_{\Sigma_L}(u)$ and $\mathcal{A}^{(1)}_{\Sigma_G}(u)$ — We begin by calculating $\int_{S'} d\sigma' - \int_S d\sigma$, in terms of the field of displacement $u(M)$ defined at all points M of a surface S. For this, we introduce a *representation* of the "deformed" surface S' by means of $u(\alpha, \beta)$, where (α, β) are the Gauss coordinates — considered as "lagrangian coordinates" of S' (Fig. 4.2(a)). We therefore write:

$$M'(\alpha, \beta) \; = \; M(\alpha, \beta) \; + \; u(\alpha, \beta) \tag{4.9}$$

If n' denotes the unit normal to S', the expression for the area $d\sigma'$ of S' — as a function of $d\sigma$ according to (4.9) — in terms of the area element $d\sigma$ of S, is deduced from $n' d\sigma' = (\frac{\partial M'}{\partial \alpha} \times \frac{\partial M'}{\partial \beta}) d\alpha \, d\beta$ by replacing M' by $M + u$. We then obtain:

$$n' d\sigma' \; = \; (n \; + \; n^{(1)} \; + \; n^{(2)}) d\sigma \tag{4.10}$$

---

[3] W represents the ***total potential energy*** of the liquid. We note that $\rho_F V g Z$ is simply the classical potential energy of a liquid in a gravitational field. The other terms, which are proportional to the area of the surfaces of separation, are due to the surface tension (*cf.* Landau & Lifchitz [115]).

where $n^{(1)}$ and $n^{(2)}$ are the respectively linear and quadratic functionals of u expressed as:

$$n^{(1)} = \frac{\frac{\partial M}{\partial \alpha} \times \frac{\partial u}{\partial \beta} + \frac{\partial u}{\partial \alpha} \times \frac{\partial M}{\partial \beta}}{\|\frac{\partial M}{\partial \alpha} \times \frac{\partial M}{\partial \beta}\|} \quad (a) \quad , \quad n^{(2)} = \frac{\frac{\partial u}{\partial \alpha} \times \frac{\partial u}{\partial \beta}}{\|\frac{\partial M}{\partial \alpha} \times \frac{\partial M}{\partial \beta}\|} \quad (b) \qquad (4.11)$$

The expression (4.10) enables $d\sigma'$ to be related to $d\sigma$. Limiting the expansion to the second order in terms of u, we obtain:

$$d\sigma' = d\sigma \left[1 + n\cdot n^{(1)} + \frac{1}{2}\left(|n^{(1)}|^2 + 2n\cdot n^{(2)} - (n\cdot n^{(1)})^2\right)\right] \qquad (4.12)$$

From (4.11) and (4.12), the first order expansion $\mathcal{A}_S^{(1)}(u)$ of the variation of area is expressed $\int_\Gamma n\cdot n^{(1)} d\sigma$. To calculate $n^{(1)}$ defined by (4.11(a)), we decompose u according to $u = u_\| + (u\cdot n)n$, (Fig. 4.2(a)), and we use the relations (4.6).
We then obtain: $\mathcal{A}_S^{(1)}(u) = \int_S (\frac{1}{R_\alpha} + \frac{1}{R_\beta}) u\cdot n d\sigma + \int_S div_s u_\| d\sigma$.
By transforming the second term of $\mathcal{A}^{(1)}$ by Stokes' formula (4.5), we obtain finally: [4]

$$\mathcal{A}_S^{(1)}(u) = \int_S (\frac{1}{R_\alpha} + \frac{1}{R_\beta}) u\cdot n d\sigma + \oint_{\partial S} u\cdot\nu\, dl \qquad (4.13)$$

$\mathcal{A}_\Gamma^{(1)}(u)$ is given by (4.13) on replacing S by Γ and ∂S by γ.
For the calculation of $\mathcal{A}_{\Sigma_L}^{(1)}(u)$ and $\mathcal{A}_{\Sigma_G}^{(1)}(u)$, we first note that, since $\mathcal{A}(\Sigma_L) + \mathcal{A}(\Sigma_G) =$ Area $(\Sigma) =$ constant, we have $\mathcal{A}_{\Sigma_G}^{(1)}(u) = -\mathcal{A}_{\Sigma_L}^{(1)}(u)$. Using next the expression (4.13) of $\mathcal{A}_S^{(1)}(u)$ applied to $S = \Sigma_L \in \Sigma$ and $\partial S = \gamma$ with u tangential to Σ, we obtain $\mathcal{A}_{\Sigma_L}^{(1)}(u) = \oint_\gamma u\cdot\nu_\Sigma\, dl$, where ν_Σ is the normal to γ in the plane tangential to Σ, external to Σ_L (Fig. 4.1).

Calculation of $V^{(1)}(u)$ and $Z^{(1)}(u)$ — The present derivation is carried out within the theory of surfaces, and we adopt here a eulerian point of view for the volume and a lagrangian point of view for the boundary. We shall express the variation of volume and of elevation Z of the center of gravity of a domain Ω_F of boundary $\partial\Omega_F$ and volume V, in terms of the displacement u of this boundary. We denote by Ω'_F the final position of Ω_F and by V', the volume of Ω'_F.
We consider the manifold — denoted $\Omega'_F - \Omega_F$ — defined parametrically from $\partial\Omega_F$ and u by: $M' = M + \lambda u$, with $M \in \partial\Omega_F$ and $0 < \lambda < 1$: $\Omega'_F - \Omega_F$ corresponds to the "volume swept by "$\partial\Omega_F$" (Fig. 4.3). We can write:

$$Z'V' - ZV = i_z\cdot\int_{\Omega'_F} \overrightarrow{OM}\, dx - i_z.\int_{\Omega_F} \overrightarrow{OM}\, dx = i_z\cdot\int_{\Omega'_F - \Omega_F} \overrightarrow{OM}\, dx \qquad (4.14)$$

Noting that $\Omega'_F - \Omega_F$ can be identified to $\partial\Omega_F \times]0,1[$ and that $dx = \det(\frac{\partial M_\lambda}{\partial\alpha}, \frac{\partial M_\lambda}{\partial\beta}, \frac{\partial M_\lambda}{\partial\lambda}) d\alpha d\beta d\lambda$, we have:

$$Z'V' - ZV = i_z\cdot\int_0^1 d\lambda \int_{\partial\Omega_F} d\alpha d\beta M_\lambda(\alpha,\beta) \det\left|\frac{\partial M_\lambda}{\partial\alpha}, \frac{\partial M_\lambda}{\partial\beta}, \frac{\partial M_\lambda}{\partial\lambda}\right| \qquad (4.15)$$

[4] This expression can also be obtained from the material derivative of the area $\int_{\partial\Omega_F} d\sigma$, *cf.* Germain [83].

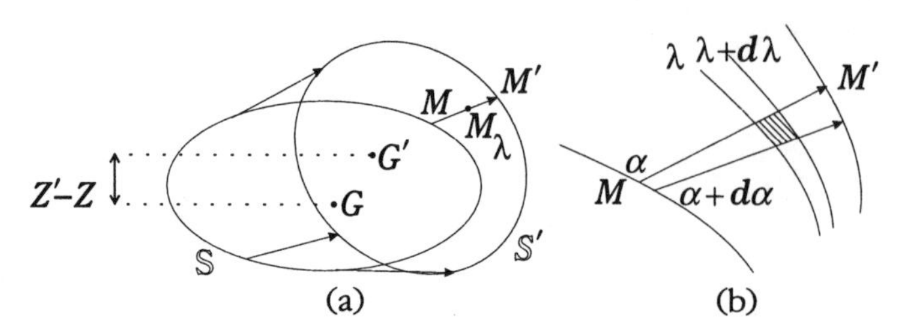

Figure 4.3: Swept volume

Replacing M_λ by its expression $M(\alpha,\beta)+\lambda u(\alpha,\beta)$ in (4.15), then retaining the term of order 1 of the development with respect to u, and finally carrying out the integration over λ, we find the following expression $(ZV)^{(1)}(u) = Z^{(1)}(u)\,V + Z\,V^{(1)}(u)$ of the first order variation in u of $Z\,V$:

$$Z^{(1)}(u)\,V + Z\,V^{(1)}(u) = \int_{\partial\Omega_F} z\,u\cdot n\,d\sigma \tag{4.16}$$

Starting with $V' - V = \int_{\Omega'_F} dx - \int_{\Omega_F} dx = \int_{\Omega'_F - \Omega_F} dx$, and using the above parametrization, we obtain: [5]

$$V^{(1)}(u) = \int_{\partial\Omega_F} u\cdot n\,d\sigma \tag{4.17}$$

Minimum of W — Applying the above results to the fluid domain Ω_F of boundary $\Sigma\cup\Gamma$, the minimization condition (4.8) yields the following condition, satisfied $\forall u$ verifying $u\cdot n = 0$ on Σ:

$$\begin{aligned}\int_\Gamma \rho_F g z u\cdot n\,d\sigma + \sigma\int_\Gamma\left(\frac{1}{R_\alpha}+\frac{1}{R_\beta}\right)u\cdot n d\sigma + \cdots\\ \cdots + \sigma\oint_\gamma u\cdot\nu_\Gamma dl + (\sigma_{13}-\sigma_{23})\oint_\gamma u\cdot\nu_\Sigma dl - \mu\int_\Gamma u\cdot n\,d\sigma = 0\end{aligned} \tag{4.18}$$

Considering first functions u vanishing on Σ, we find equation (4.3). Considering next an arbitrary u, and noting that u is colinear with ν_Σ on γ, we find the contact angle condition (4.2) (with $\nu_\Sigma\cdot\nu_\Gamma = \cos\delta$).

A study of the equilibrium capillary configurations can be found in Moiseev & Rumyantsev [135], Ekeland & Temam [62], Finn [72], Myshkis *and al* [154], Concus & Finn [42] (see also Sewell [199]).

[5] The expressions (4.16) and (4.17) can also be derived by three-dimensional lagrangian derivation. For example, for (4.16), we start with $Z'\,V' - Z\,V = i_z\cdot\int_{\Omega'_F}\overrightarrow{OM'}\,dx' - i_z\cdot\int_{\Omega_F}\overrightarrow{OM}\,dx$. We denote by J the Jacobian of the transformation, $M\to M' = M + u(M)$, it being assumed that u is defined at all points of Ω_F. After linearizing (noting that to the first order in u, $J\sim 1+divu$), and using $div(zu) = zdivu + u\cdot i_z$, we find (4.16).

4.3 Linearized free surface and contact angle conditions

This linearization is undertaken within a kinematic *representation* of the free surface by a field of normal displacement η. We propose:

1. to establish the relation between η and the eulerian fluctuation of the pressure p on Γ,

2. then to linearize the contact angle condition in the case of a moving wall.

Free surface condition

We denote by Γ' the instantaneous position of the free surface (Fig. 4.4(a)). Consider a point $M \in \Gamma$. We denote by M' the instantaneous position of a particle located at M at equilibrium. We consider also the point N' defined

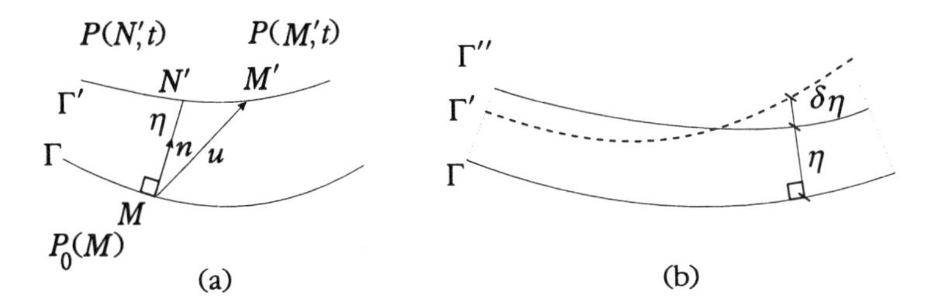

Figure 4.4: Free surface variation

geometrically as the intersection of the normal n at M to Γ with Γ', and we denote by η the distance MN'. We can thus write $N' = M + \eta(M)n$ and we introduce the variation of the pressure $p' = P(N', t) - P_0(M)$. We consider furthermore the lagrangian pressure fluctuation $p_{\mathcal{L}}(M, t) = P(M', t) - P_0(M)$, together with the eulerian fluctuation $p(M, t) = P(M, t) - P_0(M)$. We shall express p' in terms of p andy de η. The hypothesis of small motions enables us to write, to within the second order: $p_{\mathcal{L}} - p' = P(M', t) - P(N', t) \sim \nabla P \cdot \overrightarrow{N'M'} \sim \nabla P_0 \cdot (u - \eta n)$ (Fig. 4.4(a)). Using now the relation (2.7) between $p_{\mathcal{L}}$ and p, we find:

$$p' = p - \rho_F\, g\, (i_z \cdot n)\, \eta \tag{4.19}$$

From Laplace's law applied successively to $M \in \Gamma$ and to $N' \in \Gamma'$, we can, by difference, express p' — and therefore p — in terms of the variation of mean curvature of the free surface between M and N', which gives (assuming the pressure of the gas to be constant):

$$p = \rho_F g\, (i_z \cdot n)\eta + \sigma \left[\left(\frac{1}{R'_\alpha} + \frac{1}{R'_\beta}\right) - \left(\frac{1}{R_\alpha} + \frac{1}{R_\beta}\right) \right] \tag{4.20}$$

We now have to calculate the mean variation of curvature between the points M and N' in terms of η.
We shall deduce this expression by means of two derivations of the variation of area corresponding to a variation $\delta\eta$ of η. [6]
Consider the surface Γ'' — close to Γ' — corresponding to $\eta + \delta\eta$, with $\delta\eta \ll \eta$ (Fig. 4.4(b)).

1. We note that, to within the second order in $\delta\eta$, from (4.13), we have:

$$\mathcal{A}(\Gamma'') - \mathcal{A}(\Gamma') \sim \mathcal{A}^{(1)}(\delta\eta) = \int_{\Gamma'} (\frac{1}{R'_\alpha} + \frac{1}{R'_\beta})\delta\eta\, d\sigma'. \tag{4.21}$$

2. We shall now derive another expression of $\mathcal{A}(\Gamma'') - \mathcal{A}(\Gamma')$, from the expansion to the second order in η of $\mathcal{A}(\Gamma') - \mathcal{A}(\Gamma)$.

   This expansion is obtained by using (4.11) and (4.12) for a normal displacement field $u = \eta\, n$, which leads to:

$$\mathcal{A}(\Gamma') - \mathcal{A}(\Gamma) = \int_\Gamma (\frac{1}{R_\alpha} + \frac{1}{R_\beta})\eta d\sigma + \frac{1}{2}\int_\Gamma |\nabla_S \eta|^2\, d\sigma + \int_\Gamma \frac{\eta^2}{R_\alpha R_\beta}\, d\sigma \tag{4.22}$$

   Differentiating this last relation, we obtain the linearized expression in $\delta\eta$ of the variation in area between Γ' and Γ'' which can be transformed into an integral on Γ' by using the relation $d\sigma \sim d\sigma' \left[1 - \eta(\frac{1}{R_\alpha} + \frac{1}{R_\beta})\right]$ (*cf.* (4.12)). On carrying out all the calculations, we obtain:

$$\mathcal{A}(\Gamma'') - \mathcal{A}(\Gamma') = \int_{\Gamma'} \left[\nabla_S \eta \cdot \nabla_S \delta\eta - (\frac{1}{R_\alpha^2} + \frac{1}{R_\beta^2})\eta\delta\eta + (\frac{1}{R_\alpha} + \frac{1}{R_\beta})\delta\eta\right] d\sigma' \tag{4.23}$$

Integrating the first term by parts, and identifying the result obtained with (4.21), we find:

$$(\frac{1}{R'_\alpha} + \frac{1}{R'_\beta}) - (\frac{1}{R_\alpha} + \frac{1}{R_\beta}) = -(\frac{1}{R_\alpha^2} + \frac{1}{R_\beta^2})\eta - div_S \nabla_S \eta \tag{4.24}$$

From (4.20), the free surface condition can thus be written:

$$\boxed{p = \rho_F g\,(i_z \cdot n)\eta - \sigma\left[(\frac{1}{R_\alpha^2} + \frac{1}{R_\beta^2})\eta + div_S \nabla_S \eta\right] \quad \text{on } \Gamma} \tag{4.25}$$

Contact angle condition

We consider a plane Π perpendicular to γ (Fig. 4.5). We denote by M' the intersection Π with the instantaneous connecting line γ', and we let $u = \overrightarrow{MM'}$ (u is a vector in plane Π). We denote by ν_Σ and ν_Γ the geodesic normals at γ

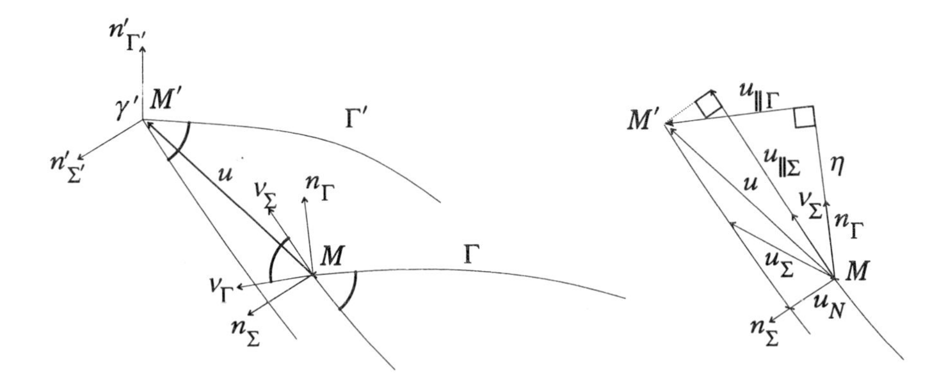

Figure 4.5: Contact angle variation

situated respectively in the tangent planes at Σ and Γ. We denote by $n'_{\Sigma'}$ the normal to the wall at M'. The contact angle condition (4.2) can then be written:

$$\boxed{n'_{\Sigma'}\cdot n'_{\Gamma'} \;=\; n_\Sigma\cdot n_\Gamma \;=\; \text{constant}} \tag{4.26}$$

We shall show that the linearization of (4.26) results in a relation between the normal displacement $\eta = u\cdot n_\Gamma$ of Γ, and the normal component $u_N = u_\Sigma\cdot n_\Sigma$ ($= u\cdot n_\Sigma$) of the displacement u_Σ of the wall.

We use the following expression for the first order variation in u of the unit normal to a surface (which can be established by means of relations (4.10), (4.11)) (Fig. 4.2(a)) :[7]

$$n' - n \sim dn = \widehat{K}(u_\|) - \nabla_S u_\perp \tag{4.27}$$

We denote respectively by $u_{\|\Gamma}, u_{\perp\Gamma} = \eta$ the projection of u onto the tangent plane at Γ, and its normal component onto n_Γ, and respectively by $u_{\|\Sigma}, u_{\perp\Sigma} = u_N$ the projection of u onto the tangent plane at Σ and its normal component onto n_Σ. $u_{\|\Gamma}$ and $u_{\|\Sigma}$ are expressed in terms of η and u_N as follows:

$$u_{\|\Gamma} \;=\; \frac{u_N + \eta\cos\delta}{\sin\delta}\nu_\Gamma \quad , \quad u_{\|\Sigma} \;=\; \frac{\eta + u_N\cos\delta}{\sin\delta}\nu_\Sigma \tag{4.28}$$

(these formulas are established by decomposing u successively in the bases n_Σ, ν_Σ and n_Γ, ν_Γ and noting that $u_{\|\Sigma}$ is colinear with ν_Σ and that $u_{\|\Gamma}$ is colinear with

[6] We note that this calculation only concerns the points $N' \in \Gamma'$ located on the normals to Γ. Using loose notation in the paragraph, we keep the notation Γ' for the set of points N'. The set of points N' actually differs from the deformed surface of an "infinitesimal strip" situated in the neighbourhood of γ which is not involved in the evaluation of the curvature variation.

[7] We note that dn is situated in the tangent plane at M, and differs from the quantity $n^{(1)}(u)$ introduced in (4.11).

ν_Γ). Differentiating (4.26), we obtain $dn_\Sigma \cdot n_\Gamma + n_\Sigma \cdot dn_\Gamma = 0$. On evaluating dn_Σ and dn_Γ in terms of u, by means of (4.27) we obtain:

$$[\widehat{K_\Sigma}(u_{\parallel\Sigma}) - \nabla_\Sigma u_{\perp\Sigma}]\cdot n_\Gamma + [\widehat{K_\Gamma}(u_{\parallel\Gamma}) - \nabla_\Gamma u_{\perp\Gamma}]\cdot n_\Sigma = 0 \tag{4.29}$$

Finally using (4.28), (4.29) leads to the following condition verified at all points of γ:

$$\boxed{\frac{\partial \eta}{\partial \nu_\Gamma} + \frac{\partial u_N}{\partial \nu_\Sigma} = -\frac{\eta + u_N \cos\delta}{\sin\delta} < K_\Sigma > + \frac{\eta\cos\delta + u_N}{\sin\delta} < K_\Gamma >} \tag{4.30}$$

where we have introduced the *curvatures* $< K_\Sigma >$ and $< K_\Gamma >$ of the sections Σ and Γ by the plane Π normal at γ defined by: [8]

$$< K_\Sigma > \sin\delta \;=\; -\widehat{K_\Sigma}(\nu_\Sigma)\cdot n_\Gamma \quad \text{et} \quad < K_\Gamma > \sin\delta \;=\; \widehat{K_\Gamma}(\nu_\Gamma)\cdot n_\Sigma \tag{4.31}$$

4.4 Harmonic response to a wall displacement u_N

We consider the harmonic response of an inviscid incompressible homogeneous liquid contained in a cavity occupying a bounded domain Ω_F, having a free surface Γ, in the presence of gravitational and surface tension forces and subject to a prescribed normal wall displacement u_N on Σ of circular frequency ω.

Boundary value problem in terms of (p, φ, η)

The boundary value problem in terms of p, φ and η, is written:

$$\boxed{\begin{array}{rcllr}
\nabla p &=& \rho_F \omega^2 \nabla\varphi & \text{in } \Omega_F & (a) \\
\Delta\varphi &=& 0 & \text{in } \Omega_F & (b) \\
\dfrac{\partial\varphi}{\partial n} &=& u_N & \text{on } \Sigma & (c) \\
\dfrac{\partial\varphi}{\partial n} &=& \eta & \text{on } \Gamma & (d) \\
p &=& \rho_F g\,(i_z\cdot n)\eta + \mathcal{L}(\eta) & \text{on } \Gamma & (e) \\
\dfrac{\partial\eta}{\partial\nu_\Gamma} &=& D\eta + E\,u_N - \dfrac{\partial u_N}{\partial\nu_\Sigma} & \text{on } \gamma & (f) \\
l(\varphi) &=& 0 & \text{with } l(1)\neq 0 & (g)
\end{array}} \tag{4.32}$$

[8] the sign $(-)$ in the definition of $< K >_\Sigma$ has been introduced so that the curvature of Σ is counted negatively when the center of curvature of Σ is situated in the liquid domain.

where $\mathcal{L}$ is defined by (*cf.* (4.25)):

$$\mathcal{L}(\eta) = -\sigma \left[(\frac{1}{R_1^2} + \frac{1}{R_2^2})\eta + div_S \nabla_S \eta \right] \quad (4.33)$$

where the functions D and E (given on γ) are defined by:

$D = \dfrac{< K >_\Gamma \cos\delta - < K_\Sigma >}{\sin\delta}$ and $E = \dfrac{< K_\Gamma > - < K >_\Sigma \cos\delta}{\sin\delta}$,

and where $l(\varphi)$ is an arbitrary linear relation for unicity of φ.

- (4.32a) results from the linearized Euler equation for displacements derived from a potential (equation (2.27)).
- (4.32b) results from the incompressibility of the fluid (equation (2.29)).
- (4.32c) is the contact condition.
- the "free surface kinematic condition" (4.32d) is obtained by letting $u = \nabla\varphi$ in $\eta = u \cdot n$.
- (4.32e) is the free surface condition (4.25), and (4.32f) is the contact angle condition (4.30) *in the presence of a moving* wall.

Solution for $\omega = 0$

We can in general verify whether the boundary value problem obtained for $\omega = 0$ is well-posed (if the equilibrium configuration is stable).
First of all we show that η^0, and then p^0, can be calculated in terms of u_N.

1. We begin by integrating (4.32b) in Ω_F, which, taking account of (4.32c) and (4.32d), yields:

$$\boxed{\int_\Sigma u_N \, d\sigma + \int_\Gamma \eta \, d\sigma = 0} \quad (4.34)$$

   which results from the invariance of the total volume of fluid.

2. Furthermore, for $\omega = 0$, (4.32a) can be written $\nabla p = 0$ which yields a constant value p^0 for p.

3. Equations (4.32ef) define a boundary value problem posed on Γ defining η^0 in terms of p^0 and u_N.

   As η^0 is linearly dependent on p^0 and u_N, we can write $\eta^0 = ap^0 + b(u_N)$, where a and $b(u_N)$ are, respectively, the solutions of the boundary value problem for $p^0 = 1, u_N = 0$ and for $p^0 = 0, u_N \neq 0$ respectively.

4. Finally, on substituting this expression of η^0 into the relation (4.34), we find $p^0 = -(\int_\Sigma u_N \, d\sigma + \int_\Gamma b(u_N) \, d\sigma)/(\int_\Gamma a \, d\sigma)$, and then $\eta^0 = ap^0 + b(u_N)$.

5. Secondly, we note that φ^0 is the solution to the Neumann problem (4.32bcdg), the condition for the existence of which results from the definition of η^0 (equation (4.34)).

Formulation in (η, φ)

The aim here is, by proceding in a way analagous to that for sloshing (chapter 3), on the one hand to eliminate p and on the other hand to arrive at a formulation which yields a well-posed boundary value problem for $\omega = 0$. Equations (4.32ag) are equivalent — according to the second definition of φ (*cf.* (2.26)) — to:

$$\boxed{\begin{array}{lr} p = \rho_F \omega^2 \varphi + \pi & (a) \\ l(\varphi) = 0 \quad \text{with } l(1) \neq 0 & (b) \end{array}} \tag{4.35}$$

On replacing p by its expression (4.35) in (4.32e), we obtain a boundary value problem in (φ, η). This formulation introduces an additional constant unknown π, which — unlike the case of sloshing (*cf.* (3.5)) — cannot be expressed explicitly in terms of φ and u_N.

$$\boxed{\begin{array}{rcllr} \Delta\varphi & = & 0 & \text{in } \Omega_F & (a) \\ \dfrac{\partial\varphi}{\partial n} & = & u_N & \text{on } \Sigma & (b) \\ \dfrac{\partial\varphi}{\partial n} & = & \eta & \text{on } \Gamma & (c) \\ l(\varphi) & = & 0 & \text{with } l(1) \neq 0 & (d) \\ \rho_F \omega^2 \varphi + \pi & = & \rho_F g \, (i_z \cdot n) \eta + \mathcal{L}(\eta) & \text{on } \Gamma & (e) \\ \dfrac{\partial\eta}{\partial\nu_\Gamma} & = & D\eta + E\, u_N - \dfrac{\partial u_N}{\partial\nu_\Sigma} & \text{on } \gamma & (f) \end{array}} \tag{4.36}$$

Solution for $\omega = 0$ — We verify that the boundary value problem (4.36) is well-posed for $\omega = 0$.

1. Integrating (4.36a) in Ω_F and taking account of (4.36bc), we find $\int_\Sigma u_N \, d\sigma + \int_\Gamma \eta \, d\sigma = 0$ (equation (4.34)).

2. Pour $\omega = 0$, (4.36e) and (4.36f) constitute a boundary value problem enabling η^0 to be calculated in terms of π^0 and u_N. We can then, using a derivation identical to that used in the formulation in p^0, η^0, φ^0 discussed previously, calculate π^0 in terms of u_N using (4.34).

3. Finally, equations (4.36abcd), uniquely define φ^0 in terms of u_N.

4.5 (η, φ) sloshing modes formulation

We are concerned with the solutions η, φ, π of the spectral problem deduced from (4.36) for $u_N = 0$. Putting $\omega^2 = \lambda$, we obtain:

$$\boxed{\begin{array}{rcllr} \Delta\varphi & = & 0 & \text{in } \Omega_F & (a) \\ \dfrac{\partial\varphi}{\partial n} & = & 0 & \text{on } \Sigma & (b) \\ \dfrac{\partial\varphi}{\partial n} & = & \eta & \text{on } \Gamma & (c) \\ l(\varphi) & = & 0 & \text{with } l(1) \neq 0 & (d) \\ \rho_F\lambda\varphi + \pi & = & \rho_F g\,(i_z\!\cdot\! n)\eta + \mathcal{L}(\eta) & \text{on } \Gamma & (e) \\ \dfrac{\partial\eta}{\partial\nu_\Gamma} & = & D\eta & \text{on } \gamma & (f) \end{array}} \tag{4.37}$$

Variational formulation in (η, φ, π)

We consider a solution η, φ, π of equations (4.37) of equations corresponding to an eigenvalue λ. We procede in two steps by the test-functions method.

1. In the first step, we introduce the subspace $\mathcal{C}_\eta$ of the smooth functions $\delta\eta$ defined on Γ. Multiplying (4.37e) (with definition (4.33) of $\mathcal{L}$) by an arbitrary test-function $\delta\eta \in \mathcal{C}_\eta$, integrating over Γ, then applying Green's formula [9] as follows:

$$\int_\Gamma div_S \nabla_S\eta\,\delta\eta\,d\sigma = -\int_\Gamma \nabla_S\eta\cdot\nabla_S\delta\eta\,d\sigma + \oint_\gamma \frac{\partial\eta}{\partial\nu_\Gamma}\delta\eta\,dl \tag{4.38}$$

   and finally, taking into account (4.37f), we obtain, $\forall\delta\eta \in \mathcal{C}_\eta$:

$$\boxed{k(\eta,\delta\eta) - \pi\int_\Gamma \delta\eta\,d\sigma - \lambda\int_\Gamma \rho_F\varphi\delta\eta\,d\sigma = 0} \tag{4.39}$$

   where we have put:

$$\boxed{\begin{aligned} k(\eta,\delta\eta) = & \int_\Gamma \rho_F g\,(i_z\!\cdot\! n)\eta\,\delta\eta\,d\sigma + \cdots \\ & \ldots + \sigma\int_\Gamma \nabla_S\eta\cdot\nabla_S\delta\eta\,d\sigma - \sigma\int_\Gamma (\frac{1}{R_1^2} + \frac{1}{R_2^2})\eta\,\delta\eta\,d\sigma - \sigma\oint_\gamma D\eta\,\delta\eta\,dl \end{aligned}} \tag{4.40}$$

2. In the second step, we introduce the space $\mathcal{C}_\varphi$ of the smooth test-functions φ in Ω_F, and we apply Green's formula (3.16).

[9] which is deduced from (4.4) and (4.5) with $u_{\|} = \nabla_S\eta$.

Next, in the integral on $\partial\Omega_F$, distinguishing the contributions related to Σ and Γ, and finally, taking account of equations (4.37b) and (4.37c), we obtain (after multiplication by ρ_F):

$$\boxed{\int_{\Omega_F} \rho_F \nabla\varphi \cdot \nabla\delta\varphi \, dx - \int_\Gamma \rho_F \eta \, \delta\varphi \, d\sigma = 0 \quad \forall \delta\varphi \in \mathcal{C}_\varphi} \tag{4.41}$$

Conversely, by going formally through the calculations, we verify that the variational properties (4.39) and (4.41), completed by (4.37d), characterize the solutions η, φ, π of the initial spectral problem.

Remark — It is important to note that, in (4.41), the test-functions $\delta\varphi$ are not subject to condition (4.37d) which φ must satisfy.
In order to establish a variational formulation where φ and $\delta\varphi$ belong to the same admissible space, we need to introduce the space $\mathcal{C}_\varphi^l$ defined by:

$$\boxed{\mathcal{C}_\varphi^l = \{\varphi \in \mathcal{C}_\varphi \mid l(\varphi) = 0, \, l(1) \neq 0\}} \tag{4.42}$$

We begin by showing that $\varphi \in \mathcal{C}_\varphi$ can be uniquely written in the form:

$$\varphi = \varphi^l + \text{constant} \quad \text{with } \varphi \in \mathcal{C}_\varphi \text{ and } \varphi^l \in \mathcal{C}_\varphi^l \tag{4.43}$$

On now applying l to (4.43) and using $l(\varphi^l) = 0$, we find that the value of the constant is $l(\varphi)/l(1)$. Conversely, (4.43) is verified by $\varphi^l = \varphi - l(\varphi)/l(1)$.
As a result, $\mathcal{C}_\varphi$ decomposes according to the direct sum:

$$\mathcal{C}_\varphi = \mathcal{C}_\varphi^l \oplus \mathbb{R} \tag{4.44}$$

Given this decomposition, (4.41) may be replaced by the set equivalent to the two variational equations obtained by successively restricting (4.41) to the class of test-functions $\delta\varphi \in \mathcal{C}_\varphi^l$ and to the constant test-functions $\delta\pi \in \mathbb{R}$, giving:

$$\boxed{\begin{array}{lr} \displaystyle\int_{\Omega_F} \rho_F \nabla\varphi \cdot \nabla\delta\varphi \, dx - \int_\Gamma \rho_F \eta \, \delta\varphi \, d\sigma = 0 \quad \forall \delta\varphi \in \mathcal{C}_\varphi^l & (a) \\ \displaystyle\delta\pi \int_\Gamma \eta \, d\sigma = 0 \quad \forall \delta\pi \in \mathbb{R} & (b) \end{array}} \tag{4.45}$$

We note that (4.45b) leads to $\int_\Gamma \eta \, d\sigma = 0$ which results from the invariancy of fluid volume.

Conclusion — The variational formulation of the boundary value spectral problem (4.37) can be stated:

find $\lambda \in \mathbb{R}^+$; $(\eta, \varphi, \pi) \in \mathcal{C}_\eta \times \mathcal{C}_\varphi^l \times \mathbb{R}$ verifying (4.39), (4.45a), (4.45b), $\forall(\delta\eta, \delta\varphi, \delta\pi) \in \mathcal{C}_\eta \times \mathcal{C}_\varphi^l \times \mathbb{R}$.

Remarks

- In the special case where $\sigma = 0$ and Γ is a horizontal surface, formulation (4.39),(4.45) constitutes an alternative formulation in (η, φ) of the problem of sloshing in the absence of surface tension (studied in chapter 3).
- The above formulation can readily be adapted to the case of two fluids (inviscid, immiscible) in contact along a surface Γ — one of the fluids may be incompressible — taking account of the coupling conditions (2.34) on Γ (§ 2.4).

4.6 Spectral problem discretized by finite elements

We denote by $\boldsymbol{\eta}$ the vector of the N_η nodal values of η, and by $\boldsymbol{\Phi}$ the vector of the N_φ nodal values of φ.
The matrices corresponding to the various bilinear forms involved in the variational formulation (4.39), (4.45a) and (4.45b) are defined by:

$$\begin{array}{|ccc|}
k(\eta,\delta\eta) & \Longrightarrow & \delta\boldsymbol{\eta}^T\boldsymbol{K}\boldsymbol{\eta} \quad (a)\\
\int_{\Omega_F}\rho_F\nabla\varphi\cdot\nabla\delta\varphi\,dx & \Longrightarrow & \delta\boldsymbol{\Phi}^T\boldsymbol{F}\boldsymbol{\Phi} \quad (b)\\
\int_\Gamma\rho_F\varphi\delta\eta\,d\sigma \Leftrightarrow \delta\boldsymbol{\eta}^T\boldsymbol{B}\boldsymbol{\Phi} & | & \int_\Gamma\rho_F\eta\,\delta\varphi\,d\sigma \Leftrightarrow \delta\boldsymbol{\Phi}^T\boldsymbol{B}^T\boldsymbol{\eta} \quad (c)\\
\pi\int_\Gamma\delta\eta\,d\sigma \Leftrightarrow \delta\boldsymbol{\eta}^T\boldsymbol{b}\pi & | & \delta\pi\int_\Gamma\eta\,d\sigma \Leftrightarrow \delta\pi\boldsymbol{b}^T\boldsymbol{\eta} \quad (d)
\end{array} \qquad (4.46)$$

where $\boldsymbol{K}(N_\eta \times N_\eta)$, $\boldsymbol{F}(N_\varphi \times N_\varphi)$ are symmetric matrices, $\boldsymbol{B}(N_\eta \times N_\varphi)$ is a coupling matrix between $\boldsymbol{\Phi}$ and $\boldsymbol{\eta}$ and $\boldsymbol{b}(N_\eta)$ is a "coupling vector" between $\boldsymbol{\eta}$ and π.

Practical choice of the constraint $l(\varphi)$ — For practical numerical purposes, it is convenient to choose for $l(\varphi)$ a cancellation condition of the nodal value of φ, e.g. the first: $\phi_1 = 0$. We denote by $\boldsymbol{\Phi}_2$ the "truncated" vector of the $N_\varphi - 1$ remaining nodal values.

Matrix equations — The matrix equations corresponding to the discretization of (4.39), (4.45a), (4.45b) are then written:

$$\begin{aligned}
\boldsymbol{K}\boldsymbol{\eta} - \boldsymbol{b}\pi &= \lambda\boldsymbol{B}_2\boldsymbol{\Phi}_2 && (a)\\
\boldsymbol{F}_{22}\boldsymbol{\Phi}_2 - \boldsymbol{B}_2^T\boldsymbol{\eta} &= 0 && (b)\\
\boldsymbol{b}^T\boldsymbol{\eta} &= 0 && (c)
\end{aligned} \qquad (4.47)$$

where the matrices $\boldsymbol{B}_2$ and $\boldsymbol{F}_{22}$ are obtained by eliminating both a column of $\boldsymbol{B}$ (and the row of $\boldsymbol{B}^T$), and the row and column of $\boldsymbol{F}$, corresponding to ϕ_1.

Elimination de Φ_2 — Since $\boldsymbol{F}_{22}$ is nonsingular (*cf.* § 3.4), (4.47b) enables $\boldsymbol{\Phi}_2$ to be eliminated in terms of $\boldsymbol{\eta}$, which, after substitution into (4.47), yields the following *symmetric matrix system*:

$$\boxed{\begin{bmatrix} \boldsymbol{K} & -\boldsymbol{b} \\ -\boldsymbol{b}^T & 0 \end{bmatrix} \begin{bmatrix} \boldsymbol{\eta} \\ \pi \end{bmatrix} = \lambda \begin{bmatrix} \boldsymbol{B}_2\boldsymbol{F}_{22}^{-1}\boldsymbol{B}_2^T & 0 \\ 0 & 0 \end{bmatrix} \begin{bmatrix} \boldsymbol{\eta} \\ \pi \end{bmatrix}} \tag{4.48}$$

Remarks

- We shall see that $\boldsymbol{B}_2\boldsymbol{F}_{22}^{-1}\boldsymbol{B}_2^T$ is a positive definite matrix.
- $\boldsymbol{K}$ is a positive matrix if the equilibrium configuration is stable, which we shall assume in what follows.
- Let us show that (4.48) has $N_\eta - 1$ positive eigenvalues. (4.48) may be written in the following equivalent form:

  $\delta\boldsymbol{\eta}^T\boldsymbol{K}\boldsymbol{\eta} = \lambda\delta\boldsymbol{\eta}^T\boldsymbol{B}_2\boldsymbol{F}_{22}^{-1}\boldsymbol{B}_2^T\boldsymbol{\eta}$, where $\boldsymbol{\eta}$ and $\delta\boldsymbol{\eta}$ *verify the constraint* $\boldsymbol{b}^T\boldsymbol{\eta} = 0$, which is equivalent to interpreting π as a Lagrange multiplier associated with this constraint.

  Under these conditions, (4.48) may be written:

  $$\boxed{\boldsymbol{K}\boldsymbol{\eta} = \lambda\boldsymbol{B}_2\boldsymbol{F}_{22}^{-1}\boldsymbol{B}_2^T\boldsymbol{\eta}, \quad \text{with } \boldsymbol{b}^T\boldsymbol{\eta} = 0} \tag{4.49}$$

  (4.49) enables it to be established that (4.48) possesses $N_\varphi - 1$ positive eigenvalues.
- For each eigenvalue $(\lambda_\alpha, \eta_\alpha)$, (4.47b) enables the value of the potential $\boldsymbol{\Phi}_2$ to be *retrieved*.
- Various additional numerical aspects (parametrization of the free surface through its vertical displacement $h = i_z n\,\eta$, numerical application of (4.48)) can be found in Morand & Ohayon [140], [149].

Practical construction of the matrices

To construct the matrices $(\boldsymbol{b}, \boldsymbol{b}^T)$ and $\boldsymbol{B}_2{}^T\boldsymbol{F}_{22}^{-1}\boldsymbol{B}_2^T$, we procede as follows:

1. We consider the auxiliary bilinear form [10] $\widehat{m}(\eta, \varphi | \delta\eta, \delta\varphi)$ — written $\widehat{m}(\cdot|\cdot)$ — which is symmetric in the exchange $(\eta, \varphi \leftrightarrow \delta\eta, \delta\varphi)$ and the matrix which results from its discretization by finite elements, which is expressed:

$$\widehat{m}(\cdot|\cdot) = -\int_{\Omega_F} \rho_F \nabla\varphi\cdot\nabla\delta\varphi\,dx + \left(\int_\Gamma \rho_F\eta\,\delta\varphi\,d\sigma + \int_\Gamma \rho_F\varphi\delta\eta\,d\sigma\right) \quad (a)$$

$$\Downarrow \tag{4.50}$$

$$\begin{bmatrix} \delta\boldsymbol{\eta}^T & \delta\boldsymbol{\Phi}^T \end{bmatrix} \begin{bmatrix} 0 & \boldsymbol{B} \\ \boldsymbol{B}^T & -\boldsymbol{F} \end{bmatrix} \begin{bmatrix} \boldsymbol{\eta} \\ \boldsymbol{\Phi} \end{bmatrix} \quad (b)$$

[10] where φ is not subject to the constraint $l(\varphi) = 0$.

The construction of the matrix $\begin{bmatrix} 0 & \boldsymbol{B} \\ \boldsymbol{B}^T & 0 \end{bmatrix}$ results from the simultaneous assembly of elementary contributions corresponding to the discretization of the symmetric bilinear form $(\int_\Gamma \rho_F \eta \delta\varphi \, d\sigma + \int_\Gamma \rho_F \varphi \delta\eta \, d\sigma)$ featuring in $\widehat{m}(\cdot|\cdot)$.

2. Distinguishing the value ϕ_1 of φ at a particular node, from the other nodal values $\boldsymbol{\Phi}_2$ of φ, the above matrix is partitioned as follows:

$$\begin{bmatrix} 0 & \boldsymbol{b} & \boldsymbol{B}_2 \\ \boldsymbol{b}^T & -F_{11} & -\boldsymbol{F}_{12} \\ \boldsymbol{B}_2^T & -\boldsymbol{F}_{12}^T & -\boldsymbol{F}_{22} \end{bmatrix} \Leftrightarrow \begin{bmatrix} \eta \\ \phi_1 \\ \boldsymbol{\Phi}_2 \end{bmatrix} \tag{4.51}$$

3. Applying the Gauss elimination algorithm (1.64) to this matrix, considering $\boldsymbol{\Phi}_2$ as "dependent degrees of freedom", we obtain the following condensed matrix:

$$\begin{bmatrix} \boldsymbol{B}_2 \boldsymbol{F}_{22}^{-1} \boldsymbol{B}_2^T & \boldsymbol{b} \\ \boldsymbol{b}^T & 0 \end{bmatrix} \tag{4.52}$$

whose submatrices are those featuring in (4.48). We note that the diagonal term corresponding to ϕ_1 is null, which is because $\boldsymbol{F}$ is of rank $N_\varphi - 1$, with a kernel reducing to the vector $\boldsymbol{\Phi}$ of components equal to 1 (*cf.* § 3.4).

Axisymmetric reservoirs

Here we examine the case of a reservoir of revolution. We work in cylindrical coordinates: z is the axis of revolution, r, θ, z denote the local coordinates. We can decompose φ, η into a Fourier series as follows:

$$\underbrace{\begin{Bmatrix} \eta \\ \varphi \\ \pi \end{Bmatrix}}_{\mathcal{C}} = \underbrace{\begin{Bmatrix} \eta_0(r,z) \\ \varphi_0(r,z) \\ \pi \end{Bmatrix}}_{\mathcal{C}_0} + \underbrace{\begin{Bmatrix} \eta_n^+(r,z) \\ \varphi_n^+(r,z) \\ 0 \end{Bmatrix}}_{\mathcal{C}_n^+} \cos n\theta + \underbrace{\begin{Bmatrix} \eta_n^-(r,z) \\ \varphi_n^-(r,z) \\ 0 \end{Bmatrix}}_{\mathcal{C}_n^-} \sin n\theta \tag{4.53}$$

We now show that the eigenvalue problem $\mathcal{P}$ splits into a series of two-dimensional problems $\mathcal{P}_n^\pm$ posed in the meridian plane, where n is a positive or null integer. We note that the constant functions π are in the category $\mathcal{C}_0$. Consequently, φ_0 is subject to the constraint $l(\varphi_0) = 0$.
But for $n \geq 1$, π is not identically null, and φ_n is not subject to the constraint $l(\varphi) = 0$.
The variational formulation corresponding to $\mathcal{P}_n^\pm$ is obtained by restricting variational formulation (4.39), (4.45a), (4.45b) to the admissible subspaces $\mathcal{C}_\eta$ and $\mathcal{C}_\varphi^l$ associated with the decomposition of $\mathcal{C}$ resulting from (4.53). [11]

[11] We note que that $\mathcal{C}_\eta$ decomposes according to $\mathcal{C}_\eta = \mathcal{C}_{\eta_0} \oplus \mathcal{C}_{\eta_n^+} \oplus \mathcal{C}_{\eta_n^-}$ and that the admissible subspace $\mathcal{C}_\varphi^l$ decomposes according to $\mathcal{C}_\varphi^l = \mathcal{C}_{\varphi_0}^l \oplus \mathcal{C}_{\varphi_n^+} \oplus \mathcal{C}_{\varphi_n^-}$.

1. **Axisymmetric modes** — the axisymmetric modes correspond to the non-degenerate eigenvalues of the three- dimensional problem. The numerical application issues from the above discussion.

2. **Non-axisymmetric modes** — For $n \geq 1$, the eigenvalues of $\mathcal{P}_n^+$ and $\mathcal{P}_n^-$ are identical (they correspond to a multiplicity 2 of the eigenvalues of the three-dimensional problem) and the eigenvectors of $\mathcal{P}_n^-$ can be deduced from those of $\mathcal{P}_n^+$ by rotation through $\pi/2n$.

   The numerical application is therefore carried out for modes of type $\mathcal{C}_n^+$, for example.

   The fact that φ_n is not subject to the constraint l and that π is identically zero, enables the initial boundary value problem of the variational formulation and the matrix structure of the discretized problem, to be simplified. We have:

   (a) the (two-dimensional) boundary value problem in η_n, φ_n is deduced from (4.37), by ignoring (4.37d) and by putting $\pi = 0$ in (4.37e) — the pressure being given by $p_n = \rho_F \omega^2 \varphi_n$ from (4.35).

   (b) The variational formulation of the (two-dimensional) problem in η_n, φ_n is deduced from (4.39) (putting $\pi = 0$) and (4.41).

   (c) The matrix equations are written:

$$\mathcal{P}_n \Longrightarrow \left\{ \begin{array}{ll} \boldsymbol{K}\boldsymbol{\eta}^n & = \lambda \boldsymbol{B}\boldsymbol{\Phi}^n \\ \boldsymbol{F}\boldsymbol{\Phi}^n & \boldsymbol{B}^T\boldsymbol{\eta}^n = 0 \end{array} \right. \tag{4.54}$$

   where $\boldsymbol{\Phi}^n$ is the vector of the N_{φ_n} nodal values of φ_n.

   (d) Unlike the case $n = 0$ where $\boldsymbol{F}$ is of rank $N_{\varphi_0} - 1$, for $n \neq 0$, $\boldsymbol{F}$ is non-singular and therefore the second equation of (4.54) enables $\boldsymbol{\Phi}^n$ to be eliminated completely, which leads to the eigenvalue problem:

$$\mathcal{P}_n \Longrightarrow \boxed{\boldsymbol{K}\boldsymbol{\eta}^n = \lambda \boldsymbol{B}\boldsymbol{F}^{-1}\boldsymbol{B}^T\boldsymbol{\eta}^n} \tag{4.55}$$

4.7 η symmetric formulation and added mass

The variational equations (4.39), (4.45ab) may be interpreted by considering (4.45b) as a constraint and π as the associated Lagrange multiplier. By forcing η to satisfy this constraint, we can eliminate π from this formulation. We therefore have to consider the space of the η admissible $\mathcal{C}_\eta^0$ defined by:

$$\boxed{\mathcal{C}_\eta^0 = \left\{ \eta \in \mathcal{C}_\eta \mid \int_\Gamma \eta \, d\sigma = 0 \right\}} \tag{4.56}$$

We then obtain the following equivalent formulation:
find λ, $\eta \in \mathcal{C}_\eta^0$, $\varphi \in \mathcal{C}_\varphi^l$ such that, $\forall \delta\eta \in \mathcal{C}_\eta^0$, $\forall \delta\varphi \in \mathcal{C}_\varphi^l$:

$$\boxed{k(\eta, \delta\eta) - \lambda \int_\Gamma \rho_F \varphi \delta\eta \, d\sigma = 0} \tag{4.57}$$

$$\boxed{\int_{\Omega_F} \rho_F \nabla\varphi \cdot \nabla\delta\varphi \, dx \; - \; \int_\Gamma \rho_F \eta \, \delta\varphi \, d\sigma \; = \; 0} \tag{4.58}$$

Elimination of φ

The variational equation (4.58) considered in isolation for a given $\eta \in \mathcal{C}_\eta^0$, characterizes the unique solution φ to the Neumann problem (4.37abc), whose existence results from the constraint on η appearing in the admissible class and the unicity of the condition $l(\varphi) = 0$ satisfied by $\varphi \in \mathcal{C}_\varphi^l$.
We furthermore verify that this solution — which is denoted φ_η — is linearly dependent on η.
The elimination consists of substituting φ_η into (4.57), which yields the following *variational formulation* in η (whose symmetry is established in what follows):
Find λ and $\eta \in \mathcal{C}_\eta^0$ such that $\forall \delta\eta \in \mathcal{C}_\eta^0$:

$$k(\eta, \delta\eta) \; = \; \lambda \int_\Gamma \rho_F \varphi_\eta \, \delta\eta \, d\sigma \tag{4.59}$$

The bilinear form $\mathcal{M}_A(\eta, \delta\eta)$ defined by:

$$\mathcal{M}_A \; = \; \int_\Gamma \rho_F \varphi_\eta \, \delta\eta \, d\sigma \tag{4.60}$$

is known as the *added mass operator*. We shall show that *this operator is symmetric.*
Consider solution $\varphi_{\delta\eta}$ to (4.58) for a given $\delta\eta$, taking as special test-function $\delta\varphi = \varphi_\eta$, which is written:

$$\int_\Gamma \rho_F \, \delta\eta \varphi_\eta \, d\sigma \; = \; \int_{\Omega_F} \rho_F \nabla\varphi_{\delta\eta} \cdot \nabla\varphi_\eta \, dx \tag{4.61}$$

which enables $\mathcal{M}_A$ to be written in the following symmetric form:

$$\boxed{\mathcal{M}_A \; = \; \int_{\Omega_F} \rho_F \nabla\varphi_\eta \cdot \nabla\varphi_{\delta\eta} \, dx} \tag{4.62}$$

This symmetric operator is obviously positive.

Properties of eigenmodes

From (4.59) and (4.60), the spectral problem in η is written:
find $\lambda \in \mathbb{R}$ and $\eta \in \mathcal{C}_\eta^0$, such that, $\forall \, \delta\eta \in \mathcal{C}_\eta^0$,

$$k(\eta, \delta\eta) \; = \; \lambda \mathcal{M}_A(\eta, \delta\eta) \tag{4.63}$$

which yields the following orthogonality properties of the eigenmodes:

$$k(\eta_\alpha, \eta_\beta) = \lambda_\alpha \mu_\alpha \delta_{\alpha\beta}, \quad \mathcal{M}_A(\eta_\alpha, \eta_\beta) = \mu_\alpha \delta_{\alpha\beta}, \tag{4.64}$$

Basis of eigenmodes — We can show, by choosing $\mathcal{C}_\eta = H^1(\Gamma)$, that $\{\eta_\alpha\}$ constitutes a hilbertian basis of $\mathcal{C}^0_\eta$.
We note that there corresponds to each eigenmode $\lambda_\alpha, \eta_\alpha$ a harmonic function φ_α which is a solution to the Neumann problem (4.58) for $\eta = \eta_\alpha$.
The study of sloshing eigenmodes under surface tension and in the presence of gravity can be found in Moiseev & Rumyantsev [135], Tong [206] and for particular geometries in Satterlee & Reynolds [197], Benjamin & Scott [15], Veldman & Vogels [209].

4.8 Comparison of eigenfrequencies

We consider the problems $\mathcal{P}$ and $\mathcal{P}'$ corresponding to the two liquid domains Ω_F and Ω'_F of the same free surface Γ, such that $\Omega'_F \subset \Omega_F$ and such that the walls Σ and Σ' are tangential at all points to γ, and furthermore have the same curvature tensors at each point of γ (Fig. 4.6). We denote respectively by n and n' the unitary normal to Σ and Σ' external to Ω_F and Ω'_F. We wish to compare the respective eigenvalues of these two problems. The demonstration is based on the first theorem of comparison (1.37), which involves the Rayleigh quotient.

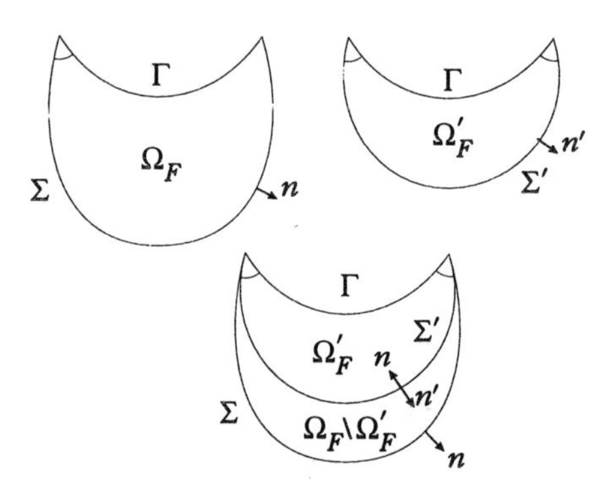

Figure 4.6: Reservoirs of various shapes

Rayleigh quotient — The variational formulation (4.59) results from the stationarity of the Rayleigh quotient (after taking account of (4.60) and (4.61)):

$$R(\eta) = \frac{k(\eta,\eta)}{\mathcal{M}_A(\eta,\eta)} \tag{4.65}$$

Physical interpretation — The numerator represents the potential energy of the liquid and the denominator corresponds to the inertia of the liquid: we know that $u^F = \nabla\varphi$ represents the displacement of the liquid and therefore, from (4.62), we have $\mathcal{M}_A(\eta, \delta\eta) = \int_{\Omega_F} \rho_F |u^F|^2 \, dx$.

Extremal property of $\mathcal{M}_A$ — We show that the quadratic form $\mathcal{M}_A(\eta,\eta)$ verifies the property:

$$\mathcal{M}_A(\eta,\eta) = \max_{\varphi\in\mathcal{C}^l_\varphi}\left\{-\int_{\Omega_F}\rho_F|\nabla\varphi|^2\,dx + 2\int_\Gamma \rho_F\eta\,\varphi\,d\sigma\right\} \tag{4.66}$$

The demonstration is based on the classical property of the solution u of $a(u,v) = <f,v>\ \forall v$, with $a(\cdot,\cdot) > 0$:
if $J(v) = -a(v,v)+2<f,v>$, u verifies $J(u) = \max_v J(v)$. Using the variational equation satisfied by u, we deduce from this that $J(u) = \max_v J(v) = a(u,u) = 2<f,u>$. The stated property is obtained by applying the above result for $a(u,v) = \int_{\Omega_F}\rho_F\nabla\varphi\cdot\nabla\delta\varphi\,dx$ and $<f,v> = \int_\Gamma\rho_F\eta\,\delta\varphi\,d\sigma$.

Remark — The above property enables us to show that the eigenvalues of the discretized problem are higher than the eigenvalues of the same rank of the continuum problem (to an approximation of a domain).

Comparison — We denote by φ_η and φ'_η the solutions of (4.58) in Ω_F and Ω'_F, for the same value of η, and by $\mathcal{M}_A$ and $\mathcal{M}'_A$ the corresponding added mass operators defined by (4.62).
The Rayleigh quotients corresponding to $\mathcal{P}$ and $\mathcal{P}'$ — defined for $\eta\in\mathcal{C}^0_\eta$ — are expressed:

$$R(\eta) = \frac{k(\eta,\eta)}{\mathcal{M}_A(\eta,\eta)}\ ,\quad R'(\eta) = \frac{k(\eta,\eta)}{\mathcal{M}'_A(\eta,\eta)} \tag{4.67}$$

Let us show that $\forall\eta\in\mathcal{C}^0_\eta$, $R(\eta)\geq R'(\eta)$. This is equivalent to demonstrating that $\mathcal{M}_A\leq\mathcal{M}'_A$, i.e. that:

$$\Delta\mathcal{M}_A = \int_{\Omega'_F}\rho_F|\nabla\varphi'_\eta|^2\,dx - \int_{\Omega_F}\rho_F|\nabla\varphi_\eta|^2\,dx \geq 0 \tag{4.68}$$

$\Delta\mathcal{M}_A$ can be written:

$$\Delta\mathcal{M}_A = \int_{\Omega'_F}\rho_F(|\nabla\varphi'_\eta|^2 - |\nabla\varphi_\eta|^2)\,dx - \int_{\Omega_F\setminus\Omega'_F}\rho_F|\nabla\varphi|^2\,dx \tag{4.69}$$

By using the identity $a^2 - b^2 = (a-b)^2 + 2b(a-b)$, $\Delta\mathcal{M}_A$ is written:

$$\int_{\Omega'_F}\rho_F|\nabla\varphi'_\eta-\nabla\varphi_\eta|^2\,dx + 2\int_{\Omega'_F}\rho_F\nabla\varphi_\eta\cdot(\nabla\varphi'_\eta-\nabla\varphi_\eta)\,dx - \int_{\Omega_F\setminus\Omega'_F}\rho_F|\nabla\varphi_\eta|^2\,dx \tag{4.70}$$

On transforming the second term by means of Green's formula (3.16), taking into account both the fact that φ_η and φ'_η are harmonic functions, and that $\partial\varphi_\eta/\partial n = \partial\varphi'_\eta/\partial n = \eta$ on Γ, we obtain $2\int_{\Omega'_F}\rho_F\nabla\varphi_\eta\cdot(\nabla\varphi'_\eta-\nabla\varphi_\eta)\,dx = -2\int_{\Sigma'}\rho_F\varphi_\eta\frac{\partial\varphi_\eta}{\partial n'}\,d\sigma = 2\int_{\Sigma'}\rho_F\varphi_\eta\frac{\partial\varphi_\eta}{\partial n}\,d\sigma$.
Applying Green's formula in $\Omega_F\setminus\Omega'_F$, this term writes $2\int_{\Omega_F\setminus\Omega'_F}\rho_F|\nabla\varphi_\eta|^2\,dx$, yielding the following expression for $\Delta\mathcal{M}_A$:

$$\Delta\mathcal{M}_A = \int_{\Omega'_F}\rho_F|\nabla\varphi'_\eta - \nabla\varphi_\eta|^2\,dx + \int_{\Omega_F\setminus\Omega'_F}\rho_F|\nabla\varphi_\eta|^2\,dx \tag{4.71}$$

which shows that $\mathcal{M}_A \leq \mathcal{M}'_A$ and, therefore, that $R(\eta) \geq R'(\eta)$.
From the first theorem of comparison, we deduce that the *nth* eigenfrequency of the "large reservoir" Ω_F is higher than the *nth* eigenfrequency of the "small reservoir" Ω'_F.
This result extends the result found in chapter 3 to the case of sloshing vibrations under surface tension (*cf.* Morand [141]).

4.9 Modal analysis of the vibratory response of the liquid

We begin by writing the variational formulation of the boundary value problem (4.36) describing the response of the liquid to a prescribed motion u_N of the wall. We then derive the variational formulation in η.
We proceed by the test-functions method. We simply repeat the derivation of the variational formulation (4.39), (4.45ab) of the eigenmodes and include the additional terms involving u_N.
As before, $\mathcal{C}_\eta$ denotes the space of the smooth functions $\delta\eta$ defined on Γ, $\mathcal{C}^l_\varphi$ the space of the φ subject to the constraint $l(\varphi) = 0$ (*cf.* (4.42)).
The variational property writes as follows. For given ω and u_N, find $\eta \in \mathcal{C}_\eta$, $\varphi \in \mathcal{C}^l_\varphi$, $\pi \in \mathbb{R}$ such that $\forall\delta\eta \in \mathcal{C}_\eta$, $\forall\delta\varphi \in \mathcal{C}^l_\varphi$, $\forall\delta\pi \in \mathbb{R}$, we have:

$$k(\eta, \delta\eta) - \pi \int_\Gamma \delta\eta \, d\sigma - \omega^2 \int_\Gamma \rho_F \varphi \delta\eta \, d\sigma = \sigma \oint_\gamma (E\, u_N - \frac{\partial u_N}{\partial \nu_\Sigma}) \delta\eta \, dl \tag{4.72}$$

$$\begin{aligned} \int_{\Omega_F} \rho_F \nabla\varphi \cdot \nabla\delta\varphi \, dx - \int_\Gamma \rho_F \eta \delta\varphi \, d\sigma &= \int_\Sigma \rho_F u_N \, \delta\varphi \, d\sigma \quad (a) \\ \delta\pi \int_\Gamma \eta \, d\sigma &= -\delta\pi \int_\Sigma u_N \, d\sigma \quad (b) \end{aligned} \tag{4.73}$$

Formulation in terms of η

As before, we derive the condensed formulation in η. To begin with, we introduce the space $\mathcal{C}^{u_N}_\eta$ defined by:

$$\mathcal{C}^{u_N}_\eta = \left\{ \eta \in \mathcal{C}_\eta \mid \int_\Gamma \eta \, d\sigma + \int_\Sigma u_N \, d\sigma = 0 \right\} \tag{4.74}$$

Under these conditions, the variational formulation (4.72), (4.73ab) may be interpreted by considering π as the Lagrange multiplier associated with the constraint appearing in (4.74). We have exactly the equivalent formulation:
for given ω and u_N, find $\eta \in \mathcal{C}^{u_N}_\eta$, $\varphi \in \mathcal{C}^l_\varphi$ such that, $\forall\delta\eta \in \mathcal{C}^0_\eta$, $\forall\delta\varphi \in \mathcal{C}^l_\varphi$, we have:

$$k(\eta, \delta\eta) - \omega^2 \int_\Gamma \rho_F \varphi \delta\eta \, d\sigma = \sigma \oint_\gamma (E\, u_N - \frac{\partial u_N}{\partial \nu_\Sigma}) \delta\eta \, dl \tag{4.75}$$

$$\int_{\Omega_F} \rho_F \nabla\varphi \cdot \nabla\delta\varphi \, dx - \int_\Gamma \rho_F \eta \delta\varphi \, d\sigma = \int_\Sigma \rho_F u_N \, \delta\varphi \, d\sigma \tag{4.76}$$

We can then eliminate φ in terms of η and u_N. The variational equation (4.76), considered alone for a given η in $\mathcal{C}_\eta^{u_N}$, characterizes the unique solution φ of the Neumann problem (4.36abc), (whose existence results from the constraint appearing in the admissible space $\mathcal{C}_\eta^{u_N}$).
We furthermore verify that this solution — which will be denoted φ_{η,u_N} — is linearly dependent on (η, u_N).
The elimination consists of substituting φ_{η,u_N} in (4.75), yielding the following *variational formulation in* η:
for given ω and u_N, find $\eta \in \mathcal{C}_\eta^{u_N}$, such that, $\forall \delta\eta \in \mathcal{C}_\eta^0$, we have:

$$k(\eta, \delta\eta) \; - \; \omega^2 \int_\Gamma \rho_F \varphi_{\eta,u_N} \delta\eta \, d\sigma \; = \; \sigma \oint_\gamma (E\, u_N - \frac{\partial u_N}{\partial \nu_\Sigma}) \delta\eta \, dl \tag{4.77}$$

By a derivation similar to that carried out to derive $\mathcal{M}_A$, we can show that $\int_\Gamma \rho_F \varphi_{\eta,u_N} \delta\eta \, d\sigma = \int_{\Omega_F} \rho_F \nabla\varphi_{\eta,u_N} \cdot \nabla\varphi_{\delta\eta,u_N} \, dx$.

Modal analysis

If η denotes the solution to (4.77), then $\eta - \overline{\eta}$ with $\overline{\eta} = (1/\text{Area}(\Gamma)) \int_\Gamma \eta \, d\sigma$, is a function of zero mean value on Γ, which belongs therefore to $\mathcal{C}_\eta^0$ defined by (4.56). Since η belongs to $\mathcal{C}_\eta^{u_N}$, we have $\int_\Gamma \eta \, d\sigma = -\int_\Sigma u_N \, d\sigma$, and therefore $\overline{\eta} = -(1/\text{Area}(\Gamma)) \int_\Sigma u_N \, d\sigma$, which is a linear function of u_N denoted $\overline{\eta}_{u_N}$. The term $\eta - \overline{\eta}_{u_N}$ is an element of $\mathcal{C}_\eta^0$, and may therefore be decomposed on the basis of eigenvectors η_α. We then apply the method of Ritz-Galerkin by seeking a solution η to (4.77) in the form:

$$\boxed{\eta \; = \; \overline{\eta}_{u_N} \; + \; \sum_{\alpha \geq 1} \kappa_\alpha \eta_\alpha} \tag{4.78}$$

Replacing η by this expression in (4.77), then putting successively $\delta\eta = \eta_1, \eta_2, \ldots$, and finally using the orthogonality relations (4.64), we obtain $(-\omega_\alpha^2 + \omega^2)\mu_\alpha \kappa_\alpha = \omega^2 \mathcal{M}_A(\overline{\eta}_{u_N}, \eta_\alpha) \; - k(\overline{\eta}_{u_N}, \eta_\alpha)$. This yields the expression for κ_α which we substitute into (4.78) to obtain finally the solution η of (4.77) in the form:

$$\boxed{\eta \; = \; \overline{\eta}_{u_N} \; - \sum_{\alpha \geq 1} \frac{1}{-\omega^2 + \omega_\alpha^2} \frac{k(\overline{\eta}_{u_N}, \eta_\alpha) - \omega^2 \mathcal{M}_A(\overline{\eta}_{u_N}, \eta_\alpha)}{\mu_\alpha}} \tag{4.79}$$

Remark — To the constant normal displacement of the free surface $\overline{\eta}_{u_N}$ there corresponds a displacement potential $\varphi_{\overline{\eta}_{u_N}, u_N}$ which is a solution to the Neumann problem $\Delta\varphi = 0|_{\Omega_F}$, $\partial\varphi/\partial n|_\Sigma = u_N$ and $\partial\varphi/\partial n|_\Gamma = \overline{\eta}_{u_N}$ of which the variational formulation is written: find φ verifying $\int_\Gamma \varphi \, d\sigma = 0$, such that $\forall \delta\varphi$ verifying $\int_\Gamma \delta\varphi \, d\sigma = 0$, we have:

$$\int_{\Omega_F} \nabla\varphi \cdot \nabla\delta\varphi \, dx \; = \; \int_\Sigma u_N \delta\varphi \, d\sigma \tag{4.80}$$

This last property results from (4.76) on replacing η by $\overline{\eta}_{u_N}$ and choosing $l(\varphi) = \int_\Gamma \varphi \, d\sigma$, as a constraint appearing in the admissible class.

4.10 Open problems

The following are examples of problems currently under development:

- analysis of problems of convergence related to discretization by finite elements and modal analysis,

- the extension of contact conditions to the case of non-constant angles of contact and the inclusion of hysteresis effects at the meniscus (*cf.* Ngan & Dussan [162]),

- deeper modeling of the condition for connecting the three-dimensional elastic medium with the free surface — the classical model of capillarity (used here) introducing a line density along the connection line γ, which is compatible with shell theory, but which introduces singularities in the case of an elastic medium,

- the numerical study of stability problems of liquids contained in rotating reservoirs in spinning satellites, taking into account the viscosity of the fluid (*cf.* Greenspan [89], El-Raheb & Wagner [63], Bauer [11], Schulkes & Cuvelier [198]).

CHAPTER 5

Hydroelastic vibrations

5.1 Introduction

We consider the vibrations of an elastic structure containing an incompressible liquid with a free surface.
An example of application to the aerospace field is the vibration of liquid propelled launch vehicles involved in studies of the *Pogo* effect [1]and in predictions of the dynamic behaviour of the launcher during flight.
We consider the approximation of a *weightless* liquid, i.e. neglecting gravity terms ($g = 0$), and surface tension terms ($\sigma = 0$) in the problem of vibrations of an incompressible inviscid liquid contained in an elastic reservoir around the equilibrium configuration.
The geometry of the liquid domain at equilibrium is characterized by the existence of a free surface Γ, which is a plane horizontal surface when surface tension effects are negligible compared to gravity effects (in the general case, Γ is not a plane surface).
We also neglect the prestress effects induced in the structure by the weight of the liquid — which we shall be analyzing in chapter 6.
We begin by establishing the variational formulation in (u, φ) of the hydroelastic response to forces applied to the structure.
We next study the spectral problem associated with "hydroelastic modes", and the matrix structure resulting from a finite element discretizion.
We show that this problem can be formulated in terms of structure displacements by introducing the concept of added mass (from continuum and discretized points of vue).

---

[1]Vibration of the launcher is accompanied by fluctuations in the liquid ergol pressures, especially at the fuel feed inlets in the propulsion unit, which induces pressure and feed rate oscillations, and hence fluctuations in the rocket engine thrust, together with oscillatory forces at the anchor points of structure lines of the launcher. These forces can in turn increase the launcher vibrations, and hence lead to an instability known as the *Pogo* effect. This phenomenon, which is a characteristic of liquid propelled launchers, involves coupling of longitudinal type vibrational modes (modes $n = 0$ in the case of an axisymmetric structure), and the unsteady model of the propulsion. For a description of the *Pogo* effect, see, e.g., Morand [146], including references.

The properties of the added mass operator will be applied to the study of the behaviour of hydroelastic natural frequencies in terms of a filling parameter.

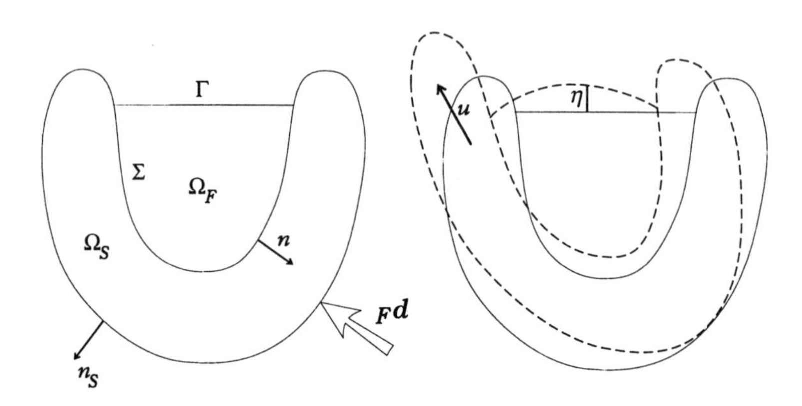

Figure 5.1: Structure containing a liquid

5.2 Variational formulation in terms of (u, φ)

Response to forces

The notation is that of Figure 5.1. The structure Ω_S is described by a field of displacement u.

For the description of the movements of the liquid, we shall use here the first definition (2.24) of the potential $\varphi = p/(\rho_F \omega^2)$, which in the present case leads as we shall see to a well-posed problem for $\omega = 0$. We denote by n the unitary normal external to the domain Ω_F occupied by the liquid.

In the harmonic state, the boundary value problem is written:

$$\begin{array}{rcll}
\Delta\varphi &=& 0 & \text{in } \Omega_F \quad (a) \\
\varphi &=& 0 & \text{on } \Gamma \quad (b) \\
\dfrac{\partial\varphi}{\partial n} &=& u \cdot n & \text{on } \Sigma \quad (c) \\
\sigma_{ij}(u)n_j^S &=& \rho_F \omega^2 \varphi n_i & \text{on } \Sigma \quad (d) \\
\sigma_{ij,j}(u) + \rho_S \omega^2 u_i &=& 0 & \text{in } \Omega_S \quad (e) \\
\sigma_{ij}(u)n_j^S &=& F_i^d & \text{on } \partial\Omega_S \setminus \Sigma \quad (f)
\end{array} \tag{5.1}$$

where φ is related to the displacement of the fluid u_F and to the pressure fluctuation p by:

$$\boxed{\begin{array}{rcll} u_F & = & \nabla\varphi & (a) \\ p & = & \rho_F\omega^2\varphi & (b) \end{array}} \quad (5.2)$$

- (5.1a) results from the condition of incompressibility of the liquid (*cf.* (2.29)).
- The free surface condition (5.1b) results from the nullity of the pressure fluctuation on the free surface in the absence of gravity (deduced from (2.33) for $g = 0$).
- (5.1c) results from the contact condition on Σ (*cf.* (2.31)).
- (5.1d) results from the action of the pressure forces exerted by the fluid on the structure $\sigma_{ij}n_j^S = -pn_i^S = pn_i$, where n denotes the normal external to Ω_F.
- (5.1e) is the elastodynamic equation (*cf.* chapter 1).
- In (5.1f), F_i^d denotes the surface density of the applied forces.

Remark — Equations (5.1abc) coincide with the boundary value problem (3.81), and therefore the potential φ introduced in this chapter coincides with φ^∞.
In other words, problem (5.1), posed for given ω and F^d, is equivalent to coupling the elastodynamic model of an elastic structure with the non-resonant behaviour model of the liquid defined by φ^∞.

Variational formulation — We apply the test-functions method. We proceed in two steps, treating in turn the equations relating to the structure "subject to the actions of fluid pressure", then the equations relating to the fluid "subject to a displacement of the wall Σ".
We first of all introduce the space C_u of smooth functions u defined in Ω_S. Multiplying equation (5.1e) by an arbitrary function $\delta u \in C_u$, then applying Green's formula (1.8), and finally, taking account of (5.1d) and (5.1f), then for $u \in C_u$ and $\forall \delta u \in C_u$, we arrive at:

$$\boxed{\int_{\Omega_S} \sigma_{ij}(u)\epsilon_{ij}(\delta u)\, dx - \omega^2 \int_{\Omega_S} \rho_S u\cdot\delta u\, dx - \omega^2 \int_\Sigma \rho_F \varphi n\cdot\delta u\, d\sigma = \int_{\partial\Omega_S\backslash\Sigma} F_i^d \delta u_i d\sigma} \quad (5.3)$$

In the second step, if C_φ denotes the space of smooth φ in Ω_F, we consider the space C_φ^* of φ verifying the constraint $\varphi = 0$ on Γ:

$$\boxed{C_\varphi^* = \{\varphi \in C_\varphi \mid \varphi = 0 \text{ on } \Gamma\}} \quad (5.4)$$

We multiply equation (5.1a) by $\delta\varphi \in C_\varphi^*$, then integrate over Ω_F; on applying Green's formula (3.17) and taking account of (5.1c), we arrive (after multiplying by ρ_F) at the following variational property verified for $\varphi \in C_\varphi^*$ and $\forall\delta\varphi \in C_\varphi^*$:

$$\boxed{-\int_{\Omega_F} \rho_F \nabla\varphi\cdot\nabla\delta\varphi\, dx + \int_\Sigma \rho_F u\cdot n\delta\varphi\, d\sigma = 0} \tag{5.5}$$

- Relations (5.3) and (5.5) constitute the variational formulation of the problem.

- (5.3) is interpreted as the variational formulation of the problem of the harmonic response of the structure to the actions of pressure by the fluid (*cf.* (1.12)) and to external forces. In the case of *slender* structures (beams, plates, shells), it is sufficient, as mentioned in chapter 1, to replace the bilinear form $\int_{\Omega_S} \sigma_{ij}(u)\epsilon_{ij}(\delta u)\, dx$ by the corresponding variational expressions [2].

- (5.5) is interpreted as the variational formulation of the harmonic response of the fluid to movement of the wall (in the absence of gravity).

Hydroelastic modes

We define the *hydroelastic modes* as the solutions of the spectral problem (set $F^d = 0$). Putting $\lambda = \omega^2$, the variational formulation of this problem is written: find λ, and $(u, \varphi) \in C_u \times C_\varphi^*$, such that $\forall\delta u \in C_u$ and $\forall\delta\varphi \in C_\varphi^*$, we have:

$$\boxed{\begin{aligned} \int_{\Omega_S} \sigma_{ij}(u)\epsilon_{ij}(\delta u)\, dx - \lambda\int_{\Omega_S} \rho_S u\cdot\delta u\, dx - \lambda\int_\Sigma \rho_F \varphi n\cdot\delta u\, d\sigma &= 0 \quad (a) \\ -\int_{\Omega_F} \rho_F \nabla\varphi\cdot\nabla\delta\varphi\, dx + \int_\Sigma \rho_F u\cdot n\delta\varphi\, d\sigma &= 0 \quad (b) \end{aligned}} \tag{5.6}$$

Solutions for $\lambda = 0$

We shall show that this problem has a solution $\lambda = 0$ of multiplicity 6, corresponding to rigid body motions of the structure - which are themselves associated with displacement (irrotational) field of the fluid.
From (5.3), we have $\int_{\Omega_S} \sigma_{ij}(u)\epsilon_{ij}(v)\, dx = 0$, of which the solution is the *rigid body displacements* $u^{\mathcal{R}} = \overrightarrow{T} + \overset{\frown}{\theta} \times \overrightarrow{OM}$ (*cf.* (1.32) and (1.33)).
To each $u^{\mathcal{R}}$ there corresponds a particular fluid displacement potential $\varphi^{\mathcal{R}}$, (unique) solution of (5.6b), or, which comes to the same thing, of equations (5.1abc).

---

[2] A shell element containing a linear interpolation of displacements enabling normal rotations to be eliminated in the calculation of hydroelastic vibrations, can be found in Ohayon & Nicolas-Vullierme [166].

Remarks

- To the rotation $\overset{\curvearrowleft}{\theta} \times \overrightarrow{OM}$ of the structure, there corresponds the potential $\varphi^{\mathcal{R}}$ solution of (5.1abc) with $\frac{\partial\varphi}{\partial n} = (\overset{\curvearrowleft}{\theta} \times \overrightarrow{OM})\cdot n$.

  To the vertical translation T of the structure there corresponds a vertical translation displacement as a whole of the liquid and a potential $\varphi = -Tz$, where z is the elevation of the point under consideration, taking as origin a point located on the free surface.

  It is important to note, on the other hand, that the movement of fluid which accompanies a horizontal translation of the structure differs from a simple translation movement as a whole. In particular, the free surface does not remain plane.

- The above formulation, which uses the first definition of φ, (2.24), yields a well-posed problem for $\omega = 0$ (this potential coincides with that obtained with the second definition (2.26) of φ taking $l(\varphi) = \int_\Gamma \varphi \, d\sigma$, which indeed yields $\pi = 0$).

5.3 Spectral problem discretized by finite elements

$\boldsymbol{U}$ denotes the vector of the N_u nodal values introduced by the finite element discretization of u. Similarly, $\boldsymbol{\Phi}$ denotes the vector of the N_φ nodal values introduced by the finite element discretization of φ.
The finite element discretization of the bilinear forms of (5.6ab) *before application of the constraints* yields the following matrices:

$$
\boxed{
\begin{array}{llll}
\int_{\Omega_S} \sigma_{ij}(u)\epsilon_{ij}(\delta u)\, dx & \Longrightarrow & \delta \boldsymbol{U}^T \boldsymbol{K} \boldsymbol{U} & (a) \\
\int_{\Omega_S} \rho_S u\cdot\delta u\, dx & \Longrightarrow & \delta \boldsymbol{U}^T \boldsymbol{M} \boldsymbol{U} & (b) \\
\int_{\Omega_F} \rho_F \nabla\varphi\cdot\nabla\delta\varphi\, dx & \Longrightarrow & \delta \boldsymbol{\Phi}^T \boldsymbol{F} \boldsymbol{\Phi} & (c) \\
\int_\Sigma \rho_F \varphi n\cdot\delta u\, d\sigma \Longrightarrow \delta \boldsymbol{U}^T \boldsymbol{C} \boldsymbol{\Phi} & \Big| & \int_\Sigma \rho_F \delta\varphi u\cdot n\, d\sigma \Longrightarrow \delta \boldsymbol{\Phi}^T \boldsymbol{C}^T \boldsymbol{U} & (d)
\end{array}
}
\qquad (5.7)
$$

Under these conditions, the matrix equations which discretize equations (5.6ab) are written:

$$
\begin{aligned}
\boldsymbol{K}\boldsymbol{U} - \lambda \boldsymbol{M}\boldsymbol{U} - \lambda \boldsymbol{C}_* \boldsymbol{\Phi}_* &= 0 \quad (a) \\
\boldsymbol{F}_* \boldsymbol{\Phi}_* - \boldsymbol{C}_*^T \boldsymbol{U} &= 0 \quad (b)
\end{aligned}
\qquad (5.8)
$$

given that:

- $\boldsymbol{K}$ and $\boldsymbol{M}$ are the symmetric positive $(N_u \times N_u)$ matrices of *stiffness* and *mass* for the structure. For a structure not subject to clamping conditions, $\boldsymbol{K}$ is of rank $N_u - 6$ and $\boldsymbol{M}$ is non-singular.

- $\boldsymbol{C}$ is a rectangular "coupling" matrix ($N_u \times N_\varphi$). We note from (5.7c) that the coupling term only involves the values of u (and of φ) on Σ, which means that $\boldsymbol{C}$ and $\boldsymbol{C}^T$ only couple the degrees of freedom relative to the nodes on Σ.

- $\boldsymbol{F}$ is a positive symmetric matrix ($N_\varphi \times N_\varphi$) of rank $N_\varphi - 1$ (*cf.* § 3.4).

- When the constraint $\varphi = 0$ (and $\delta\varphi = 0$) on Γ, appearing in the admissible class $\mathcal{C}_\varphi^*$, is taken into account, this results in the cancellation of the nodal values of φ and $\delta\varphi$. We shall denote by $\boldsymbol{\Phi}_*$ and $\delta\boldsymbol{\Phi}_*$ the N_{φ_*} component vectors so defined. Consequently:

  - the square matrix $\boldsymbol{F}_*(N_{\varphi_*} \times N_{\varphi_*})$ is obtained by eliminating the corresponding rows and columns of $\boldsymbol{F}$. The matrix $\boldsymbol{F}_*$ is thus non-singular.
  - The coupling matrices $\boldsymbol{C}_*$ and $\boldsymbol{C}_*^T$ are obtained by eliminating respectively the corresponding columns and rows of $\boldsymbol{C}$ and $\boldsymbol{C}^T$.

Remarks

- Equations (5.8) are written:

$$\begin{bmatrix} \boldsymbol{K} & 0 \\ \boldsymbol{C}_*{}^T & -\boldsymbol{F}_* \end{bmatrix} \begin{bmatrix} \boldsymbol{U} \\ \boldsymbol{\Phi}_* \end{bmatrix} = \lambda \begin{bmatrix} \boldsymbol{M} & \boldsymbol{C}_* \\ 0 & 0 \end{bmatrix} \begin{bmatrix} \boldsymbol{U} \\ \boldsymbol{\Phi}_* \end{bmatrix} \tag{5.9}$$

  The direct solution of these *unsymmetric* equations requires specific algorithms for the extraction of eigenvalues. We can in fact deal with symmetric equations; (5.8b) has simply to be multiplied by $-\lambda$, yielding:

$$\begin{bmatrix} \boldsymbol{K} & 0 \\ 0 & 0 \end{bmatrix} \begin{bmatrix} \boldsymbol{U} \\ \boldsymbol{\Phi}_* \end{bmatrix} = \lambda \begin{bmatrix} \boldsymbol{M} & \boldsymbol{C}_* \\ \boldsymbol{C}_*{}^T & -\boldsymbol{F}_* \end{bmatrix} \begin{bmatrix} \boldsymbol{U} \\ \boldsymbol{\Phi}_* \end{bmatrix} \tag{5.10}$$

  The latter equations correspond to the discretization of the variational formulation in (u, φ) obtained by considering (5.6a) and the variational equation (5.6b) multiplied by $-\lambda$.

  The *symmetric* matrix equations (5.10) have N_{φ_*} *zero non physical unwanted eigenvalues* artificially introduced by the multiplication by λ (we note that the spectrum of positive eigenvalues is not changed).

- The hydroelastic problem can also be formulated by describing the fluid by p (which is related to φ by (5.2b)).

  Replacing φ by p in the boundary value problem (5.1) yields the following variational formulation:

  find λ and $(u, p) \in \mathcal{C}_u \times \mathcal{C}_p^*$, such that $\forall(\delta u, \delta p) \in \mathcal{C}_u \times \mathcal{C}_p^*$ we have:

$$\begin{aligned} \int_{\Omega_S} \sigma_{ij}(u)\epsilon_{ij}(\delta u)\,dx - \lambda \int_{\Omega_S} \rho_S u\cdot\delta u\,dx - \int_\Sigma p n\cdot\delta u\,d\sigma &= 0 \\ -\int_{\Omega_F} \nabla p\cdot\nabla\delta p\,dx + \lambda \int_\Sigma \rho_F u\cdot n\delta p\,d\sigma &= 0 \end{aligned} \tag{5.11}$$

where $\mathcal{C}_p^* = \{p \mid p|_\Gamma = 0\}$.

The discretization of (5.11) involves the matrices (5.7) (replacing φ by p and keeping the identical interpolations for p and φ). If $\boldsymbol{P}_*$ denotes the vector of the nodal values of p corresponding to the nodes not located on the free surface, the matrix equations are written:

$$\begin{bmatrix} \boldsymbol{K} & -\frac{1}{\rho_F}\boldsymbol{C}_* \\ 0 & \frac{1}{\rho_F}\boldsymbol{F}_* \end{bmatrix} \begin{bmatrix} \boldsymbol{U} \\ \boldsymbol{P}_* \end{bmatrix} = \lambda \begin{bmatrix} \boldsymbol{M} & 0 \\ \boldsymbol{C}_*^T & 0 \end{bmatrix} \begin{bmatrix} \boldsymbol{U} \\ \boldsymbol{P}_* \end{bmatrix} \tag{5.12}$$

We note that these matrix equations are not symmetric.

In what follows, to avoid the difficulties of solving two-field problems directly, we shall show that the problem is equivalent to a "reduced" symmetric problem in u having no non physical unwanted zero eigenvalue.

5.4 Formulation of the spectral problem in u: added mass

$\boldsymbol{\Phi}_*$ can be eliminated from equations (5.8): since $\boldsymbol{F}_*$ is non-singular, we have from (5.8b):

$$\boldsymbol{\Phi}_* = \boldsymbol{F}_*^{-1}\boldsymbol{C}_*^T\boldsymbol{U} \tag{5.13}$$

Replacing $\boldsymbol{\Phi}_*$ by this expression in (5.8), we have:

$$\boxed{\boldsymbol{K}\boldsymbol{U} = \lambda(\boldsymbol{M} + \boldsymbol{M}_A)\boldsymbol{U}} \tag{5.14}$$

where $\boldsymbol{M}_A$ is the *added mass matrix* defined by:

$$\boxed{\boldsymbol{M}_A = \boldsymbol{C}_*\boldsymbol{F}_*^{-1}\boldsymbol{C}_*^T} \tag{5.15}$$

- $\boldsymbol{M}_A$ is a *positive symmetric* matrix.

  The associated quadratic form, which is written $\boldsymbol{U}^T\boldsymbol{M}_A\boldsymbol{U} = \boldsymbol{X}^T\boldsymbol{F}_*^{-1}\boldsymbol{X}$, with $\boldsymbol{X} = \boldsymbol{C}_*^T\boldsymbol{U}$, is indeed positive since $\boldsymbol{F}_*$ is positive.

- The *symmetric* matrix equation with eigenvalues (5.14) only involves the unknowns $\boldsymbol{U}$ relating to the structure. Solution of this equation yields the eigenvalues λ_α and eigenvectors $\boldsymbol{U}_\alpha$. Relation (5.13) enables $\boldsymbol{\Phi}_{*\alpha}$ to be "retrieved" for each solution (which enables the pressure on each node of the mesh to be calculated by means of (5.2b)).

- For a mathematical study of the spectrum and aerospace applications, using an approximative expression of the gravity terms [3] on the wall, with

[3] These terms are discussed more thoroughly in chapter 6.

u, φ and the displacement of the free surface η as unknowns *cf.* Boujot [23], Berger, Boujot & Ohayon [17]. A mathematical study of the spectrum in the case of shells can be found in Aganovic [2]. Applications in the nuclear domain are found in Gibert [84], Gupta & Hutchinson [90], and in biomechanics in Holmes [95], [96]. The case of systems having a network type symmetry has been studied in Conca, Planchard, Thomas & Vanninathan [41].

Remarks

- If we denote by $\boldsymbol{F}^d$ the vector of the applied forces (resulting from the discretization of $\int_{\partial\Omega_S \setminus \Sigma} F_i^d \delta u_i \, d\sigma$), the discretization of equations (5.3) and (5.5), after elimination of $\boldsymbol{\Phi}_*$ yields:

$$\boxed{\boldsymbol{KU} - \omega^2(\boldsymbol{M} + \boldsymbol{M}_A)\boldsymbol{U} = \boldsymbol{F}^d} \tag{5.16}$$

- In the transient state, we similarly establish that:

$$\boxed{(\boldsymbol{M} + \boldsymbol{M}_A)\ddot{\boldsymbol{U}} + \boldsymbol{KU} = \boldsymbol{F}^d(t)} \tag{5.17}$$

Solution of these equations by modal analysis, by means of the hydroelastic mode solutions to (5.14), which verify the obvious orthogonality relations ${\boldsymbol{U}_\alpha}^T(\boldsymbol{M} + \boldsymbol{M}_A)\boldsymbol{U}_\beta = \delta_{\alpha\beta}\mu_\alpha$, and ${\boldsymbol{U}_\alpha}^T\boldsymbol{K}\boldsymbol{U}_\beta = \delta_{\alpha\beta}\lambda_\alpha\mu_\alpha$, is carried out from the methods described in chapter 1 (*cf.* (1.68) and (1.69)).

Calculation of M_A

We show that $\boldsymbol{M}_A$ may be deduced, by *static condensation*, from the matrix $\widehat{\boldsymbol{M}}$ which discretizes the "auxiliary" bilinear form $\widehat{m}(u, \varphi|\delta u, \delta\varphi)$ by:

$$\widehat{m}(u, \varphi|\delta u, \delta\varphi) = -\int_{\Omega_F} \rho_F \nabla\varphi \cdot \nabla\delta\varphi \, dx + \int_\Sigma \rho_F \varphi n \cdot \delta u \, d\sigma + \int_\Sigma \rho_F \delta\varphi n \cdot u \, d\sigma \tag{5.18}$$

Taking into account (5.7c) and (5.7d), $\widehat{\boldsymbol{M}}$ is expressed:

$$\widehat{\boldsymbol{M}} = \begin{bmatrix} 0 & \boldsymbol{C} \\ \boldsymbol{C}^T & -\boldsymbol{F} \end{bmatrix} \tag{5.19}$$

We see that the matrix $\boldsymbol{M}_A = \boldsymbol{C}_* {\boldsymbol{F}_*}^{-1} {\boldsymbol{C}_*}^T$ (equation (5.15)) may be deduced from $\widehat{\boldsymbol{M}}$ by applying the constraint $\varphi = 0$ on Γ, then eliminating the remaining $\boldsymbol{\Phi}_*$ by means of the condensation algorithm (1.64).
In practice the calculation of $\boldsymbol{M}_A$ only requires the mesh of the fluid domain and of the "wetted" surface Σ of the structure: the coupling term (5.7d) only involves the values of u on Σ.

- Calculation of $\widehat{\boldsymbol{M}}$ involves two types of finite elements: classical corresponding to the discretization of $\int_{\Omega_F} \rho_F \nabla\varphi \cdot \nabla\delta\varphi \, dx$ — which involves a mesh of Ω_F — and those corresponding to the discretization of $\int_\Sigma \rho_F(\varphi n \cdot \delta u + \delta\varphi n \cdot u) \, d\sigma$ which require a mesh of Σ.

  The matrix $\begin{bmatrix} 0 & \boldsymbol{C} \\ \boldsymbol{C}^T & 0 \end{bmatrix}$ is thus directly obtained by the assembly of *square symmetric* matrices corresponding to each finite element of Σ. There is therefore no need to construct separately the rectangular matrices $\boldsymbol{C}$ and $\boldsymbol{C}^T$.

- We note that the added mass matrix is a full matrix which only couples the nodal values of $\boldsymbol{U}$ on Σ.

- In practice there are various elimination algorithms enabling $\boldsymbol{M}_A$ to be constructed (frontal elimination, ...)

Remark — The elimination, by substitution, of $\boldsymbol{P}_*$ in (5.12), also gives (5.14).

Continuum-based interpretation of the added mass

For a given $u \cdot n$ on Σ — denoted here u_N — the variational equation (5.6b), which is written:
find $\varphi \in \mathcal{C}_\varphi^*$ such that, $\forall \delta\varphi \in \mathcal{C}_\varphi^*$:

$$\boxed{\int_{\Omega_F} \rho_F \nabla\varphi \cdot \nabla\delta\varphi \, dx = \int_\Sigma \rho_F u_N \delta\varphi \, d\sigma} \tag{5.20}$$

characterizes the unique solution φ to the Neumann-Dirichlet problem (5.1abc):

$$\boxed{\begin{array}{rcll} \Delta\varphi & = & 0 \quad \text{in } \Omega_F & (a) \\ \varphi & = & 0 \quad \text{on } \Gamma & (b) \\ \dfrac{\partial\varphi}{\partial n} & = & u_N \quad \text{on } \Sigma & (c) \end{array}} \tag{5.21}$$

It is readily verified that this solution, denoted φ_{u_N}, is linearly dependent on u_N.
The matrix elimination of $\boldsymbol{\Phi}_*$, which consists of replacing $\boldsymbol{\Phi}_*$ in (5.8a) by its expression (5.13) in terms of $\boldsymbol{U}$, is equivalent at the continuum based level to considering φ in (5.6) as the functional φ_{u_N} previously defined.
The spectral problem can thus be stated:
find $\lambda \in \mathbb{R}$ and $u \in \mathcal{C}_u$, such that, $\forall \delta u \in \mathcal{C}_u$:

$$\boxed{\int_{\Omega_S} \sigma_{ij}(u)\epsilon_{ij}(\delta u) \, dx = \lambda \left(\int_{\Omega_S} \rho_S u \cdot \delta u \, dx + \mathcal{M}_A(u_N, \delta u_N) \right)} \tag{5.22}$$

where the bilinear form $\mathcal{M}_A(u_N, \delta u_N)$, called the *hydroelastic added mass operator* [4] is defined by:

$$\mathcal{M}_A(u_N, \delta u_N) = \int_\Sigma \rho_F \varphi_{u_N} \delta u_N \, d\sigma \tag{5.23}$$

We show that this operator is *symmetric*: consider the solution $\varphi_{\delta u_N}$ of (5.20) for a given δu_N, taking as special test-fuction $\delta\varphi = \varphi_{u_N}$. After multiplication by ρ_F, we obtain the identity:

$$\int_\Sigma \rho_F \delta u_N \varphi_{u_N} \, d\sigma = \int_{\Omega_F} \rho_F \nabla\varphi_{\delta u_N} \cdot \nabla\varphi_{u_N} \, dx \tag{5.24}$$

which enables $\mathcal{M}_A$ to be written in the following symmetric form:

$$\boxed{\mathcal{M}_A(u_N, \delta u_N) = \int_{\Omega_F} \rho_F \nabla\varphi_{\delta u_N} \cdot \nabla\varphi_{u_N} \, dx} \tag{5.25}$$

The matrix $\boldsymbol{M}_A$ may thus be interpreted as a discretization by finite elements of this operator [5].

Spectrum of eigenvalues — In the form (5.25) it is obvious that $\mathcal{M}_A$ is positive. Given the symmetry and positivity properties of $\mathcal{M}_A$, formulation (5.22) enables us to establish that the hydroelastic problem has a denumerable infinite sequence of positive eigenvalues.

Orthogonality of the hydroelastic modes — Consider two distinct solutions $(\lambda_\alpha, u_\alpha)$ and (λ_β, u_β) of (5.22). We denote by $\varphi_{(u_N)_\alpha}$ — written φ_α — and by $\varphi_{(u_N)_\beta}$ written φ_β, the corresponding values of φ, i.e. the solutions of (5.20) for $u = u_\alpha$ and $u = u_\beta$. We have the following orthogonality relations — resulting from (5.22) and of (5.25):

$$\begin{aligned} \int_{\Omega_S} \rho_S u_\alpha \cdot u_\beta \, dx + \int_{\Omega_F} \rho_F \nabla\varphi_\alpha \cdot \nabla\varphi_\beta \, dx &= \delta_{\alpha\beta}\mu_\alpha & (a) \\ \int_{\Omega_S} \sigma_{ij}(u_\alpha)\epsilon_{ij}(u_\beta) \, dx &= \delta_{\alpha\beta}\omega_\alpha^2 \mu_\alpha & (b) \end{aligned} \tag{5.26}$$

We note that $\mu_\alpha = \int_{\Omega_S} \rho_S |u_\alpha|^2 dx + \int_{\Omega_F} \rho_F |u_\alpha^F|^2 dx$ where $u_\alpha^F = \nabla\varphi_\alpha$ is the displacement field of the liquid.

Physical interpretation of the added mass operator

The quadratic form associated with $\mathcal{M}_A$ corresponds to the inertia of the fluid: we know that $u_F = \nabla\varphi$ represents the displacement of the fluid, and therefore, from (5.25), we have:

$$\boxed{\mathcal{M}_A(u_N, u_N) = \int_{\Omega_F} \rho_F |u_F|^2 \, dx} \tag{5.27}$$

[4] $\mathcal{M}_A$ coincides with $\mathcal{M}_B^\infty$ defined in (3.103).

[5] The matrix discretizing $\mathcal{M}_A$ may be calculated by integral equations; *cf.* e.g., Khabbaz [109].

The elimination of the internal variables in the fluid in terms of the (normal) displacement of Σ, means that within the hypothesis of an inviscid incompressible liquid, and neglecting gravity terms, the kinetic energy of the fluid only depends on the normal velocity of the wall Σ.

Extremal property of the added mass operator

We show that $\mathcal{M}_A(u_N, u_N)$ verifies the following extremal property:

$$\boxed{\mathcal{M}_A(u_N, u_N) = \max_{\varphi \in \mathcal{C}^*_\varphi} \left\{ -\int_{\Omega_F} \rho_F |\nabla\varphi|^2 \, dx + 2\int_\Sigma \rho_F u_N \varphi \, d\sigma \right\}} \tag{5.28}$$

The demonstration is based on the classical property of the solution u of $a(u,v) = <f,v> \quad \forall v$, avec $a(\cdot,\cdot) > 0$:
if $J(v) = -a(v,v) + 2 <f,v>$, u verifies $J(u) = \max_v J(v)$. Using the variational equation satisfied by u, we deduce that $J(u) = \max_v J(v) = a(u,u) = 2 <f,u>$. The stated property is obtained by applying the previous result for $a(u,v) = \int_{\Omega_F} \rho_F \nabla\varphi \cdot \nabla\delta\varphi \, dx$ and $< f, v >= \int_\Sigma \rho_F u_N \, \delta\varphi \, d\sigma$.

Rayleigh quotient

The variational property (5.22) — taking into account (5.24) and (5.25) — results from the stationarity of the Rayleigh quotient $R(u)$ defined for $u \in \mathcal{C}_u$ by:

$$\boxed{R(u) = \frac{\int_{\Omega_S} \sigma_{ij}(u)\epsilon_{ij}(u) \, dx}{\int_{\Omega_S} \rho_S |u|^2 \, dx + \mathcal{M}_A(u_N, u_N)}} \tag{5.29}$$

- The numerator represents the potential energy of deformation of the structure.
- The denominator is the sum of two terms: the first corresponds classically to the inertia of the structure, and the second, as we have seen previously, to the inertia of the liquid.

Overestimation of the eigenvalues of the discretized problem

The discretization by finite elements consists of introducing the spaces of finite dimension $\mathcal{C}^h_u \subset \mathcal{C}_u$ and $\mathcal{C}^{*h}_\varphi \subset \mathcal{C}^*_\varphi$.
We show first of all that for a given discretization u^h of u, the discretization of φ has the effect of "underestimating the added mass".
The discretized added mass operator $\mathcal{M}^h_A$ results from (5.28) restricted to $\mathcal{C}^{*h}_\varphi$, and consequently, for $u^h \in \mathcal{C}^h_u$, we have the inequality:

$$\mathcal{M}^h_A(u^h_N, u^h_N) < \mathcal{M}_A(u^h_N, u^h_N) \tag{5.30}$$

Secondly, we define the two Rayleigh coefficients $R(u^h)$ and $R^h(u_h)$ calculated respectively with $\mathcal{M}_A(u^h_N, u^h_N)$ and $\mathcal{M}^h_A(u^h_N, u^h_N)$.

We thus have $R^h(u^h) > R(u^h)$ (from (5.30)). From the comparison theorem (1.37), the eigenvalues λ^h of the discretized problem are greater than the eigenvalues of the same rank of the continuum problem (to within domain approximation).
In other words, the overestimation of the eigenvalues results from the cumulative effect of the discretizations
— of φ: which results in an underestimation of $\mathcal{M}_A$ (which tends to increase the Rayleigh quotient).
— of u: which, by restricting the admissible class $\mathcal{C}_u$, also tends to increase the eigenvalues.

Remarks

- There exists another variational characterization of the added mass of an inviscid incompressible fluid, as follows:

$$\boxed{\begin{array}{c} \mathcal{M}_A(u_N, u_N) = \min\limits_{\mathcal{C}_{u_F}} \displaystyle\int_{\Omega_F} |u_F|^2 \, dx \\ \text{with} \quad \mathcal{C}_{u_F} = \{u_F \mid div u_F|_{\Omega_F} = 0 \; ; \; u_F{\cdot}n|_\Sigma = u_N\} \end{array}} \tag{5.31}$$

  This property can be used for an alternative construction of $\mathcal{M}_A$, the complication in this case arising on the one hand when we take into account the constraint $div\, u_F = 0$ at the discretized level (*cf.* Girault & Raviart [85], Oden & Reddy [164], Pironneau [187]), and on the other hand, from the vectorial nature of the field describing the fluid. This approach is equivalent to eliminating u_F in terms of u_N in the following "displacements" formulation. Consider $\mathcal{C}_{u,u_F} = \{(u, u_F) \mid div u_F = 0 \text{ in } \Omega_S, (u - u_F){\cdot}n = 0 \text{ on } \Sigma\}$. This formulation consists of finding $\lambda \in \mathbb{R}$, and $(u, u_F) \in \mathcal{C}_{u,u_F}$ such that $\forall(\delta u, \delta u_F) \in \mathcal{C}_{u,u_F}$, we have:

$$\int_{\Omega_S} \sigma_{ij}(u)\epsilon_{ij}(\delta u)\, dx = \lambda \left(\int_{\Omega_S} \rho_S u{\cdot}\delta u\, dx + \int_{\Omega_F} \rho_F u_F{\cdot}\delta u_F\, dx \right) \tag{5.32}$$

- The calculation of $\mathcal{M}_A$ by discretization of (5.31) could enable numerical bracketing of the added mass operator.

5.5 Axisymmetric structures

We work in cylindrical coordinates: z is the axis of revolution, u_r, u_z, u_θ are the components of u in local coordinates (r, z, θ)(*cf.* chapter 1). We can decompose u, φ into a Fourier series as follows:

$$\underbrace{\begin{Bmatrix} u_r \\ u_\theta \\ u_z \\ \varphi \end{Bmatrix}}_{\mathcal{C}} = \underbrace{\begin{Bmatrix} U_0^+(r,z) \\ 0 \\ W_0^+(r,z) \\ \varphi_0^+(r,z) \end{Bmatrix}}_{\mathcal{C}_0^+} + \underbrace{\begin{Bmatrix} 0 \\ V_0^-(r,z) \\ 0 \\ 0 \end{Bmatrix}}_{\mathcal{C}_0^-} + \underbrace{\begin{Bmatrix} U_n^+(r,z)\cos n\theta \\ V_n^+(r,z)\sin n\theta \\ W_n^+(r,z)\cos n\theta \\ \varphi_n^+(r,z)\cos n\theta \end{Bmatrix}}_{\mathcal{C}_n^+} + \underbrace{\begin{Bmatrix} U_n^-(r,z)\sin n\theta \\ -V_n^-(r,z)\cos n\theta \\ W_n^-(r,z)\sin n\theta \\ \varphi_n^-(r,z)\sin n\theta \end{Bmatrix}}_{\mathcal{C}_n^-} \tag{5.33}$$

The classification of the hydroelastic modes is the same as that discussed in chapter 1, (*cf.* § 1.7).

1. We note that $\varphi = 0$ on the axis of revolution, except for modes of index $n = 0$. Similarly, we have $p = \rho_F \omega^2 \varphi = 0$.

2. Modes of the type $\mathcal{C}_0^-$ (axisymmetric torsion) do not depend on the liquid (since the normal displacement of the structure is identically null in this case).

3. Since the eigenvalues show a degeneracy of 2 for $n \geq 1$, the numerical application is carried out for modes of the type $\mathcal{C}_n^+$ for example (*cf.* § 1).

5.6 Comparison of eigenfrequencies

We wish to compare the hydroelastic eigenfrequencies of a given reservoir for different filling rates.
Consider the geometrical configuration $\mathcal{P}$ of a reservoir Ω_S, $(\Omega_F, \Sigma, \Gamma)$, and the geometrical configuration $\mathcal{P}'$ of a "less filled" reservoir Ω_S, $(\Omega'_F \subset \Omega_F, \Sigma' \subset \Sigma, \Gamma')$ (Fig. 5.2). In the general case, $\mathcal{P}'$ corresponds to a lower filling rate of the inclined reservoir Ω_S (Fig. 5.2). A special case consists of considering two horizontal filling levels of the same reservoir.

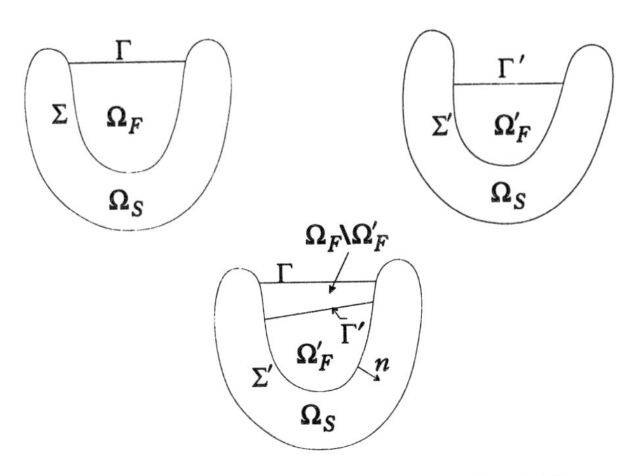

Figure 5.2: Geometric configurations $\mathcal{P}$ and $\mathcal{P}'$

We denote by φ_{u_N} and φ'_{u_N} solutions (5.21) corresponding to $\mathcal{P}$ and $\mathcal{P}'$, and by $\mathcal{M}_A$ and $\mathcal{M}'_A$ the corresponding added mass operators defined by (5.25).
The method consists of applying the first comparison theorem (1.37) to the problems having identical admissible classes.
In the present case, we therefore use the "condensed" formulation in u, whose admissible space is made up of the class $\mathcal{C}_u$ of structure displacements and whose Rayleigh quotient is given by (5.29).

The Rayleigh quotients are expressed:

$$\left.\begin{aligned} R(u) &= \frac{\int_{\Omega_S} \sigma_{ij}(u)\epsilon_{ij}(u)\,dx}{\int_{\Omega_S} \rho_S|u|^2\,dx + \mathcal{M}_A(u_N, u_N)} \\ R'(u) &= \frac{\int_{\Omega_S} \sigma_{ij}(u)\epsilon_{ij}(u)\,dx}{\int_{\Omega_S} \rho_S|u|^2\,dx + \mathcal{M}'_A(u_N, u_N)} \end{aligned}\right\}, \quad \text{with} \quad u \in \mathcal{C}_u \tag{5.34}$$

We show that:

$$R(u) \leq R'(u) \quad \forall u \in \mathcal{C}_u \tag{5.35}$$

This is equivalent to showing that, for the same u_N, $\mathcal{M}_A \geq \mathcal{M}'_A$, i.e. that:

$$\Delta\mathcal{M}_A = \int_{\Omega_F} \rho_F|\nabla\varphi_{u_N}|^2\,dx - \int_{\Omega'_F} \rho_F|\nabla\varphi'_{u_N}|^2\,dx \geq 0 \tag{5.36}$$

$\Delta\mathcal{M}_A$ may be written:

$$\Delta\mathcal{M}_A = \int_{\Omega_F\backslash\Omega'_F} \rho_F|\nabla\varphi_{u_N}|^2\,dx + \int_{\Omega'_F} \rho_F\left(|\nabla\varphi_{u_N}|^2 - |\nabla\varphi'_{u_N}|^2\right)\,dx \tag{5.37}$$

Using the identity $a^2 - b^2 = (a-b)^2 + 2b(a-b)$, $\Delta\mathcal{M}_A$ is written:

$$\begin{aligned} \Delta\mathcal{M}_A = &\int_{\Omega_F\backslash\Omega'_F} \rho_F|\nabla\varphi_{u_N}|^2\,dx + \cdots \\ \cdots + &\int_{\Omega'_F} \rho_F|\nabla\varphi_{u_N} - \nabla\varphi'_{u_N}|^2\,dx + 2\int_{\Omega'_F} \rho_F\nabla\varphi'_{u_N}\cdot(\nabla\varphi_{u_N} - \nabla\varphi'_{u_N})dx \end{aligned} \tag{5.38}$$

We show that the last term is null.
We note that, from Green's formula (3.16), we have:

$$\int_{\Omega'_F} \nabla\varphi'\cdot(\nabla\varphi - \nabla\varphi')\,dx = -\int_{\Omega'_F} \varphi'\Delta(\varphi-\varphi')\,dx + \int_{\partial\Omega'_F} \varphi'(\frac{\partial\varphi}{\partial n} - \frac{\partial\varphi'}{\partial n})\,d\sigma \tag{5.39}$$

The first integral is null since φ and φ' satisfy (5.21a). The contribution to Γ' of the second term is null as φ' satisfies (5.21b). The contribution on Σ' is null as φ and φ' verify (5.21c), for the same value of u_N. Therefore, $\Delta\mathcal{M}_A$ is written:

$$\Delta\mathcal{M}_A = \int_{\Omega_F\backslash\Omega'_F} \rho_F|\nabla\varphi_{u_N}|^2\,dx + \int_{\Omega'_F} \rho_F|\nabla\varphi_{u_N} - \nabla\varphi'_{u_N}|^2\,dx \tag{5.40}$$

From the first comparison theory, if λ_n and λ'_n respectively denote the *nth* eigenvalues of the problems $\mathcal{P}$ and $\mathcal{P}'$, we have:

$$\boxed{\lambda_n \leq \lambda'_n} \tag{5.41}$$

- This inequality means that the natural hydroelastic frequencies increase when the mass of liquid decreases (we have seen that for sloshing, the situation is different) (*cf.* Morand [141]).
- The inequality (5.41) enables the hydroelastic frequencies to be bracketed: for the domain of liquids verifying $\Omega'_F \subset \Omega_F \subset \Omega''_F$, we have the two-sided estimation $\lambda''_n \leq \lambda_n \leq \lambda'_n$, $\forall n$.

Derivative of eigenvalues in terms of the height h of liquid

We consider the geometric configuration $\mathcal{P}(h)$ corresponding to a height h of liquid (domain $\Omega_F(h)$, free surface $\Gamma(h)$, (Fig. 5.3)). We assume that the eigenvalues are *simple*, and we consider the *nth* natural hydroelastic mode $\lambda_n(h), u_n(h), \varphi_n(h)$. To $\varphi_n(h)$ corresponds the vertical displacement $\eta_n = \frac{\partial \varphi_n}{\partial z}|_\Gamma$ of the free surface.

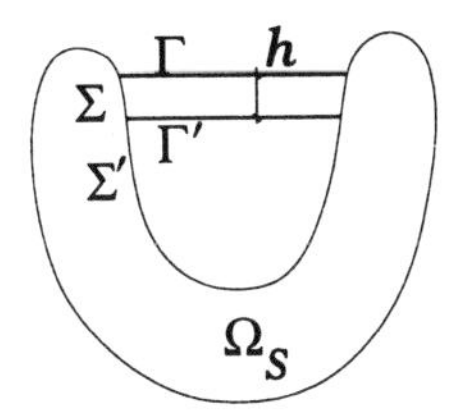

Figure 5.3: Variation in the height of the liquid

We shall establish the following result:

$$\boxed{\frac{1}{\lambda_n}\frac{\partial \lambda_n}{\partial h} = -\frac{\int_\Gamma \rho_F \eta_n^2}{\mu_n}} \tag{5.42}$$

Referring to the Rayleigh quotient (5.29), taking into account the fact that only $\mathcal{M}_A$ depends on h, the derivative of the nth. eigenvalue with respect to h is expressed (*cf.* chapter 1):

$$\frac{1}{\lambda_n}\frac{\partial \lambda_n}{\partial h} = -\frac{1}{\mu_n}\frac{\partial \mathcal{M}_A(u_N, u_N)}{\partial h} \tag{5.43}$$

The demonstration is based on the calculation of the derivative of $\mathcal{M}_A$ with respect to the parameter h.
We call the infinitesimal variation of the liquid height δh, and the corresponding variation of $\mathcal{M}_A$, $\delta\mathcal{M}_A$, which we shall evaluate from the expression of the finite variation $\Delta\mathcal{M}_A$ given by (5.40), for two successive configurations $(\Omega_F, \Sigma, \Gamma)$ and $(\Omega'_F \subset \Omega_F, \Sigma' \subset \Sigma, \Gamma')$.
We begin by showing that the second term of (5.40) is of order $(\delta h)^2$.
Assuming the continuity of φ and $\nabla\varphi$, we have $\nabla(\varphi - \varphi') \sim \delta h$ and thus, we have $|\nabla(\varphi - \varphi')|^2 \sim \delta h^2$.
In the second step, we transform the integral on $\Omega_F \setminus \Omega'_F$ by means of Green's formula, taking into account equations (5.21) satisfied by φ, which yields:

$$\delta\mathcal{M}_A \sim -\int_{\Gamma'} \rho_F \varphi \frac{\partial \varphi}{\partial z}\, d\sigma + \int_{\Sigma\setminus\Sigma'} \rho_F \varphi\, u_N\, d\sigma \tag{5.44}$$

The second term is in h^2 since, on the one hand, the size of $\Sigma \setminus \Sigma'$ is of the order h, and on the other hand $\varphi|_{\Gamma'} \sim -\frac{\partial\varphi}{\partial z}|_\Gamma h$ (noting that $\varphi|_\Gamma = 0$).

Finally, replacing $\varphi|_{\Gamma'}$ by this latter expression in the first term, we arrive at $\delta \mathcal{M}_A \sim h \int_{\Gamma'} \rho_F (\frac{\partial \varphi}{\partial z})^2 \, d\sigma$ which, replacing $\frac{\partial \varphi}{\partial z}$ by η and replacing (to the order h) the integral on Γ' by an integral Γ, is written:

$$\frac{\partial \mathcal{M}_A}{\partial h} = \int_\Gamma \rho_F \eta^2 \, d\sigma \tag{5.45}$$

Substituting this expression into (5.43) we find the stated result (5.42).
For a given configuration h whose modal characteristics are known, formula (5.42) enables the slope at this point to be calculated from the curves of frequency as a function of h (*cf.* Morand [136] and Morand & Ohayon [139] for application of the Ritz method to the case of finite perturbations).

Case of systems presenting symmetries — We can show that systems with symmetries, having multiple eigenvalues, can be reduced to simple eigenvalue problems.
As an example, the case of axisymmetric structures containing liquids can be reduced to the case of a simple eigenvalue, provided we consider the behaviour of eigenvalues corresponding to modes of the given type $\mathcal{C}_n^+$.

5.7 Open problems

Problems under development include deeper study of local problems of connection of reservoirs to flexible pipes in the presence of flow of compressible viscous liquids ("outflow").

Normally, the coupling between the reservoir and the feed line involves the modal pressure in the liquid calculated at the reservoir-pipe interface and assuming the reservoir to be sealed at the pipe inlet.

We should also indicate that pipe structures are generally modelled by beam models, whereas reservoirs are modeled by shell models and, therefore, we have in general the problem of junctions in multi-structures occurring in launchers (*cf.* Ciarlet [39]).

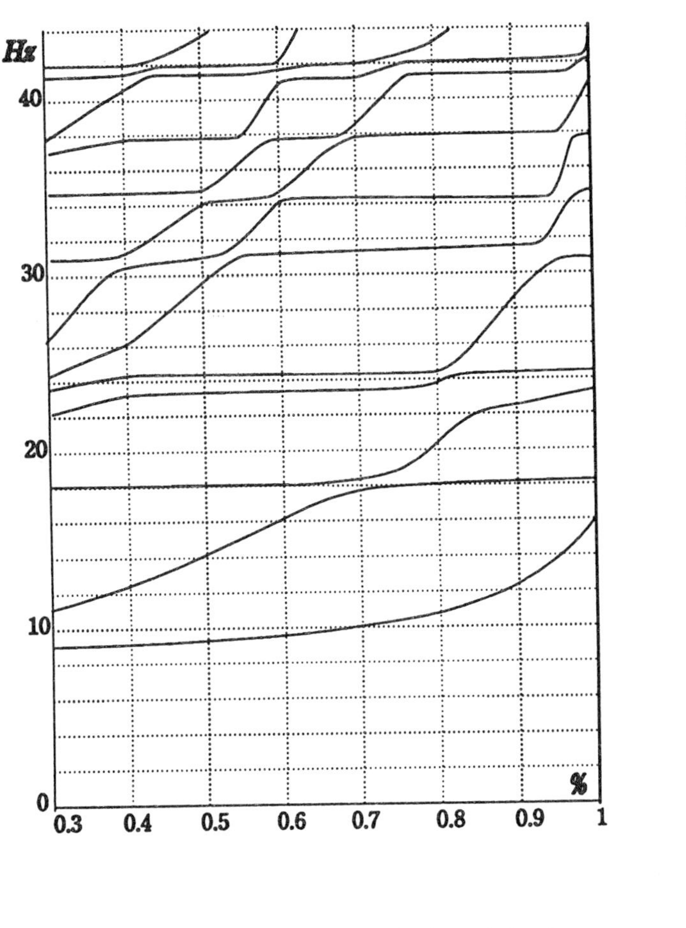

Upper part (payload and supports of satellites)

Third stage

Second stage

First stage

The above diagram shows the evolution of the frequencies of the axisymmetric hydroelastic vibration modes ($n = 0$) of the launcher Ariane 4 as a function of the combustion rate of the first stage (liquid filling ratio).

Note the increasing of the frequencies (*cf.*§(5.6)) and the curve veering phenomenon (*cf.*§(1.8)).

The diagram opposite shows the first hydroelastic mode at end of flight (frequency 15Hz), represented in a meridian plane (equilibrium and deformed configuration for an arbitrary normalization).

The structure of the launcher is modelled by shells of revolution discretized by finite elements, and the liquids are modelled by an added mass matrix calculated by finite elements (*cf.*§(5.4)); only deformations of the shells are shown here, with liquids shown at their equilibrium level).

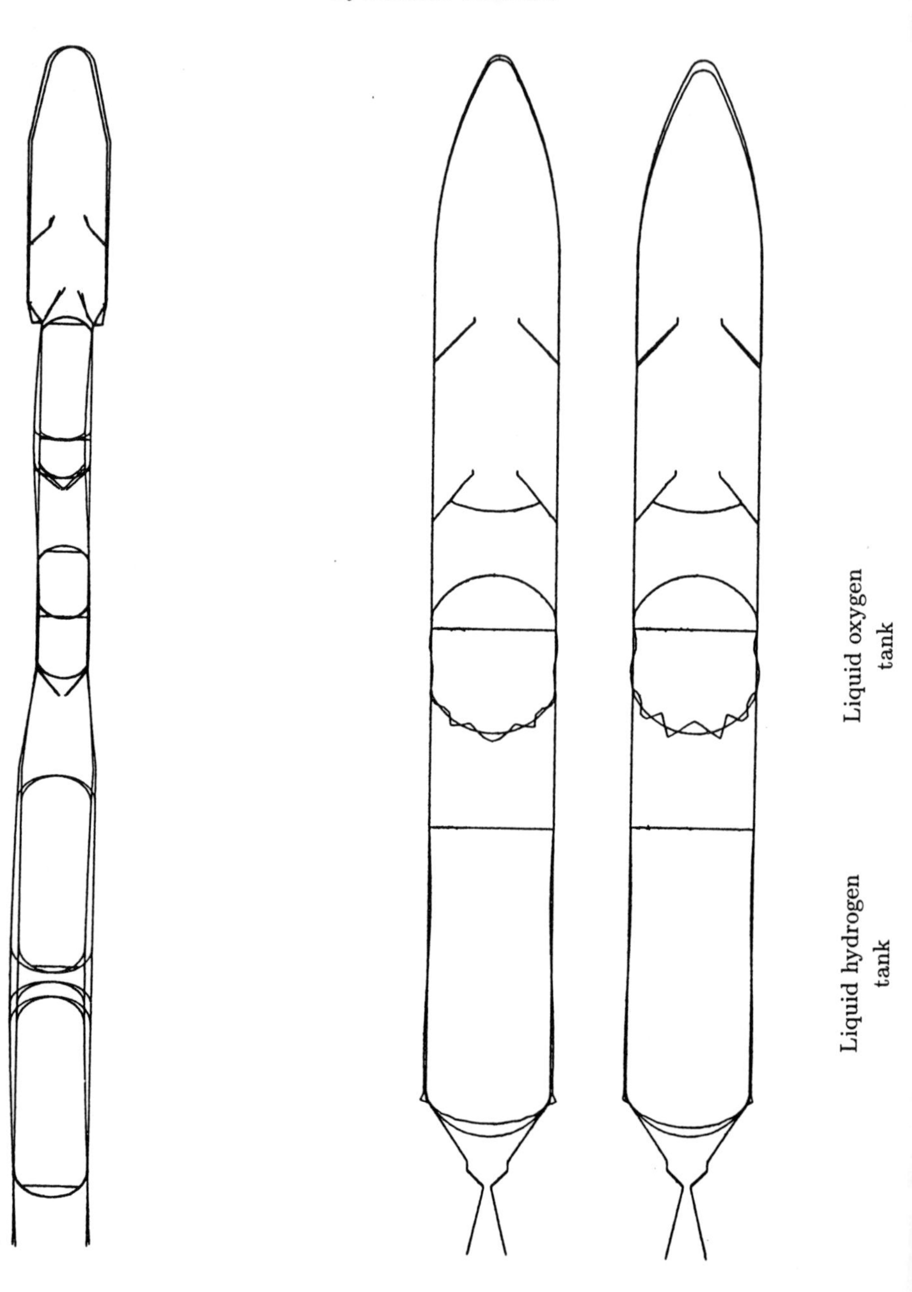

Structural modal shape of the fourth hydroelastic mode $n = 1$ of the launcher Ariane 4 in an end-of-flight configuration of the first stage (note the similarity to a beam bending mode).

Structural modal shapes of two hydroelastic modes $n = 0$ with close frequencies (47.5 and 47.7Hz) of the launcher Ariane 5, after separation of the solid propergol stages. These modal shapes appear as a combination in phase and out of phase of a mode of the liquid oxygen tank and a mode of the liquid hydrogen tank (*cf.* §(1.8), page 16).

CHAPTER 6

Hydroelastic vibrations under gravity

6.1 Introduction

In chapter 5, we considered the problem of vibrations of an elastic structure containing an incompressible liquid with a free surface, neglecting the effects of gravity. Within that approximation, we saw that the coupled problem can be formulated in terms of displacements u of the structure, after introducing an added mass operator.

Here we are interested in coupled vibrations of an elastic structure containing an incompressible liquid, taking into consideration the effects of gravity. [1]

An example of the application, in the aerospace domain, is the attitude control of liquid propelled launchers which requires the simultaneous treatment of sloshing and the flexural vibrations of the launcher.

Note that gravity is involved as a restoring force on the free surface and through the prestress effect induced at equilibrium in the structure.

First we establish the linearized equations of the response of a structure subjected to a pressure and to given forces in the presence of a gravitational field, together with the corresponding variational formulation which introduces the *symmetric elastogravity* stiffness operator.

We next establish a non-symmetric formulation in (u, φ) of the coupled problem, which will be used in chapter 9 for analysis of this problem by modal substructuring.

We shall show that introduction of an additional unknown η describing the displacement of the free surface enables us to arrive at a *symmetric* formulation of this problem, for the purpose of direct calculation of the response to given forces by finite elements.

---

[1] A general treatment of these problems may be found in Abramson [1] (*cf.* also Ma [126] and Crolet & Ohayon [50] for various applications.)

6.2 Structure subject to a pressure field and to gravity

The notation is that of Fig. 6.1.

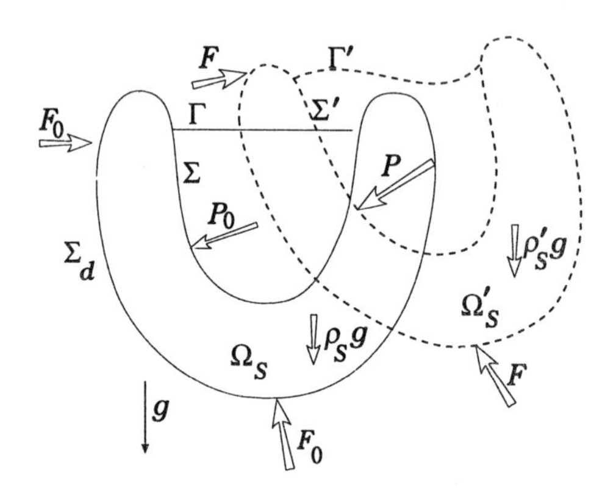

Figure 6.1: Equilibrium and deformed configurations

Configuration	equilibrium	instant t
Solid (*resp.* liquid) particle position	M_S (*resp.* M_F)	M'_S (*resp.* M'_F)
Structure domain (*resp.* liquid)	Ω_S (*resp.* Ω_F)	Ω'_S (*resp.* Ω'_F)
Liquid free surface	Γ	Γ'
Structure-liquid contact surface	$\Sigma = \partial\Omega_F \setminus \Gamma$	$\Sigma' = \partial\Omega'_F \setminus \Gamma'$
Normal external to the structure	n	n'
Area element (physical)	$d\sigma$	$d\sigma'$
Mass density of structure	ρ_S	ρ'_S
Cauchy stress tensor	σ^0_{ij}	σ_{ij}
Pressure actions of liquid	$P_0\, d\sigma$	$P\, d\sigma'$
External forces	$F_0\, d\sigma$	$F\, d\sigma'$

To linearize the equations, we proceed as follows:

We write the dynamic equations in the current state Ω'_S and in the equilibrium state Ω_S, in the form of the principle of virtual power.

We then carry out a "lagrangian transport by reciprocal mappings" on the principle of virtual power in Ω'_S, in order to arrive at a formulation of the problem in the equilibrium domain Ω_S.

Finally, we establish the equations of small movements by linearization.

Principle of virtual power

Current configuration

For any "kinematically admissible virtual velocity field" $v(M')$ defined in Ω'_S we have:

$$\begin{aligned}\int_{\Omega'_S} \sigma_{ij} v_{i,j}\, dx' + \int_{\Omega'} \rho'_S \gamma' \cdot v\, dx' = \\ \cdots - \int_{\Sigma'} (P\, n'\, d\sigma') \cdot v - \int_{\Omega'_S} \rho'_S g v_z\, dx' + \int_{\partial\Omega'_S \setminus \Sigma'} (F\, d\sigma') \cdot v\end{aligned} \qquad (6.1)$$

where γ' is the acceleration vector in the current state, $v_z = v \cdot i_z$ is the vertical component of v, and where F denotes the surface density of the given external forces on $\Sigma'_d = \partial\Omega'_S \setminus \Sigma'$.

Equilibrium configuration

For all kinematically admissible $v^0(M)$ defined in Ω_S, we have:

$$\int_{\Omega_S} \sigma^0_{ij} v^0_{i,j}\, dx = - \int_{\Sigma} (P_0\, n\, d\sigma) \cdot v^0 - \int_{\Omega_S} \rho_S g \cdot v^0_z\, dx + \int_{\partial\Omega_S \setminus \Sigma} (F_0\, d\sigma) \cdot v^0 \qquad (6.2)$$

where P_0 is the external pressure exerted by the fluid on Σ (which verifies the equation $\nabla P_0 = -\rho_F g i_z$ (2.2) in the fluid domain), $v^0_z = v^0 \cdot i_z$ the vertical component of v^0 and F_0 the surface density of the *external forces.* [2]

Lagrangian transport in Ω_S

In order to compare (6.1) and (6.2), we "shift" equation (6.1) by reciprocal mapping onto the equilibrium domain (Fig. 6.2). We introduce the displacement u^S of the particle located at M_S at equilibrium according to: $M'_S = M_S + u^S(M_S, t)$, and we introduce the reciprocal mappings G^* in Ω_S of a quantity G defined in Ω'_S by the relation:

$$G^*(M_S, t) = G(M'_S, t) \quad \text{with } M'_S = M_S + u^S(M_S, t),\ \forall M_S \in \Omega_S \qquad (6.3)$$

The reciprocal mappings of the surface and volume integrals are then written respectively:

$$\begin{aligned}\int_{\Omega'_S} G(M'_S)\, dx' &= \int_{\Omega_S} G^*(M_S)\, (dx')^* & (a)\\ \int_{\Sigma'_S} G(M'_S)\, n' \cdot d\sigma' &= \int_{\Sigma_S} G^*(M_S)\, (n'\, d\sigma')^* & (b)\end{aligned} \qquad (6.4)$$

where $(dx')^* = J\, dx$ is the reciprocal mapping of the volume element dx', and J denotes the Jacobian of the transformation $M_S \to M'_S$ ($J = \det F$ with $dM'_S = F dM_S$), and where $(n'\, d\sigma')^*$ is the reciprocal mapping of the "oriented element of area" the expression for which is discussed below.
With these notations, each term (6.1) is transformed in the following way:

---

[2]These forces balance the forces of gravity on the structure and of the liquid pressure on $\Sigma_d = \partial\Omega_S \setminus \Sigma$.

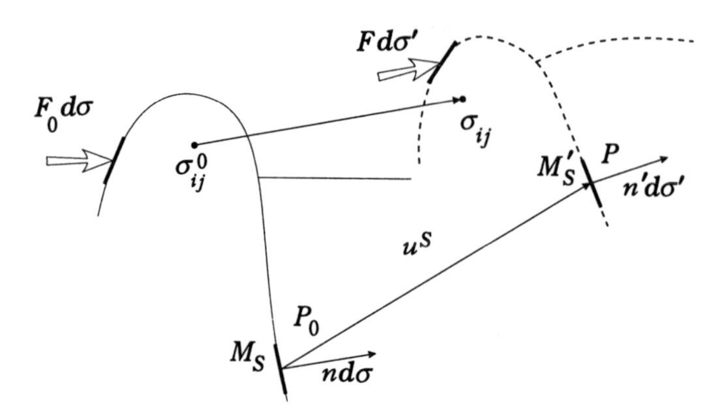

Figure 6.2: Lagrangian transport

1. The first term becomes:

$$\int_{\Omega'_S} \sigma_{ij} v_{i,j}\, dx' = \int_{\Omega_S} \theta_{ij} v^*_{i,j}\, dx \tag{6.5}$$

where θ_{ij} is the first Piola-Kirchhoff stress tensor (non-symmetric Boussinesq tensor) defined by:

$$\theta_{ik} = J\sigma^*_{ij} \frac{\partial x_k}{\partial x'_j} \tag{6.6}$$

This result is obtained by using $(dx')^* = Jdx$ and noting that $(v_{i,j})^* = \dfrac{\partial v^*_i}{\partial x_k} \dfrac{\partial x_k}{\partial x'_j}$.

2. The second term becomes:

$$\int_{\Omega'_S} \rho'_S \gamma' \cdot v\, dx' = \int_{\Omega_S} \rho_S \frac{\partial^2 u^S}{\partial t^2} \cdot v^*\, dx \tag{6.7}$$

This result comes from the equality $(\rho'_S\, dx')^* = \rho_S dx$ (which results from the conservation of mass), and from $\gamma'^* = \dfrac{\partial^2 u^S(M_S, t)}{\partial t^2}$.

3. The first term of the second member of (6.1) is written:

$$\int_{\Sigma'} (Pv)^* \cdot (n'\, d\sigma') = \int_{\Sigma} P^* v^* \cdot (n\, d\sigma)^* \tag{6.8}$$

where $P^*(M_S, t) = P(M'_S, t)$ is the instantaneous pressure at the solid particle located at M_S at equilibrium .

$P^*(M_S, t)$ is therefore the reciprocal mapping (on Σ) of the trace on Σ' of $P(M'_S, t)$.

4. The second term of the second member is written:

$$\int_{\Omega'_S} \rho'_S g v_z \, dx' = \int_{\Omega_S} \rho_S g v_z^* \, dx \tag{6.9}$$

This result arises from the fact that $(\rho'_S dx')^* = \rho_S dx$ (conservation of mass), and also noting that g is constant.

5. The last term of the second member is written:

$$\int_{\partial\Omega'_S \setminus \Sigma'} (F \, d\sigma') \cdot v = \int_{\partial\Omega_S \setminus \Sigma} (F \, d\sigma)^* \cdot v^* \tag{6.10}$$

We can now compare the dynamic equations in Ω_S with the equilibrium conditions (6.2). We *choose* $v^0 = v^*$ in (6.2), i.e. we take, for v^0, the reciprocal mapping of v. The reciprocal mappings of (6.1) and equation (6.2) are written respectively:

$$\begin{aligned}
\int_{\Omega_S} \theta_{ij} v_{i,j}^* dx + \int_{\Omega_S} \rho_S \frac{\partial^2 u^S}{\partial t^2} \cdot v^* dx &= \cdots \\
\cdots - \int_\Sigma P^* v^* \cdot (n d\sigma)^* - \int_{\Omega_S} \rho_S g v_z^* dx + \int_{\Sigma_d} (F d\sigma)^* \cdot v^* &\quad (a) \\
\int_{\Omega_S} \sigma_{ij}^0 v_{i,j}^* dx = - \int_\Sigma P_0 v^* \cdot (n d\sigma) - \int_{\Omega_S} \rho_S g v_z^* dx + \int_{\Sigma_d} (F_0 d\sigma) \cdot v^* &\quad (b)
\end{aligned} \tag{6.11}$$

A general treatment of non-linear elasticity may be found in Germain [82], [83], Marsden & Hughes [129], Ciarlet [37], [38].

Linearization

We begin by writing the difference between equations (6.11a) and (6.11b), which yields: [3]

$$\begin{aligned}
\int_{\Omega_S} (\theta_{ij} - \sigma_{ij}^0) v_{i,j}^* dx + \int_{\Omega_S} \rho_S \frac{\partial^2 u^S}{\partial t^2} \cdot v^* dx &= \\
- \int_\Sigma [P^* (n \, d\sigma)^* - P_0 (n \, d\sigma)] \cdot v^* + \int_{\Sigma_d} [(F \, d\sigma)^* - (F_0 \, d\sigma)] \cdot v^*
\end{aligned} \tag{6.12}$$

1. **Linearization of the given forces term** — We introduce the surface force density $\widetilde{F}$ on Σ_d defined by:

$$\widetilde{F} \, d\sigma = (F \, d\sigma)^* - (F_0 \, d\sigma) \tag{6.13}$$

where $\widetilde{F}$ represents the variation of the applied forces with respect to the equilibrium state, *assumed to be given* (infinitesimal).

[3] We note that the volumic terms due to gravity are not involved.

2. **Linearization of the stress terms** — For a *hyperelastic medium* the linearization of $\theta_{ij} - \sigma_{ij}^0$ yields:

$$\theta_{ij} - \sigma_{ij}^0 = u_{i,h}^S \sigma_{hj}^0 + a_{ijkh}\epsilon_{kh}(u^S) \tag{6.14}$$

where $\epsilon_{kh}(u) = (u_{k,h} + u_{h,k})/2$ is the linear part of the deformation tensor, and a_{ijkh} are the usual coefficients of elasticity (which coincide with those defined in the natural state for the small prestresses σ_{ij}^0, see for example Mandel [128] vol. 2, Marsden & Hughes [129]).

3. **Linearization of the pressure term** — To begin with, we let:

$$p^S(M_S, t) = P^*(M_S, t) - P_0(M_S) \tag{6.15}$$

p^S is the (lagrangian) pressure fluctuation of the particle located at $M_S \in \Sigma$ at equilibrium. We next consider the development to the first order in u^S of the oriented area element $n\, d\sigma$:

$$(n\, d\sigma)^* = n\, d\sigma + n_1(u^S)\, d\sigma \tag{6.16}$$

where $n_1(u^S)$ is a term which is linearly dependent on u^S and which will be explained later. Under these conditions, we have:

$$P^*(n\, d\sigma)^* - P_0(n\, d\sigma) = P_0 n_1(u^S)\, d\sigma + p^S\, d\sigma \tag{6.17}$$

Under these conditions, the linearization of (6.12) yields:

$$\begin{aligned} &\int_{\Omega_S} a_{ijkh}\epsilon_{kh}(u^S)\epsilon_{ij}(v^*)\, dx + \int_{\Omega_S} \sigma_{hj}^0 u_{i,h}^S v_{i,j}^*\, dx + \int_{\Omega_S} \rho_S \frac{\partial^2 u^S}{\partial t^2} \cdot v^* dx + \cdots \\ &\cdots + \int_\Sigma P_0\, n_1(u^S) \cdot v^*\, d\sigma + \int_\Sigma p^S\, v^* \cdot n d\sigma = \int_{\Sigma_d} \widetilde{F} \cdot v^*\, d\sigma \end{aligned} \tag{6.18}$$

Remarks — The first term coincides with the mechanical stiffness term discussed previously.
The second term arises from consideration of the prestresses in the structure at equilibrium.

Expression of p^S in terms of p and u^S

In (6.18), $\int_\Sigma p^S v^* \cdot n\, d\sigma$ results from the pressure actions of the liquid on Σ.
The basic equations of the liquid involve the eulerian fluctuation p (*cf.* (2.5)); we shall show that p^S can be expressed in terms of p and u^S.
We consider a point M of Σ, and the liquid and solid particles in contact at equilibrium at that point ($M_S = M_F = M$). The respective instantaneous positions of these particles, denoted M'_S and M'_F, are thus given by (Fig. 6.3):

$$M \in \Sigma \quad \left\{ \begin{array}{lcll} M'_S & = & M + u^S(M,t) & (a) \\ M'_F & = & M + u^F(M,t) & (b) \end{array} \right. \tag{6.19}$$

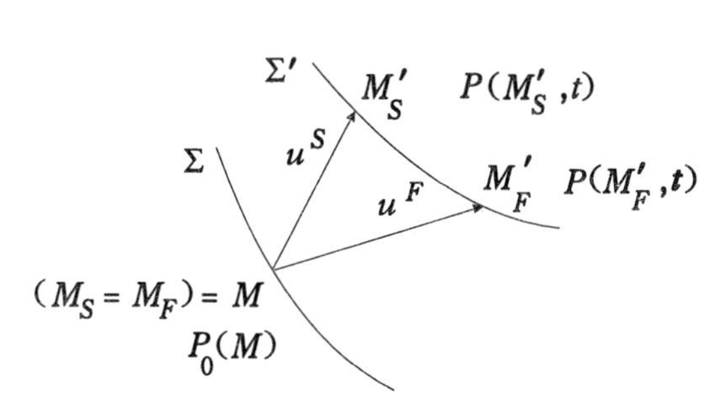

Figure 6.3: Liquid-structure contact

We consider the trace of the lagrangian fluctuation of pressure $p_{\mathcal{L}}$ on Σ given by (*cf.* (2.4)):

$$p_{\mathcal{L}}(M,t) = P(M'_F,t) - P_0(M) \quad \text{with } M'_F = M + u^F(M,t) \tag{6.20}$$

We then calculate $p_{\mathcal{L}} - p^S$ by difference from (6.15) and (6.20). Next we perform the expansion limited to the first order of $p_{\mathcal{L}} - p^S$, which yields, using $\nabla P_0 = -\rho_F g i_z$:

$$p_{\mathcal{L}}(M) - p^S(M) \sim \nabla P_0 \cdot (u^F - u^S) = -\rho_F g\, i_z \cdot (u^F - u^S) \tag{6.21}$$

Finally, using the relation $p_{\mathcal{L}} = p - \rho_F g i_z \cdot u^F$ between lagrangian and eulerian pressure fluctuations of the liquid (relation 2.7), we obtain:

$$\boxed{p^S = p - \rho_F\, g\, i_z \cdot u^S \quad \text{on } \Sigma} \tag{6.22}$$

Variational formulation

On substituting (6.22) into (6.18), and replacing P_0 by its value $P_0 = -\rho_F g z$, (z denoting the elevation of $M \in \Sigma$ with respect to an origin point located on Γ), we obtain the following variational property relating the displacement u^S to the pressure p of the liquid:

$$\begin{aligned} &\underbrace{\int_{\Omega_S} a_{ijkh}\epsilon_{kh}(u^S)\epsilon_{ij}(v^*)\,dx}_{k_E} + \underbrace{\int_{\Omega_S} \sigma^0_{hj} u^S_{i,h} v^*_{i,j}\,dx}_{k_G} + \int_{\Omega_S} \rho_S \frac{\partial^2 u^S}{\partial t^2} \cdot v^* dx + \cdots \\ &\underbrace{-\rho_F g \int_\Sigma \left[z n_1(u^S) \cdot v^* + i_z \cdot u^S\, v^* \cdot n \right] d\sigma}_{k_\Sigma} = \int_{\Sigma_d} \widetilde{F} \cdot v^*\, d\sigma - \int_\Sigma p\, v^* \cdot n\, d\sigma \end{aligned} \tag{6.23}$$

Symmetric elastogravity operator

The first member of (6.23) involves, as well as the inertia term, the *elastogravity* operator $\widehat{k}(u^S, v^*)$ defined as the sum of the following three contributions:

1. k_E is the usual elastic stiffness *symmetric* bilinear form,

2. k_G is the so-called "geometric rigidity" *symmetric* bilinear form which involves the prestresses σ^0_{ij} originating both from gravity terms and from any internal pressurization of the reservoirs,

3. k_Σ is a bilinear form which is a function of the displacements on Σ, of which we shall establish the symmetry (for the linearization and establishing of k_Σ, *cf.* Morand [148]). [4]

In the case of *slender* structures (beams, plates, shells), the bilinear forms k_E and k_G have simply to be replaced by the corresponding variational expressions.

Symmetry of k_Σ — We use here the abbreviated notation u, v for the trace of (u^S, v^*) on Σ. We denote by (u_z, v_z) the vertical components of u and v, and by u_N and v_N the components of u and v normal to Σ.
We thus write:

$$\boxed{k_\Sigma(u,v) \;=\; -\rho_F g \int_\Sigma [z\, n_1(u)\cdot v \;+\; u_z v_N]\; d\sigma} \tag{6.24}$$

We have to show that $k_\Sigma(u,v) = k_\Sigma(v,u)$, for all vector pairs (u,v) defined on Σ.
The symmetry condition $k_\Sigma(u,v) = k_\Sigma(v,u)$ is written:

$$\int_\Sigma z\,[n_1(u)\cdot v - n_1(v)\cdot u]\; d\sigma \;=\; \int_\Sigma [v_z\, u_N - u_z\, v_N]\; d\sigma \tag{6.25}$$

We transform the first member $\int_\Sigma z\,[n_1(u)\cdot v - n_1(v)\cdot u]\; d\sigma$ of (6.25) by using the formalism of the external differential forms on the differential manifolds.
We denote by dM the 1-form $i_x dx + i_y dy + i_z dz$, by $\wedge$ the exterior (or wedge) product and by $d\omega$ the derivation of the differential form ω.
Given two differential 1-forms ω_1 and ω_2 with vector values in R^3, and a bilinear mapping $\mathcal{B}$ of $R^3 \times R^3 \longrightarrow R^3$, we denote by $\omega_1 \wedge_\mathcal{B} \omega_2$ the exterior product of ω_1 and ω_2 in the sense of $\mathcal{B}$.
We can then define the "oriented element of area" $n\, d\sigma = (1/2) dM \wedge_\times dM$ where $\times$ denotes the vectorial product in R^3 oriented.

1. From (6.16), and noting that according to (6.3) $dM' = dM + du$, we have $n_1(u)\, d\sigma = dM \wedge_\times du$, to the first order in u. The first member — denoted I — of (6.25) is then written:

$$I \;=\; \int_\Sigma z[(dM \wedge_\times du)\cdot v - (dM \wedge_\times dv)\cdot u] \tag{6.26}$$

[4] Another derivation of k_Σ may be found in Debongnie [55]. An approximate consideration of the contribution of gravity terms on Σ has yielded a non symmetric bilinear form of the type $\int_\Sigma \rho_F\, g i_z\cdot u^F\, n\cdot\delta u^S\, d\sigma$, replaced by the symmetric bilinear form $\int_\Sigma \rho_F g i_z\cdot nn\cdot u^S n\cdot\delta u^S\, d\sigma$ (*cf.* Tong [206]).

Putting $dM = i_x dx + i_y dy + i_z dz$ and $du = i_x du_x + i_y du_y + i_z du_z$ in the preceding expression, we establish the property:

$$d\{(u \times v)\cdot dM\} = (dM \wedge_\times dv)\cdot u - (dM \wedge_\times du)\cdot v \tag{6.27}$$

2. On writing $zd\{(u \times v)\cdot dM\} = d[z\{(u \times v)\cdot dM\}] - dz \wedge \{(u \times v)\cdot dM\}$, then applying Stokes' formula, and noting that $z = 0$ on the boundary $\partial\Sigma = \gamma$ of Σ, I is written:

$$I = \int_\Sigma dz \wedge \{(u \times v)\cdot dM\} \tag{6.28}$$

3. By calculating I in terms of the components (u, v), we find the following expression for the first member of (6.25):

$$I = \int_\Sigma [(u_y v_z - u_z v_y)\, dz \wedge dx - (u_z v_x - u_x v_z)\, dy \wedge dz] \tag{6.29}$$

4. Finally, on calculating the second member $\int_\Sigma [v_z\, u\cdot n - u_z\, v\cdot n]\, d\sigma$ of (6.25) in terms of the components of u and v, we verify that the result obtained coincides with (6.29), which completes the demonstration.

We can thus write k_Σ in the following explicit *symmetric* form:

$$\boxed{\begin{aligned} k_\Sigma(u, v) = -\frac{1}{2}\rho_F g \Big\{ & \int_\Sigma [z\, n_1(u)\cdot v + u_z\, v_N]\, d\sigma + \cdots \\ & \int_\Sigma [z\, n_1(v)\cdot u + v_z\, u_N]\, d\sigma \Big\} \end{aligned}} \tag{6.30}$$

Remark

The elastogravity operator has been used in geophysics for the study of vibrations of the Earth involving the internal fluid masses and the gravity-induced prestresses (*cf.* Dahlen [51], Valette [208]).

Variational formulation in the harmonic case

We arrive at the harmonic case by letting $u^S(M, t) = u(M) \cos \omega t$.
If we denote by $\mathcal{C}_u$ the space of the admissible u, and use the previous notation, the variational formulation (6.23) in the harmonic state, for $u \in \mathcal{C}_u$ and $\forall \delta u \in \mathcal{C}_u$ is written [5]:

$$\boxed{\widehat{k}(u, \delta u) - \omega^2 \int_{\Omega_S} \rho_S u\cdot\delta u\, dx - \int_\Sigma p\, \delta u\cdot n^F\, d\sigma = \int_{\Sigma_d} \widetilde{F}\cdot\delta u\, d\sigma} \tag{6.31}$$

where $n^F (= -n)$ is the external unitary normal to the domain Ω_F, and where we have let:

$$\widehat{k} = (k_E + k_G + k_\Sigma) \tag{6.32}$$

[5] we have replaced v by the notation δu.

6.3 (u, φ) unsymmetric formulation

We obtain the variational formulation of the coupled problem [6], if we recall:

1. the variational formulation (3.65) of the response of the liquid to a "deformation of the wall" established in chapter 3, (*cf.* 3.65),

2. and the variational formulation (6.31) of the response of the structure "subject to the pressure p of the liquid" — on replacing p by its expression $p = \rho_F \omega^2 \varphi - (\rho_F g/\text{Area}(\Gamma)) \int_\Sigma u \cdot n^F \, d\sigma$ (*cf.* 3.9).

The variational formulation of the coupled problem is thus written:
For given ω and $\widetilde{F}$, find $(u, \varphi) \in \mathcal{C}_u \times \mathcal{C}^*_\varphi$ such that: $\forall(\delta u, \delta\varphi) \in \mathcal{C}_u \times \mathcal{C}^*_\varphi$, we have:

$$\boxed{\begin{array}{ll} k^0(u, \delta u) - \omega^2 \int_{\Omega_S} \rho_S u \cdot \delta u \, dx - \omega^2 \int_\Sigma \rho_F \varphi \delta u \cdot n^F \, d\sigma = \int_{\Sigma_d} \widetilde{F} \cdot \delta u \, d\sigma & (a) \\ \int_{\Omega_F} \rho_F \nabla\varphi \cdot \nabla\delta\varphi \, dx - \omega^2 \int_\Gamma \frac{\rho_F}{g} \varphi \delta\varphi \, d\sigma - \int_\Sigma \rho_F u \cdot n^F \delta\varphi \, d\sigma = 0 & (b) \end{array}} \tag{6.33}$$

where we have let:

$$\boxed{k^0 = \widehat{k}(u, \delta u) + \mathcal{K}^0_B(u, \delta u)} \tag{6.34}$$

with:

$$\boxed{\mathcal{K}^0_B(u, \delta u) = \frac{\rho_F g}{\text{Area}(\Gamma)} \left(\int_\Sigma u \cdot n^F \, d\sigma \right) \left(\int_\Sigma \delta u \cdot n^F \, d\sigma \right)} \tag{6.35}$$

where $\mathcal{K}^0_B$ coincides with the operator introduced in (3.113).

Matrix structure of the discretized problem

We denote by $\boldsymbol{U}$ the vector N_u of the nodal values of u, and by $\boldsymbol{\Phi}$ the vector of the N_φ nodal values of φ. The matrices corresponding to the various bilinear and linear forms involved in (6.33) are defined by:

$$\boxed{\begin{array}{lcll} \int_{\Omega_S} a_{ijkh} \epsilon_{kh}(u) \epsilon_{ij}(\delta u) \, dx & \Longrightarrow & \delta \boldsymbol{U}^T \boldsymbol{K}_E \boldsymbol{U} & (a) \\ \int_{\Omega_S} \sigma^0_{ij} u_{l,i} \delta u_{l,j} \, dx & \Longrightarrow & \delta \boldsymbol{U}^T \boldsymbol{K}_G \boldsymbol{U} & (b) \\ k_\Sigma(u, \delta u) & \Longrightarrow & \delta \boldsymbol{U}^T \boldsymbol{K}_\Sigma \boldsymbol{U} & (c) \\ \frac{\rho_F g}{\text{Area}(\Gamma)} \left(\int_\Sigma u \cdot n^F \, d\sigma \right) \left(\int_\Sigma \delta u \cdot n^F \, d\sigma \right) & \Longrightarrow & \delta \boldsymbol{U}^T \boldsymbol{K}^0_B \boldsymbol{U} & (d) \\ \int_{\Omega_S} \rho_S u \cdot \delta u \, dx & \Longrightarrow & \delta \boldsymbol{U}^T \boldsymbol{M} \boldsymbol{U} & (e) \\ \int_{\Sigma_d} \widetilde{F} \cdot \delta u d\sigma & \Longrightarrow & \delta \boldsymbol{U}^T \widetilde{\boldsymbol{F}} & (f) \end{array}} \tag{6.36}$$

[6] An analytical study, starting from the local equations of the problem, may be found in Rapoport [190].

with, in addition

$$\boxed{\int_\Sigma \rho_F \varphi \delta u \cdot n^F \, d\sigma \Rightarrow \delta \boldsymbol{U}^T \boldsymbol{C} \boldsymbol{\Phi} \;\Big|\; \int_\Sigma \rho_F \delta\varphi u \cdot n^F \, d\sigma \Rightarrow \delta \boldsymbol{\Phi}^T \boldsymbol{C}^T \boldsymbol{U}} \tag{6.37}$$

Finally, discretization of (6.33b) yields the following matrices: [7]

$$\begin{array}{|lcll|} \displaystyle\int_{\Omega_F} \rho_F \nabla\varphi \cdot \nabla\delta\varphi \, dx & \Longrightarrow & \delta\boldsymbol{\Phi}^T \boldsymbol{F} \boldsymbol{\Phi} & (a) \\ \displaystyle\int_\Gamma \frac{\rho_F}{g} \varphi \delta\varphi \, d\sigma & \Longrightarrow & \delta\boldsymbol{\Phi}^T \boldsymbol{S} \boldsymbol{\Phi} & (b) \end{array} \tag{6.38}$$

Consequently, the matrix $\boldsymbol{K}^0$ discretizing k^0 defined in (6.34) is written:

$$\boldsymbol{K}^0 = \boldsymbol{K}_E + \boldsymbol{K}_G + \boldsymbol{K}_\Sigma + \boldsymbol{K}_B^0 \tag{6.39}$$

In discretized form, the variational formulation (6.33) is written:

$$\begin{bmatrix} \boldsymbol{K}^0 & 0 \\ -\boldsymbol{C}^T & \boldsymbol{F} \end{bmatrix} \begin{bmatrix} \boldsymbol{U} \\ \boldsymbol{\Phi} \end{bmatrix} - \omega^2 \begin{bmatrix} \boldsymbol{M} & \boldsymbol{C} \\ 0 & \boldsymbol{S} \end{bmatrix} \begin{bmatrix} \boldsymbol{U} \\ \boldsymbol{\Phi} \end{bmatrix} = \begin{bmatrix} \widetilde{\boldsymbol{F}} \\ 0 \end{bmatrix} \tag{6.40}$$

where $\boldsymbol{\Phi}$ is subject to the discretized constraint $\int_\Gamma \varphi \, d\sigma = 0$.

- **Construction of $\boldsymbol{K}_B^0$** — We start from relations (6.37): replacing ρ_F by 1 and $\boldsymbol{\Phi}$ by the vector $\mathbf{1}$ of components equal to 1, and introducing the vector $\boldsymbol{C}_1 = \boldsymbol{C}\mathbf{1}$ we see that $\boldsymbol{K}_B^0 = \frac{\rho_F g}{\text{Area}(\Gamma)} \boldsymbol{C}_1 {\boldsymbol{C}_1}^T$, which we can obtain by the condensation algorithm (1.65) (with $\alpha = \text{Area}(\Gamma)/\rho_F g$).

- For $\omega = 0$, the problem is well-posed: it describes the static response of a structure containing a liquid with weight.

- The preceding formulation is *unsymmetric*. The matrix equations (6.40) can nevertheless be symmetrized by means of various algebraic manipulations. In particular, in chapter 9 we shall establish a procedure for symmetrization by modal projection.

- In what follows, we shall derive a *symmetric variational formulation*, enabling *direct numerical solution* by finite elements.

6.4 Symmetric variational formulation in (u, η) and added mass

We shall show that introducing the additional variable $\eta = \left.\dfrac{\partial\varphi}{\partial z}\right|_\Gamma$ enables a symmetric variational formulation of the coupled problem to be established. The introduction of this new variable concerns only the formulation of the response of the liquid to a displacement of the wall, which we now examine.

[7] We note that the matrices $\boldsymbol{F}$ and $\boldsymbol{S}$ differ from those used in (3.67) in the respective multiplying factors ρ_F and ρ_F/g.

Introduction of the supplementary unknown η

We start with equations (3.1). Expressing p by means of relation (3.4), which we recall is:

$$\boxed{\begin{array}{rclr} p & = & \rho_F \omega^2 \varphi + \pi & (a) \\ l(\varphi) & = & 0 \quad \text{with } l(1) \neq 0 & (b) \end{array}} \qquad (6.41)$$

equations (3.1) can be written in the form:

$$\boxed{\begin{array}{rcllr} \Delta\varphi & = & 0 & \text{in } \Omega & (a) \\ \dfrac{\partial\varphi}{\partial z} & = & u \cdot n^F & \text{on } \Sigma & (b) \\ \dfrac{\partial\varphi}{\partial n^F} & = & \eta & \text{on } \Gamma & (c) \\ l(\varphi) & = & 0 & \text{with } l(1) \neq 0 & (d) \\ \rho_F \omega^2 \varphi + \pi & = & \rho_F g \eta & \text{on } \Gamma & (e) \end{array}} \qquad (6.42)$$

Remark on the constraint $l(\varphi)$ — Two ways are open.

1. The first is to choose $l(\varphi) = \int_\Gamma \varphi \, d\sigma$, which then enables the unknown constant π to be eliminated from the formulation of this problem. In numerical applications, this constraint results in a linear relation imposed on the degrees of freedom $\mathbf{\Phi}$.

2. The second solution — which we pursue here — is to keep the constant π as an unknown, and to keep $l(\varphi)$ arbitrary. It is then sufficient, at the discretized level, to cancel any nodal unknown φ (this approach has already been used in chapter 4).

Response of the liquid: formulation in (η, φ)

In the first step, we introduce the space $\mathcal{C}_\eta$ of the functions $\delta\eta$. Multiplying (6.42e) by a test-function $\delta\eta \in \mathcal{C}_\eta$, then integrating over Γ, we obtain, $\forall \delta\eta \in \mathcal{C}_\eta$:

$$\boxed{\int_\Gamma \rho_F g \, \eta \delta\eta \, d\sigma \; - \; \pi \int_\Gamma \delta\eta \, d\sigma \; - \; \omega^2 \int_\Gamma \rho_F \, \varphi \delta\eta \, d\sigma \; = \; 0} \qquad (6.43)$$

In the second step, we introduce the space $\mathcal{C}_\varphi$ of the test-functions $\delta\varphi$ smooth in Ω_F. We multiply (6.42a) by $\delta\varphi \in \mathcal{C}_\varphi$ and integrate in Ω_F. We next use Green's formula (3.16), distinguishing in the integral on $\partial\Omega_F$ the relative contributions to Σ and Γ. Finally, taking account of (6.42b) and (6.42c), we obtain (after multiplication by ρ_F), $\forall \delta\varphi \in \mathcal{C}_\varphi$:

$$\int_{\Omega_F} \rho_F \nabla\varphi \cdot \nabla\delta\varphi \, dx - \int_\Gamma \rho_F \, \eta \delta\varphi \, d\sigma - \int_\Sigma \rho_F \, u \cdot n^F \delta\varphi \, d\sigma = 0 \tag{6.44}$$

Conversely, going formally through the calculations, we verify that the variational properties (6.43) and (6.44) completed by the constraint (6.42d) "restore" the boundary value problem (6.42).
We note that in (6.44), the test-functions $\delta\varphi \in \mathcal{C}_\Phi$ are not subject to the constraint (6.42d) imposed to φ.
In order to establish a variational formulation where φ and $\delta\varphi$ belong to the same admissible space, we have (*cf.* also § 4.5) to introduce the space $\mathcal{C}^l_\varphi$ defined by:

$$\mathcal{C}^l_\varphi = \{\varphi \in \mathcal{C}_\varphi \mid l(\varphi) = 0, \text{ with } l(1) \neq 0\} \tag{6.45}$$

Remark — We note that any $\varphi \in \mathcal{C}_\varphi$ may be written in a unique way in the form:

$$\varphi = \varphi^l + \text{constant}, \quad \varphi^l \in \mathcal{C}^l_\varphi \tag{6.46}$$

On applying l to (6.46) and using $l(\varphi^l) = 0$, we then find the value of the constant: constant $= l(\varphi)/l(1)$. Conversely, (6.46) is verified by $\varphi^l = \varphi - l(\varphi)/l(1)$. Consequently, $\mathcal{C}_\varphi$ decomposes according to:

$$\mathcal{C}_\varphi = \mathcal{C}^l_\varphi \oplus \mathbb{R} \tag{6.47}$$

In view of this decomposition, (6.44) may be replaced by the equivalent set of the two variational equations obtained by successively restricting (6.44) to the class of test-functions $\delta\varphi \in \mathcal{C}^l_\varphi$, and to the constant test-functions $\delta\pi \in \mathbb{R}$, yielding:

$$\boxed{\begin{aligned} \int_{\Omega_F} \rho_F \nabla\varphi \cdot \nabla\delta\varphi \, dx - \int_\Gamma \rho_F \eta \, \delta\varphi \, d\sigma - \int_\Sigma \rho_F u \cdot n^F \delta\varphi \, d\sigma &= 0 \quad (a) \\ \delta\pi \int_\Gamma \eta \, d\sigma + \delta\pi \int_\Sigma u \cdot n^F \, d\sigma &= 0 \quad (b) \end{aligned}} \tag{6.48}$$

We note that (6.48b) leads to $\int_\Gamma \eta \, d\sigma + \int_\Sigma u \cdot n^F \, d\sigma = 0$ which results from the invariance of the fluid volume.

Response of the structure

Replacing p by its expression (6.41) in terms of φ and π in (6.31), we obtain the following variational property of the response of the structure:
For given $u \in \mathcal{C}_u$, and $\forall \delta u \in \mathcal{C}_u$:

$$\boxed{\begin{aligned} &\widehat{k}(u, \delta u) - \omega^2 \int_{\Omega_S} \rho_S u \cdot \delta u \, dx - \omega^2 \int_\Sigma \rho_S \varphi \, \delta u \cdot n^F d\sigma - \pi \int_\Sigma \delta u \cdot n^F \, d\sigma = \cdots \\ &\cdots = \int_{\Sigma_d} \widetilde{F} \cdot \delta u \, d\sigma \end{aligned}} \tag{6.49}$$

$(\boldsymbol{u},\boldsymbol{\eta},\boldsymbol{\varphi})$ variational formulation of the coupled problem

Recapitulating the above results, we arrive at the following variational formulation of the coupled problem:
For given ω and $\widetilde{F}$, find $(u,\eta,\varphi,\pi) \in \mathcal{C}_u \times \mathcal{C}_\eta \times \mathcal{C}^l_\varphi \times \mathbb{R}$ verifying (6.49), (6.43) and (6.48ab), $\forall(\delta u,\delta\eta,\delta\varphi,\delta\pi) \in \mathcal{C}_u \times \mathcal{C}_\eta \times \mathcal{C}^l_\varphi \times \mathbb{R}$.

Discretization

We denote by $\boldsymbol{\eta}$ the vector of the N_η nodal values of η. In addition to the matrices already introduced, we define:

$$\boxed{\int_\Gamma \rho_F g \eta \delta\eta \, d\sigma \implies \delta\boldsymbol{\eta}^T \boldsymbol{K}_\eta \boldsymbol{\eta}} \tag{6.50}$$

together with the coupling terms:

$$\begin{array}{rcl|rcll}
\pi \displaystyle\int_\Sigma \delta u \cdot n^F & \Rightarrow & \delta\boldsymbol{U}^T \boldsymbol{c}\pi & \delta\pi \displaystyle\int_\Sigma u\cdot n^F & \Rightarrow & \delta\pi \boldsymbol{c}^T \boldsymbol{U} & (a) \\
\displaystyle\int_\Gamma \rho_F \varphi \delta\eta \, d\sigma & \Rightarrow & \delta\boldsymbol{\eta}^T \boldsymbol{B}\boldsymbol{\Phi} & \displaystyle\int_\Gamma \rho_F \eta \, \delta\varphi \, d\sigma & \Rightarrow & \delta\boldsymbol{\Phi}^T \boldsymbol{B}^T \boldsymbol{\eta} & (b) \\
\pi \displaystyle\int_\Gamma \delta\eta \, d\sigma & \Rightarrow & \delta\boldsymbol{\eta}^T \boldsymbol{b}\pi & \delta\pi \displaystyle\int_\Gamma \eta \, d\sigma & \Rightarrow & \delta\pi \boldsymbol{b}^T \boldsymbol{\eta} & (c)
\end{array} \tag{6.51}$$

where $\boldsymbol{c}(N_u)$ is a coupling vector between $\boldsymbol{U}$ and π, $\boldsymbol{B}(N_\eta \times N_\varphi)$ a coupling matrix between $\boldsymbol{\Phi}$ and $\boldsymbol{\eta}$, and $\boldsymbol{b}(N_\eta)$ is a coupling vector.

Practical choice of the constraint imposed to φ — For numerical applications, it is convenient to choose as the constraint l imposed at φ, a condition for cancellation of a nodal value of φ, for example the first: $\phi_1 = 0$. We denote by $\boldsymbol{\Phi}_2$ the "truncated" vector of the $N_\varphi - 1$ remaining nodal values.

Matrix equations — The matrix equations coresponding to the discretization of (6.49), (6.43), and (6.48ab) are then written:

$$\begin{aligned}
\widehat{\boldsymbol{K}}\boldsymbol{U} - \boldsymbol{c}\pi - \omega^2 \boldsymbol{M}\boldsymbol{U} - \omega^2 \boldsymbol{C}_2\boldsymbol{\Phi}_2 &= \widetilde{\boldsymbol{F}} && (a) \\
\boldsymbol{K}_\eta\boldsymbol{\eta} - \boldsymbol{b}\pi - \omega^2 \boldsymbol{B}_2\boldsymbol{\Phi}_2 &= 0 && (b) \\
\boldsymbol{F}_{22}\boldsymbol{\Phi}_2 - \boldsymbol{C}_2^T\boldsymbol{U} - \boldsymbol{B}_2^T\boldsymbol{\eta} &= 0 && (c) \\
\boldsymbol{c}^T\boldsymbol{U} + \boldsymbol{b}^T\boldsymbol{\eta} &= 0 && (d)
\end{aligned} \tag{6.52}$$

where the matrices $\boldsymbol{C}_2$, $\boldsymbol{B}_2$ and $\boldsymbol{F}_{22}$ are obtained by eliminating, on the one hand the column of $\boldsymbol{C}$ and that of $\boldsymbol{B}$ (together with the row of $\boldsymbol{C}^T$ and that of $\boldsymbol{B}^T$), and on the other hand the row and column of $\boldsymbol{F}$ which correspond to ϕ_1, and where we have let:

$$\widehat{\boldsymbol{K}} = \boldsymbol{K}_E + \boldsymbol{K}_G + \boldsymbol{K}_\Sigma \tag{6.53}$$

Elimination of Φ_2 and the added mass matrix

We begin by noting that the matrix F_{22} is nonsingular, since F is of rank $N_\varphi - 1$ (*cf.* § 3.4).
Equation (6.52c) thus enables Φ_2 to be eliminated in terms of U and η, which, after substitution into (6.52a) and (6.52b) leads to the following *symmetric matrix system*:

$$\left\{ \left[\begin{array}{cc|c} \widehat{K} & 0 & -c \\ 0 & K_\eta & -b \\ \hline -c^T & -b^T & 0 \end{array} \right] - \omega^2 \left[\begin{array}{cc|c} M + C_2 F_{22}^{-1} C_2^T & C_2 F_{22}^{-1} B_2^T & 0 \\ B_2 F_{22}^{-1} C_2^T & B_2 F_{22}^{-1} B_2^T & 0 \\ \hline 0 & 0 & 0 \end{array} \right] \right\} \left[\begin{array}{c} U \\ \eta \\ \hline \pi \end{array} \right] = \left[\begin{array}{c} \widetilde{F} \\ 0 \\ \hline 0 \end{array} \right] \tag{6.54}$$

1. Relation (6.52c) enables Φ_2 to be "retrieved", followed by the value of p by means of (6.41).

2. Equation (6.54) may be written in the following equivalent form:

$$\left\{ \begin{bmatrix} \widehat{K} & 0 \\ 0 & K_\eta \end{bmatrix} - \omega^2 \begin{bmatrix} M + C_2 F_{22}^{-1} C_2^T & C_2 F_{22}^{-1} B_2^T \\ B_2 F_{22}^{-1} C_2^T & B_2 F_{22}^{-1} B_2^T \end{bmatrix} \right\} \begin{bmatrix} U \\ \eta \end{bmatrix} = \begin{bmatrix} \widetilde{F} \\ 0 \end{bmatrix} \tag{6.55}$$

   with the constraint $c^T U + b^T \eta = 0$

   In this form, we see in (6.54) that π is interpreted as the Lagrange multiplier associated with the constraint (6.52d).

3. We note that the fluid is involved, on the one hand through a "free surface stiffness" matrix K_η, due to gravity, and on the other hand through the added mass matrix M_A, coupling U and η, which is expressed:

$$M_A = \begin{bmatrix} C_2 F_{22}^{-1} C_2^T & C_2 F_{22}^{-1} B_2^T \\ B_2 F_{22}^{-1} C_2^T & B_2 F_{22}^{-1} B_2^T \end{bmatrix} \tag{6.56}$$

   We shall later verify that M_A is *positive, definite*.

Practical application

In practice, to construct the matrices (c, c^T), (b, b^T) and M_A, we proceed as follows:

1. We consider the following auxiliary bilinear form [8] $\widehat{m}(u, \eta, \varphi | \delta u, \delta \eta, \delta \varphi)$ — abbreviated to $\widehat{m}(\cdot|\cdot)$ — which is symmetric in the exchange

[8] where φ is not subject to the constraint $l(\varphi) = 0$.

$(u, \eta, \varphi \leftrightarrow \delta u, \delta\eta, \delta\varphi)$:

$$\widehat{m}(\cdot|\cdot) = -\int_{\Omega} \rho_F \nabla\varphi \cdot \nabla\delta\varphi \, dx + \left(\int_{\Sigma} \rho_F u \cdot n^F \, \delta\varphi \, d\sigma + \int_{\Sigma} \rho_F \delta u \cdot n^F \, \varphi \, d\sigma \right) + \cdots$$
$$\cdots + \left(\int_{\Gamma} \rho_F \eta \delta\varphi \, d\sigma + \int_{\Gamma} \rho_F \varphi \, \delta\eta \, d\sigma \right) \tag{6.57}$$

which in discretized form is written:

$$\begin{bmatrix} \delta \boldsymbol{U}^T & \delta \boldsymbol{\eta}^T & \delta \boldsymbol{\Phi}^T \end{bmatrix} \begin{bmatrix} 0 & 0 & \boldsymbol{C} \\ 0 & 0 & \boldsymbol{B} \\ \boldsymbol{C}^T & \boldsymbol{B}^T & -\boldsymbol{F} \end{bmatrix} \begin{bmatrix} \boldsymbol{U} \\ \boldsymbol{\eta} \\ \boldsymbol{\Phi} \end{bmatrix} \tag{6.58}$$

The matrix $\begin{bmatrix} 0 & \boldsymbol{C} \\ \boldsymbol{C}^T & 0 \end{bmatrix}, \begin{bmatrix} 0 & \boldsymbol{B} \\ \boldsymbol{B}^T & 0 \end{bmatrix}$ are constructed by simultaneous assembly of the elementary contributions corresponding to the discretization of the symmetric bilinear coupling forms featuring in $\widehat{m}(\cdot|\cdot)$.

2. Distinguishing the value ϕ_1 of φ in a particular node from other values $\boldsymbol{\Phi}_2$ of φ, the matrix (6.58) is partitioned as follows:

$$\begin{bmatrix} 0 & 0 & \boldsymbol{c} & \boldsymbol{C}_2 \\ 0 & 0 & \boldsymbol{b} & \boldsymbol{B}_2 \\ \boldsymbol{c}^T & \boldsymbol{b}^T & -F_{11} & -\boldsymbol{F}_{12} \\ \boldsymbol{C}_2^T & \boldsymbol{B}_2^T & -\boldsymbol{F}_{12}^T & -\boldsymbol{F}_{22} \end{bmatrix} \Leftrightarrow \begin{bmatrix} \boldsymbol{U} \\ \boldsymbol{\eta} \\ \phi_1 \\ \boldsymbol{\Phi}_2 \end{bmatrix} \tag{6.59}$$

3. Applying the Gauss elimination algorithm (1.64) to this matrix, considering $\boldsymbol{\Phi}_2$ as "dependent degrees of freedom", we obtain the following condensed matrix:

$$\left[\begin{array}{cc|c} \boldsymbol{C}_2\boldsymbol{F}_{22}^{-1}\boldsymbol{C}_2^T & \boldsymbol{C}_2\boldsymbol{F}_{22}^{-1}\boldsymbol{B}_2^T & \boldsymbol{c} \\ \boldsymbol{B}_2\boldsymbol{F}_{22}^{-1}\boldsymbol{C}_2^T & \boldsymbol{B}_2\boldsymbol{F}_{22}^{-1}\boldsymbol{B}_2^T & \boldsymbol{b} \\ \hline \boldsymbol{c}^T & \boldsymbol{b}^T & 0 \end{array}\right] \tag{6.60}$$

of which the various submatrices are those in (6.54).

We note the nullity of the diagonal term corresponding to ϕ_1, which arises because $\boldsymbol{F}$ is of rank $N_\varphi - 1$, and that the kernel of $\boldsymbol{F}$ is of dimension 1, spanned by the vector $\boldsymbol{\Phi}$ of components equal to 1.

Continuum-based added mass operator

The matrix equations (6.54), (6.55) and the matrix $\boldsymbol{M}_A$ defined by (6.56), may be interpreted as resulting from the discretization of a variational property in (u, η).

The variational equations (6.49), (6.43) and (6.48ab) can indeed be interpreted by considering (6.48b) as a constraint, and π as the associated Lagrange multiplier. By forcing (u, η) to satisfy the constraint, we can eliminate π from this formulation. We are thus led to considering the space $(\mathcal{C}_u \times \mathcal{C}_\eta)^*$ of the admissible (u, η) defined by:

$$(\mathcal{C}_u \times \mathcal{C}_\eta)^* = \left\{ u \in \mathcal{C}_u,\ \eta \in \mathcal{C}_\eta \mid \int_\Gamma \eta \, d\sigma + \int_\Sigma u \cdot n^F \, d\sigma = 0 \right\} \tag{6.61}$$

We then obtain the equivalent formulation:
For given ω and $\widetilde{F}$, find $(u, \eta) \in (\mathcal{C}_u \times \mathcal{C}_\eta)^*$ and $\varphi \in \mathcal{C}^l_\varphi$ such that $\forall\, (\delta u, \delta\eta) \in (\mathcal{C}_u \times \mathcal{C}_\eta)^*$, and $\forall\, \delta\varphi \in \mathcal{C}^l_\varphi$, we have:

$$\begin{aligned}
&\widehat{k}(u, \delta u) + k_\eta(\eta, \delta\eta) - \omega^2 \int_{\Omega_S} \rho_S u \cdot \delta u \, dx + \cdots \\
&\qquad \cdots - \omega^2 \int_\Sigma \rho_F \varphi \delta u \cdot n^F \, d\sigma - \omega^2 \int_\Gamma \rho_F \varphi \delta\eta \, d\sigma = \int_{\Sigma_d} \widetilde{F} \cdot \delta u \, d\sigma \qquad (a) \\
&\int_{\Omega_F} \rho_F \nabla\varphi \cdot \nabla\delta\varphi \, dx - \int_\Gamma \rho_F \eta \, \delta\varphi \, d\sigma - \int_\Sigma \rho_F u \cdot n^F \delta\varphi \, d\sigma = 0 \qquad (b)
\end{aligned} \tag{6.62}$$

Elimination of φ

The variational equation (6.62b), considered in isolation for a given $(u, \eta) \in (\mathcal{C}_u \times \mathcal{C}_\eta)^*$, characterizes the unique solution φ to the Neumann problem (6.42abcd) (whose existence arises from the constraint on (u, η) featuring in the admissible class $(\mathcal{C}_u \times \mathcal{C}_\eta)^*$, and whose unicity arises from the condition $l(\varphi) = 0$ satisfied by $\varphi \in \mathcal{C}^l_\varphi$).

We furthermore verify that this solution — which will be denoted $\varphi_{u,\eta}$ — is linearly dependent on the couple (u, η).

The elimination consists of substituting $\varphi_{u,\eta}$ into (6.62a), which yields the following variational formulation in (u, η) (whose symmetry is established in what follows):

$$\begin{aligned}
&\widehat{k}(u, \delta u) + k_\eta(\eta, \delta\eta) - \omega^2 \int_{\Omega_S} \rho_S u \cdot \delta u \, dx + \cdots \\
&\quad \cdots - \omega^2 \int_\Sigma \rho_F \varphi_{u,\eta} \, \delta u \cdot n^F \, d\sigma - \omega^2 \int_\Gamma \rho_F \varphi_{u,\eta} \, \delta\eta \, d\sigma = \int_{\Sigma_d} \widetilde{F} \cdot \delta u \, d\sigma
\end{aligned} \tag{6.63}$$

We give the name *added mass operator* to the bilinear form $\mathcal{M}_A(u, \eta | \delta u, \delta\eta)$ defined by:

$$\mathcal{M}_A = \int_\Sigma \rho_F \varphi_{u,\eta} \delta u \cdot n^F \, d\sigma + \int_\Gamma \rho_F \varphi_{u,\eta} \delta\eta \, d\sigma \tag{6.64}$$

We show that *this operator is symmetric.*
Consider the solution $\varphi_{\delta u,\delta\eta}$ to (6.62b) for given $\delta u, \delta\eta$, using as special test-functions $\delta\varphi = \varphi_{u,\eta}$, which is written:

$$\int_{\Sigma} \rho_F \delta u \cdot n^F \varphi_{u,\eta}\, d\sigma + \int_{\Gamma} \rho_F \delta\eta \varphi_{u,\eta}\, d\sigma = \int_{\Omega_F} \rho_F \nabla\varphi_{\delta u,\delta\eta} \cdot \nabla\varphi_{u,\eta}\, dx \tag{6.65}$$

which enables $\mathcal{M}_A$ to be written in the following *symmetric* form:

$$\boxed{\mathcal{M}_A = \int_{\Omega_F} \rho_F \nabla\varphi_{u,\eta} \cdot \nabla\varphi_{\delta u,\delta\eta}\, dx} \tag{6.66}$$

Remark — This expression shows that $\mathcal{M}_A$ is a symmetric bilinear form, positive definite (since φ is subject to the constraint $l(\varphi) = 0$).
From (6.63) and (6.64), the problem in (u, η) is thus written:
for given ω and $\widetilde{F}$, find $(u, \eta) \in (\mathcal{C}_u \times \mathcal{C}_\eta)^*$ such that, $\forall(\delta u, \delta\eta) \in (\mathcal{C}_u \times \mathcal{C}_\eta)^*$, we have:

$$\boxed{\widehat{k}(u, \delta u) + k_\eta(\eta, \delta\eta) - \omega^2\Big(\int_{\Omega_S} \rho_S u \cdot \delta u\, dx + \mathcal{M}_A(u\, \eta | \delta u\, \delta\eta)\Big) = \int_{\Sigma_d} \widetilde{F} \cdot \delta u\, d\sigma} \tag{6.67}$$

6.5 Conclusion and open problems

- The case $\widetilde{F} = 0$ corresponds to the case of *dead loading* and leads to a spectral problem describing the vibrations of the free system. [9]

  We recall that this problem involves the prestress terms k_G and k_Σ (*cf.* (6.32) and (6.23)) not necessarily positive definite. If the equilibrium configuration is unstable, this problem can have negative eigenvalues.

  This is the reason why the modal analysis of the dynamic response to forces $\widetilde{F}$ is carried out, as we shall see in chapter 9, with a dynamic substructuring technique which uses both the *hydroelastic modes* calculated for $g = 0$ (studied in chapter 5), and the *sloshing modes* (studied in chapter 3).

- Problems under development include the study of non-linear vibrations such as the modelling of finite amplitude effects (*cf.* open problems, chapter 9).

[9] In the case of a launcher in flight, the "equilibrium" configuration is defined, for a given instant, by "freezing" the parameters (mass, thrust etc.) of the system. The equilibrium conditions at a point bound to the center of gravity — around which the vibrations are defined — take into account the thrust and the aerodynamic forces. $\widetilde{F}$ corresponds, in this case, to the variation of the thrust vector due to slewing of the propelling nozzle (attitude control), or else to a thrust variation in transient state (ignition or extinction of the engine).

CHAPTER 7

Acoustic cavity modes

7.1 Introduction

Here, we are concerned with the harmonic vibrations of a compressible fluid filling a cavity.
Examples include acoustic resonance calculations in payloads of launcher, in rocket engine combustion chambers, in aircraft fuselages and in automobiles, etc.
As a first step, we establish the variational formulation of the harmonic response of a compressible fluid to the displacement of a wall.
In the second step, we study the spectral problem associated with "acoustic modes", together with the corresponding matrix equations.
Next, we carry out the modal analysis of the problem of prescribed displacements, which leads to the introduction of the dynamic impedance operators of a fluid.
Finally we transpose the above results to the case of a compressible liquid with a free surface, neglecting gravity terms.

7.2 Harmonic response to a wall displacement u_N

We consider the harmonic response of an inviscid compressible fluid contained in a cavity occupying a bounded domain Ω_F, and subject to an prescribed normal displacement of the wall u_N on Σ, of circular frequency ω (Fig. 7.1). We assume the effects of gravity to be negligible, and the fluid *homogeneous*.

Equations in terms of (p, φ)

We use the basic equations (2.27), (2.30), (2.31), established in chapter 2:

$$\boxed{\begin{array}{rcllr} \nabla p &=& \rho_F\omega^2\nabla\varphi & \text{in } \Omega_F & (a)\\ p &=& -\rho_F c^2\Delta\varphi & \text{in } \Omega_F & (b)\\ \dfrac{\partial\varphi}{\partial n} &=& u_N & \text{on } \Sigma = \partial\Omega_F & (c)\\ l(\varphi) &=& 0 & \text{with } l(1)\neq 0 & (d) \end{array}} \qquad (7.1)$$

where $l(\varphi)$ is an arbitrary linear constraint enabling the unicity of φ to be ensured.

- (7.1a) results from the linearized Euler equation for displacements deriving from a potential (equation (2.27)).
- (7.1b) results from the fluid constitutive law (equation (2.30)).
- (7.1c) is the wall condition (equation (2.31) where we have let $u_N = u^S{\cdot}n$))

Solution for $\omega = 0$

For $\omega = 0$, and a given u_N, (7.1a) is written $\nabla p = 0$, yielding a constant value p_0 for p which can be calculated in terms of u_N. We first show that there exists a relation between p and u_N (independent of ω). Integrating (7.1b) on Ω_F and taking into account (7.1c) we indeed obtain:

$$\int_{\Omega_F} p\,dx = -\rho_F c^2 \int_\Sigma u_N\,d\sigma \qquad (7.2)$$

This relation enables the value of p_0 to be determined:

$$p^0 = -\frac{\rho_F c^2}{Vol(\Omega_F)} \int_\Sigma u_N\,d\sigma \qquad (7.3)$$

p^0 is physically interpreted as the variation of pressure accompanying a static deformation of the wall of amplitude u_N.

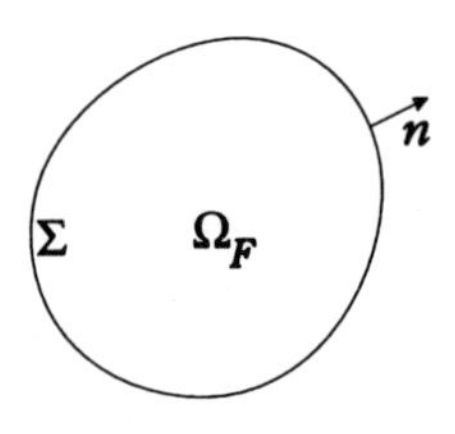

Figure 7.1: Geometrical configuration

Boundary value problem in φ

The aim here is both to eliminate p and to arrive at a formulation yielding a well-posed boundary value problem for $\omega = 0$. We know that equations (7.1ad) are equivalent — in accordance with the second definition of φ (2.26) — to:

$$\boxed{\begin{array}{ll} p = \rho_F \omega^2 \varphi + \pi & (a) \\ l(\varphi) = 0 \quad \text{with } l(1) \neq 0 & (b) \end{array}} \tag{7.4}$$

Relation between p, φ and u_N

Substituting (7.4a) into (7.2), we find:

$$\pi = -\frac{\rho_F \omega^2}{Vol(\Omega_F)} \int_{\Omega_F} \varphi \, dx - \frac{\rho_F c^2}{Vol(\Omega_F)} \int_\Sigma u_N \, d\sigma \tag{7.5}$$

from which we obtain the following expression of p in terms of φ and u_N:

$$p = \rho_F \omega^2 (\varphi - \frac{1}{Vol(\Omega_F)} \int_{\Omega_F} \varphi \, dx) - \frac{\rho_F c^2}{Vol(\Omega_F)} \int_\Sigma u_N \, d\sigma \tag{7.6}$$

We note that φ and $\varphi +$ constant give the same value for p.
In what follows we let:

$$\boxed{l(\varphi) = \int_{\Omega_F} \varphi dx} \tag{7.7}$$

Substituting (7.6) into (7.1b), we obtain the following boundary value problem in φ:

$$\boxed{\begin{array}{rcll} \Delta\varphi + \dfrac{\omega^2}{c^2}\varphi - \dfrac{1}{Vol(\Omega_F)} \displaystyle\int_\Sigma u_N \, d\sigma & = & 0 \quad \text{in } \Omega_F & (a) \\ \dfrac{\partial \varphi}{\partial n} & = & u_N \quad \text{on } \Sigma & (b) \\ \displaystyle\int_{\Omega_F} \varphi \, dx & = & 0 & (c) \end{array}} \tag{7.8}$$

knowing that from (7.6), (7.8c) and (7.3), p is given by:

$$\boxed{p = \rho_F \omega^2 \varphi - \frac{\rho_F c^2}{Vol(\Omega_F)} \int_\Sigma u_N d\sigma \quad (= \rho_F \omega^2 \varphi + p^0)} \tag{7.9}$$

Solution φ^0 for $\omega = 0$

We show that the boundary value problem in φ for $\omega = 0$ is well-posed.

We consider the solution φ^0 of (7.8) for $\omega = 0$:

$$\begin{array}{|rcllr|}\hline \Delta\varphi^0 & = & \dfrac{1}{Vol(\Omega_F)}\displaystyle\int_\Sigma u_N\, d\sigma & \text{in } \Omega_F & (a) \\ \dfrac{\partial\varphi^0}{\partial n} & = & u_N & \text{on } \Sigma & (b) \\ \displaystyle\int_{\Omega_F}\varphi^0\, dx & = & 0 & & (c) \\ \hline\end{array} \qquad (7.10)$$

This is a well-posed Neumann problem, the condition for whose existence is trivially verified, and whose unicity is ensured by (7.10c).
To φ^0, there corresponds a displacement $u^F = \nabla\varphi^0$.

Remarks on the definition of φ — If instead of the "second definition" (2.26) of φ, we use the "first definition" (2.24) $\varphi = p/(\rho_F\omega^2)$ — which is equivalent to replacing equation (7.1a) in (7.1) by $p = \rho_F\omega^2\varphi$, and ignoring equation (7.1d) — the corresponding problem in φ is written:

$$\begin{array}{|rcllr|}\hline \Delta\varphi + \dfrac{\omega^2}{c^2}\varphi & = & 0 & \text{in } \Omega_F & (a) \\ \dfrac{\partial\varphi}{\partial n} & = & u_N & \text{on } \Sigma & (b) \\ \hline\end{array} \qquad (7.11)$$

- (7.11a) coincides with the classic Helmholtz equation in linear acoustics.
- For $\omega \neq 0$, equation (7.11) are equivalent to (7.8), (7.9) (they yield the same value of p and of $u^F = \nabla\varphi$).
- For $\omega = 0$, problem (7.11) is ill-posed: this is a Neumann problem whose condition for existence is not verified, since in general $\int_\Sigma u_N\, d\sigma \neq 0$.

  We note, on the other hand, that use of the potential defined in (2.26) yields equation (7.8a) which differs from the classic Helmholtz equation (7.11a) in the presence of a "non local" term $-(1/Vol(\Omega_F))\int_\Sigma u_N\, d\sigma$. It is the presence of this term which ensures the condition for existence of the Neumann problem obtained for $\omega = 0$.

Remark on the boundary value problem in p —Taking the divergence of (7.1a), and using (7.1b), we obtain:

$$\begin{array}{|rcllr|}\hline \Delta p + \dfrac{\omega^2}{c^2}p & = & 0 & \text{in } \Omega_F & (a) \\ \dfrac{\partial p}{\partial n} & = & \rho_F\omega^2 u_N & \text{on } \Sigma & (b) \\ \hline\end{array} \qquad (7.12)$$

- (7.12a) coincides with the classic Helmholtz equation in p in linear acoustics.
- For $\omega \neq 0$, we can verify that equation (7.12) are equivalent to (7.8, 7.9) (same value of p).

- For $\omega = 0$, problem (7.12) is badly-posed since the u_N given no longer features in the equations (it yields an indeterminate constant value for p (the physical value being given by (7.3)).

7.3 φ variational formulation of the spectral problem and discretization

Spectral boundary value problem

The solutions to the spectral problem deduced from (7.8) for $u_N = 0$ (fixed wall condition) are called *acoustic modes*. Writing:

$$\lambda = \frac{\omega^2}{c^2} \tag{7.13}$$

the spectral boundary value problem is written as follows:
find λ and $\varphi \neq 0$ such that:

$$\begin{aligned} \Delta\varphi + \lambda\varphi &= 0 \quad \text{in } \Omega_F & (a) \\ \frac{\partial\varphi}{\partial n} &= 0 \quad \text{on } \Sigma & (b) \\ \int_{\Omega_F} \varphi\, dx &= 0 & (c) \end{aligned} \tag{7.14}$$

Remark

- To begin with, we verify that $\lambda = 0$ is not a solution to (7.14). For $\lambda = 0$, equations (7.14ab) have the solution $\varphi =$ constant — which, from (7.14c), is null.

- We now consider the spectral problem *reduced to equations* (7.14ab) (i.e. ignoring (7.14c)).

  1. This problem has the solution $\lambda = 0, \varphi =$ constant.
  2. On the other hand, we show that the solutions to this problem for $\lambda \neq 0$ naturally satisfy condition (7.14c), and consequently coincide with the solutions to (7.14abc): integrating (7.14a) then applying Stokes' formula, and finally taking account of (7.14b), we find $\lambda \int_{\Omega_F} \varphi\, dx = 0$, which, for $\lambda \neq 0$, yields $\int_{\Omega_F} \varphi\, dx = 0$.

Variational formulation

In accordance with the above remark, and for purposes of simplification, we shall be concerned with solutions to the spectral boundary value problem (7.14),

ignoring condition (7.14c), which only differs from those of (7.14abc) in the presence of the non physical unwanted solution $\lambda = 0, \varphi =$ constant .
We proceed formally by the test-functions method. We introduce the space $\mathcal{C}$ of the test-functions φ smooth in Ω_F. We multiply (7.14a) par $\delta\varphi \in \mathcal{C}$, then integrate in the domain Ω_F, apply Green's formula (3.16), and finally take into account the boundary condition (7.14b). We then obtain the variational property of the boundary value problem (7.14) which is stated:
find $\lambda \in \mathbb{R}, \varphi \in \mathcal{C}$ such that, $\forall\delta\varphi \in \mathcal{C}$:

$$\boxed{\int_{\Omega_F} \nabla\varphi\cdot\nabla\delta\varphi\, dx - \lambda \int_{\Omega_F} \varphi\delta\varphi\, dx = 0} \tag{7.15}$$

- We note that this formulation involves the two *bilinear symmetric* forms $\int_{\Omega_F} \nabla\varphi\cdot\nabla\delta\varphi\, dx$ and $\int_{\Omega_F} \varphi\delta\varphi\, dx$ on $\mathcal{C} \times \mathcal{C}$.

- The eigenvalues λ are positive: putting $\delta\varphi = \varphi$ in (7.15), we see that $\lambda = \int_{\Omega_F} |\nabla\varphi|^2\, dx / \int_{\Omega_F} \varphi^2\, dx$ which shows that λ is the quotient of two positive numbers.

Converse — We show that, if λ and φ verify (7.15), λ and φ satisfy equations (7.14ab). Going formally through the calculations by means of Green's formula (3.16), we find:

$$-\int_{\Omega_F} (\Delta\varphi + \lambda\varphi)\, \delta\varphi\, dx + \int_{\partial\Omega_F} \delta\varphi \frac{\partial\varphi}{\partial n}\, d\sigma = 0 \quad \forall\delta\varphi \in \mathcal{C} \tag{7.16}$$

To retrieve the local equations, we proceed in two steps:

1. On choosing test-functions $\delta\varphi$ vanishing on $\partial\Omega_F = \Sigma$, (7.16) reduces to the first term, whose nullity leads to satisfaction of (7.14a).

2. We refer to (7.16) and consider the test-functions $\delta\varphi \in \mathcal{C}$, $\delta\varphi \neq 0$ on $\partial\Omega_F$. The first term is identically null since we have just established that φ satisfies (7.14a). The nullity of the remaining integral then leads to equation (7.14b).

Remark — The space $\mathcal{C}$ required in the variational formulation (7.16) is the Sobolev space $H^1(\Omega_F)$.

Orthogonality properties

We consider two solutions $(\lambda_\alpha, \varphi_\alpha)$ and $(\lambda_\beta, \varphi_\beta)$ of (7.15), with $\lambda_\alpha \neq \lambda_\beta$.
We show that:

$$\boxed{\begin{array}{rcll} \displaystyle\int_{\Omega_F} \varphi_\alpha\varphi_\beta\, dx &=& 0 & (a) \\ \displaystyle\int_{\Omega_F} \nabla\varphi_\alpha\cdot\nabla\varphi_\beta\, dx &=& 0 & (b) \end{array}} \tag{7.17}$$

Applying the variational property (7.15) to φ_α (with $\delta\varphi = \varphi_\beta$) and φ_β (with $\delta\varphi = \varphi_\alpha$), we obtain:

$$\boxed{\begin{array}{ll} \int_{\Omega_F} \nabla\varphi_\alpha \cdot \nabla\varphi_\beta \, dx = \lambda_\alpha \int_{\Omega_F} \varphi_\alpha \varphi_\beta \, dx & (a) \\ \int_{\Omega_F} \nabla\varphi_\beta \cdot \nabla\varphi_\alpha \, dx = \lambda_\beta \int_{\Omega_F} \varphi_\beta \varphi_\alpha \, dx & (b) \end{array}} \quad (7.18)$$

then subtracting term by term, and using the symmetry of the bilinear forms, we obtain $(\lambda_\alpha - \lambda_\beta) \int_{\Omega_F} \varphi_\alpha \varphi_\beta \, dx = 0$.
Consequently, $\int_{\Omega_F} \varphi_\alpha \varphi_\beta \, dx = 0$, from which, referring to (7.18), $\int_{\Omega_F} \nabla\varphi_\alpha \cdot \nabla\varphi_\beta \, dx = 0$, which concludes the demonstration.

Discussion

- (7.17ab) constitutes the orthogonality of the eigensolutions.

- We have seen that all solutions $(\lambda_\alpha \neq 0, \varphi_\alpha)$ satisfy $\int_{\Omega_F} \varphi_\alpha \, dx = 0$.

  We recall that φ = constant is a solution of (7.14ab) (and therefore of (7.15)) for $\lambda = 0$ (unwanted solution mentioned previously).

  Consequently, $\int_{\Omega_F} \varphi_\alpha \, dx = 0$ is here interpreted as a special case of the orthogonality condition (7.17a) of the solutions $\lambda_\alpha > 0, \varphi_\alpha$ and of the *non physical unwanted mode* $(\lambda = 0, \varphi = \text{constant})$.

- It is possible to eliminate this unwanted solution by adding condition (7.14c) to equations (7.14ab), which is equivalent to restricting the variational formulation (7.15) to the admissible class of functions of $\mathcal{C}$ of *zero mean value in* Ω_F.

  We emphasize that this type of constraint, consisting of an orthogonality relation on an eigensolution — corresponding to a zero eigenvalue in the present case — results in the elimination of this solution in the modified spectral problem, *without changing the other solutions*.

  In numerical applications, formulation (7.15) can therefore be used.

  It will be noticed that this latter formulation is equivalent to the variational formulation of the spectral boundary value problem in p obtained on putting $u_N = 0$ in (7.12)— which does include the non physical unwanted solution $\omega = 0$.

Energy interpretation

We consider an eigensolution $(\lambda_\alpha > 0, \varphi_\alpha)$. We denote by $u_\alpha = \nabla\varphi_\alpha$ the corresponding displacement of the fluid, and by $p_\alpha = -\rho_F c^2 div u_\alpha = \rho_F \omega_\alpha^2 \varphi_\alpha$ the corresponding pressure (in conformity with (2.15) and (7.9)).

From (7.15) and for $\delta\varphi = \varphi_\alpha$ we have:

$$\int_{\Omega_F} |\nabla\varphi_\alpha|^2 \, dx = \frac{\omega_\alpha^2}{c^2} \int_{\Omega_F} \varphi_\alpha^2 \, dx \tag{7.19}$$

We show that this relation results from the conservation of total mechanical energy of the fluid during a harmonic oscillation, i.e. the sum of the kinetic and potential energies.
We therefore have to express the kinetic and potential energies in terms of φ_α.

Kinetic energy — Knowing that the instantaneous displacement is written $u^F(M,t) = u_\alpha(M)\cos\omega_\alpha t$, the kinetic energy $E_C = \frac{1}{2}\int_{\Omega_F} \rho_F |\frac{\partial u^F}{\partial t}|^2 \, dx$ is expressed:

$$E_C = \left(\frac{\omega_\alpha^2}{2} \int_{\Omega_F} \rho_F |u_\alpha|^2 \, dx\right) \sin^2\omega_\alpha t = \left(\frac{\omega_\alpha^2}{2} \int_{\Omega_F} \rho_F |\nabla\varphi_\alpha|^2 \, dx\right) \sin^2\omega_\alpha t \tag{7.20}$$

Potential energy — This is written classically (using relation (2.15)):

$$E_P = -\frac{1}{2}\int_{\Omega_F} p \, div u^F \, dx = \rho_F c^2 \int_{\Omega_F} |div u^F|^2 \, dx = \frac{1}{2\rho_F c^2} \int_{\Omega_F} p^2 \, dx \tag{7.21}$$

from which, for the oscillation under consideration:

$$E_P = \left(\frac{1}{2\rho_F c^2} \int_{\Omega_F} p_\alpha^2 \, dx\right) \cos^2(\omega_\alpha t) \tag{7.22}$$

Conservation of total energy — From (7.19), the factors $\sin^2\omega_\alpha t$ in E_C and of $\cos^2\omega_\alpha t$ in E_P are identical, which leads to a constant value of $E_C + E_P$ during an oscillation.

Generalized mass and rigidity

1. We call *generalized modal mass* [1] μ_α the quantity:

$$\boxed{\mu_\alpha = \int_{\Omega_F} \rho_F |u_\alpha|^2 \, dx = \int_{\Omega_F} \rho_F |\nabla\varphi_\alpha|^2 \, dx} \tag{7.23}$$

2. We call *generalized modal rigidity* γ_α the quantity:

$$\boxed{\gamma_\alpha = \int_{\Omega_F} \frac{p_\alpha^2}{\rho_F c^2} \, dx = \frac{\rho_F \omega_\alpha^4}{c^2} \int_{\Omega_F} \varphi_\alpha^2 \, dx} \tag{7.24}$$

((γ_α = twice the maximum potential energy)

[1] By analogy with the definition of the generalized mass of a structural mode of vibration (*cf.* § 1.6).

From (7.19), μ_α and γ_α are thus related by:

$$\boxed{\gamma_\alpha = \mu_\alpha \omega_\alpha^2} \tag{7.25}$$

With these definitions,the orthogonality relations (7.17) can be stated indifferently:

$$\begin{array}{c}
\boxed{\begin{aligned}
\int_{\Omega_F} \frac{p_\alpha p_\beta}{\rho_F c^2}\, dx &= \delta_{\alpha\beta}\omega_\alpha^2 \mu_\alpha \quad (a) \\
\int_{\Omega_F} \rho_F u_\alpha \cdot u_\beta \, dx &= \delta_{\alpha\beta}\mu_\alpha \qquad (b)
\end{aligned}} \\
\Updownarrow \\
\boxed{\begin{aligned}
\int_{\Omega_F} \varphi_\alpha \varphi_\beta \, dx &= \delta_{\alpha\beta} \frac{c^2}{\rho_F} \frac{\mu_\alpha}{\omega_\alpha^2} \quad (c) \\
\int_{\Omega_F} \nabla\varphi_\alpha \cdot \nabla\varphi_\beta \, dx &= \delta_{\alpha\beta} \frac{\mu_\alpha}{\rho_F} \qquad (d)
\end{aligned}}
\end{array} \tag{7.26}$$

Normalization — The eigenvectors φ_α are defined to within a multiplying factor C. Taking $C = 1/\sqrt{\mu_\alpha}$, we define an eigenvector (defined to within a factor of ± 1) of generalized mass equal to 1.

Matrix equations of the discretized problem

Discretization by finite elements of the symmetric bilinear forms involved in (7.15) yields the following matrices:

$$\boxed{\begin{aligned}
\int_{\Omega_F} \nabla\varphi \cdot \nabla\delta\varphi \, dx &\Longrightarrow \delta\boldsymbol{\Phi}^T \boldsymbol{F} \boldsymbol{\Phi} \qquad (a) \\
\int_{\Omega_F} \varphi \, \delta\varphi \, dx &\Longrightarrow \delta\boldsymbol{\Phi}^T \boldsymbol{S} \boldsymbol{\Phi} \qquad (b)
\end{aligned}} \tag{7.27}$$

The matrix form of (7.15) is then written:

$$\boxed{\boldsymbol{F}\boldsymbol{\Phi} = \lambda \boldsymbol{S}\boldsymbol{\Phi}} \tag{7.28}$$

Practical application

- ($\lambda = 0$, $\varphi =$ constant) being a solution to (7.15), the matrix $\boldsymbol{F}(N \times N)$ is of rank $N - 1$ (§ 3.4).

- In conformity with the previous discussion (*cf.* § 2), the presence of the solution $\lambda = 0$ does not perturb the spectrum of positive eigenvalues.

- **Axisymmetric cavities** — We use the decomposition (3.41) of φ into a Fourier series which yields a series of distinct two-dimensional problems.

7.4 Modal analysis of the vibratory response of a fluid

Variational formulation of the response to a prescribed displacement u_N

We consider a given movement of the wall of normal amplitude u_N on Σ.
We start from equations (7.8) and introduce the admissible space $\mathcal{C}$ of the smooth functions defined in Ω_F and $\mathcal{C}^*$ the subspace of $\mathcal{C}$ made up of functions of zero meanvalue in Ω_F:

$$\mathcal{C}^* = \left\{\varphi \in \mathcal{C} \mid \int_{\Omega_F} \varphi \, dx = 0\right\} \tag{7.29}$$

Multiplying (7.8a) by a test-function $\delta\varphi \in \mathcal{C}^*$ and integrating in Ω_F, then applying Green's formula (3.16), and taking into account (7.8b), we obtain the variational formulation of the problem (7.8):
for given ω and u_N, find $\varphi \in \mathcal{C}^*$ such that $\forall \delta\varphi \in \mathcal{C}^*$ we have:

$$\boxed{\int_{\Omega_F} \nabla\varphi \cdot \nabla\delta\varphi \, dx - \frac{\omega^2}{c^2} \int_{\Omega_F} \varphi\delta\varphi \, dx = \int_{\Sigma} u_N \delta\varphi \, d\sigma} \tag{7.30}$$

Converse — We have to verify by going formally through the calculations that any solution φ of (7.30) is a solution of the boundary value problem (7.8).
We begin by noting that we can write (7.30) in an equivalent form obtained by taking as test-function in (7.30) the function $\delta\varphi \in \mathcal{C}^*$ defined from any function $\psi \in \mathcal{C}$ by $\delta\varphi = \psi - \frac{1}{Vol(\Omega_F)} \int_{\Omega_F} \psi \, dx$, which yields, for $\varphi \in \mathcal{C}^*$, $\forall \psi \in \mathcal{C}$:

$$\begin{aligned} \int_{\Omega_F} \nabla\varphi \cdot \nabla\psi - \frac{\omega^2}{c^2} \int_{\Omega_F} \varphi\psi \, dx = \\ \int_{\Sigma} u_N \psi \, d\sigma - \int_{\Omega_F} \left(\frac{1}{Vol(\Omega_F)} \int_{\Sigma} u_N \, d\sigma \right) \psi \, dx \end{aligned} \tag{7.31}$$

We then retrieve equations (7.8) by applying Green's formula to the first integral and proceeding by the classical method.

- The discretization of (7.30) is carried out as follows: if $\boldsymbol{U}$ denotes the vector of nodal values of u, (7.30) is written in the following discretized form:

$$\boxed{\boldsymbol{F\Phi} - \frac{\omega^2}{c^2} \boldsymbol{S\Phi} = \boldsymbol{C}^T \boldsymbol{U} \quad \text{with } \boldsymbol{L}^T \boldsymbol{\Phi} = 0} \tag{7.32}$$

  where $\delta\boldsymbol{\Phi}^T \boldsymbol{C}^T \boldsymbol{U}$ discretizes $\int_\Sigma u_N \delta\varphi \, d\sigma$, and where $\boldsymbol{L}^T \boldsymbol{\Phi} = 0$ discretizes the constraint $\int_{\Omega_F} \varphi \, dx = 0$.

In what follows, property (7.30) is used for the modal analysis of the problem, which leads to the consideration of two alternative solutions.

First modal decomposition of φ

Since the $\varphi_\alpha \in \mathcal{C}^*$ form a complete basis of $\mathcal{C}^*$, the solution $\varphi \in \mathcal{C}^*$ to (7.30) may be expanded in the form:

$$\varphi = \sum_{\alpha \geq 1} \tau_\alpha \varphi_\alpha \tag{7.33}$$

where $\{\tau_\alpha\}$ constitutes a system of generalized modal coordinates. Substituting (7.33) in (7.30) and putting sucessively $\delta\varphi = \varphi_1, \varphi_2, \ldots$, and using orthogonality relations (7.26cd), we obtain:

$$\boxed{(-\omega^2 + \omega_\alpha^2)\mu_\alpha \tau_\alpha = \omega_\alpha^2 \int_\Sigma \rho_F \, \varphi_\alpha u_N \, d\sigma} \tag{7.34}$$

then:
$$\varphi = \sum_{\alpha \geq 1} \frac{\omega_\alpha^2}{-\omega^2 + \omega_\alpha^2} \frac{\int_\Sigma \rho_F \, \varphi_\alpha u_N \, d\sigma}{\mu_\alpha} \varphi_\alpha \tag{7.35}$$

Remark — The above expression truncated to n modes yields an approximate solution to the problem.

Second modal decomposition of φ

This decomposition uses solution φ^0 of (7.10).

Introduction of φ^0 — The variational formulation of (7.10) (deduced for example from (7.30) by letting $\omega = 0$) is written, for $\varphi^0 \in \mathcal{C}^*$, $\forall \delta\varphi \in \mathcal{C}^*$:

$$\int_{\Omega_F} \nabla\varphi^0 \cdot \nabla\delta\varphi \, dx = \int_\Sigma u_N \delta\varphi \, d\sigma \tag{7.36}$$

Conjugate relations between φ^0 and φ_α

1. Putting $\delta\varphi = \varphi^0$ in equation (7.15) verified by φ_α, and $\delta\varphi = \varphi_\alpha$ in (7.36), we find the *conjugate relation* between φ^0 and φ_α:

$$\boxed{\frac{\omega_\alpha^2}{c^2} \int_{\Omega_F} \rho_F \varphi_\alpha \varphi^0 \, dx = \int_\Sigma \varphi_\alpha \rho_F u_N \, d\sigma} \tag{7.37}$$

2. Putting $\delta\varphi = \varphi_\alpha$ in (7.36) we find the *second conjugate relation*:

$$\boxed{\int_{\Omega_F} \rho_F \nabla\varphi^0 \cdot \nabla\varphi_\alpha \, dx = \int_\Sigma \rho_F \varphi_\alpha u_N \, d\sigma} \tag{7.38}$$

Modal decomposition — We put:

$$\boxed{\varphi = \varphi^0 + \underline{\varphi}} \tag{7.39}$$

Considering the variational properties (7.30) and (7.36), we find by substraction (for the same $\delta\varphi \in \mathcal{C}^*$), that $\underline{\varphi}$ verifies:

$$\int_{\Omega_F} \nabla\underline{\varphi}\cdot\nabla\delta\varphi\, dx - \frac{\omega^2}{c^2}\int_{\Omega_F} \underline{\varphi}\delta\varphi\, dx = \frac{\omega^2}{c^2}\int_{\Omega_F} \varphi^0_{u_N}\delta\varphi\, dx \tag{7.40}$$

$\underline{\varphi}$ may thus be expanded on the basis formed by eigenvectors φ_α of (7.15) according to:

$$\underline{\varphi} = \sum_{\alpha\geq 1} \kappa_\alpha\varphi_\alpha \tag{7.41}$$

Substituting this expression into (7.39) and taking into account (7.26cd) and (7.37), we obtain:

$$\boxed{(-\omega^2 + \omega_\alpha^2)\kappa_\alpha\mu_\alpha = \omega^2\int_\Sigma \rho_F\varphi_\alpha u_N\, d\sigma} \tag{7.42}$$

which from (7.39) yields the following expression for φ:

$$\varphi = \varphi^0 + \sum_{\alpha\geq 1}\frac{\omega^2}{-\omega^2+\omega_\alpha^2}\frac{\int_\Sigma \rho_F\varphi_\alpha u_N\, d\sigma}{\mu_\alpha}\varphi_\alpha \tag{7.43}$$

Remark — It is readily verified that the ratio of norms of the respective terms of rank n (in the sense of L^2) is equal to $(\omega/\omega_n)^2$. The latter expansion therefore converges faster than (7.35).

7.5 Impedance operator of the fluid

We have seen that to a displacement u_N de Σ, there corresponds, according to (7.9), a pressure fluctuation $p = \rho_F\omega^2\varphi + p^0$.
In what follows, we shall be concerned with the *dynamic component* $p_{dyn} = \rho_F\omega^2\varphi$ of p on Σ. From (7.8), φ is linearly dependent on u_N. This solution will be denoted φ_{u_N}. We then consider the virtual work of dynamic pressure forces $\int_\Sigma p_{dyn}\delta u_N\, d\sigma$. Replacing p_{dyn} by $\rho_F\omega^2\varphi$, we define the following bilinear form in $(u_N, \delta u_N)$: $\omega^2\int_\Sigma \rho_F\varphi_{u_N}\delta u_N\, d\sigma$.

Symmetry properties — We show that:

$$\int_\Sigma \rho_F\varphi_{u_N}\delta u_N\, d\sigma = \int_\Sigma \rho_F\varphi_{\delta u_N} u_N\, d\sigma \tag{7.44}$$

Putting $\delta\varphi = \varphi_{\delta u_N}$ in variational property (7.30) of φ_{u_N}, we obtain (after multiplying by ρ_F) the relation:

$$\int_{\Omega_F} \rho_F\nabla\varphi_{u_N}\cdot\nabla\varphi_{\delta u_N}\, dx - \frac{\omega^2\rho_F}{c^2}\int_{\Omega_F}\varphi_{u_N}\varphi_{\delta u_N}\, d\sigma = \int_\Sigma \rho_F\varphi_{\delta u_N}u_N\, d\sigma \tag{7.45}$$

which exposes the stated symmetry property. For a given ω, relation (7.44) enables the bilinear symmetric form $\mathcal{M}^\omega_{Ac}(u_N, \delta u_N)$, known as the acoustic impedance operator, or "dynamic mass" of the fluid, to be defined as follows:

$$\mathcal{M}^\omega_{Ac}(u_N, \delta u_N) = \int_\Sigma \rho_F\varphi_{u_N}\delta u_N\, d\sigma \tag{7.46}$$

Mass operator $\mathcal{M}^0_{Ac}$

For $\omega = 0$, (7.46) defines the bilinear form $\mathcal{M}^0_{Ac}$: [2]

$$\boxed{\mathcal{M}^0_{Ac}(u_N, \delta u_N) = \int_\Sigma \rho_F \varphi^0_{u_N} \delta u_N \, d\sigma = \int_{\Omega_F} \rho_F \nabla\varphi^0_{u_N} \cdot \nabla\varphi^0_{\delta u_N} \, dx} \tag{7.47}$$

Discretization of $\mathcal{M}^0_{Ac}$ — We have to express φ^0 in terms of u_N, with a view to calculating the discretized form of (7.47).
We can proceed, as we did in chapter 3, for the calculation of the hydrostatic mass matrix $\boldsymbol{M}^0_B$, by a technique of change of variables using the matrix $\boldsymbol{H}$ defined from the constraint $\boldsymbol{L}^T\boldsymbol{\Phi} = 0$ (*cf.* (3.97)). This is equivalent to putting $\boldsymbol{\Phi} = \boldsymbol{H}\boldsymbol{\Phi}'$ (where Φ' is an N-component vector).

Another method is to start from the *modified* boundary value problem (7.10) [3]. We introduce the space $\mathcal{C}$ of the smooth functions in Ω_F and the space $\mathcal{C}^l \subset \mathcal{C}$ of the functions verifying the constraint $l(\varphi) = 0$. Applying the test-functions method to this problem, we arrive at the variational property of $\varphi \in \mathcal{C}^l$, verified for all $\delta\varphi \in \mathcal{C}$:

$$\int_{\Omega_F} \nabla\varphi \cdot \nabla\delta\varphi \, dx = \int_\Sigma u_N \delta\varphi \, d\sigma - \frac{1}{Vol(\Omega_F)} \left(\int_\Sigma u_N \, d\sigma\right)\left(\int_{\Omega_F} \delta\varphi \, dx\right) \tag{7.48}$$

We note that (7.48) is trivially verified by $\delta\varphi$ = constant. The variational formulation may therefore be restricted to $\mathcal{C}^l$. [4]
In discretized form, taking the condition for cancellation of the first nodal value ϕ_1 of $\boldsymbol{\Phi} = (\phi_1, \boldsymbol{\Phi}_*)$, equation (7.48) becomes:

$$\delta\boldsymbol{\Phi}^T \boldsymbol{F}\boldsymbol{\Phi} = \delta\boldsymbol{\Phi}^T \left(\boldsymbol{C}^T\boldsymbol{U} - \boldsymbol{R}^T\boldsymbol{U}\right) \quad \text{with } \phi_1 = 0, \quad \text{and } \delta\phi_1 = 0 \tag{7.49}$$

where $\delta\boldsymbol{\Phi}^T \boldsymbol{R}^T\boldsymbol{U}$ discretizes $\frac{1}{Vol(\Omega_F)} \left(\int_\Sigma u_N \, d\sigma\right)\left(\int_{\Omega_F} \delta\varphi \, dx\right)$.
We deduce from this $\boldsymbol{\Phi}_* = \boldsymbol{F}_*^{-1}\left(\boldsymbol{C}_*^T - \boldsymbol{R}_*^T\right)\boldsymbol{U}$, where the matrices $\boldsymbol{C}_*$, $\boldsymbol{R}_*$, and $\boldsymbol{F}_*$ are obtained by eliminating, on the one hand, the row of $\boldsymbol{C}^T$ and of $\boldsymbol{R}^T$, and on the other hand the row and column of $\boldsymbol{F}$, which correspond to ϕ_1. From (7.27), the quadratic form corresponding to $\mathcal{M}^0_{Ac}$ is written in discretized form $\rho_F \boldsymbol{\Phi}^T\boldsymbol{F}\boldsymbol{\Phi} = \rho_F \boldsymbol{\Phi}_*^T \boldsymbol{F}_* \boldsymbol{\Phi}_*$, where $\boldsymbol{\Phi}_*$ is a function of $\boldsymbol{U}$.
By letting $\boldsymbol{E}_* = \boldsymbol{C}_* - \boldsymbol{R}_*$, we finally arrive at the following expression of $\boldsymbol{M}^0_{Ac}$:

$$\boldsymbol{M}^0_{Ac} = \rho_F \boldsymbol{E}_* \boldsymbol{F}_*^{-1} \boldsymbol{E}_*^T \tag{7.50}$$

[2] This operator describes the *quasi-static* inertial behaviour of the fluid — also called *pneumatic* in the case of a gas — (limiting behaviour for $\omega = 0$).

[3] We note that in the harmonic boundary value problem, obtained on replacing condition (7.10c) — which here acts as unicity condition — by any condition of the type $l(\varphi) = 0$ (7.8), any such substitution must be accompanied by the following modification of (7.8c) which acts as from the general relation (7.6) between p, φ and u_N: $\Delta\varphi + \frac{\omega^2}{c^2}(\varphi - \frac{1}{Vol(\Omega_F)}\int_{\Omega_F} \varphi \, dx) - \frac{1}{Vol(\Omega_F)}\int_\Sigma u_N \, d\sigma = 0$.

[4] All $\varphi \in \mathcal{C}$ may be uniquely written in the form $\varphi = \varphi^l$ + constant, with $\varphi^l \in \mathcal{C}^l$. We then see that $\varphi^l = \varphi - l(\varphi)/l(1)$ (with $l(1) \neq 0$).

We note that this matrix may be obtained by applying the Gauss condensation algorithm (1.64) to the matrix $\begin{bmatrix} 0 & \boldsymbol{E}_* \\ \boldsymbol{E}_*{}^T & \boldsymbol{F}_* \end{bmatrix}$.

Extremal property of $\mathcal{M}^0_{Ac}$ — We can show that the variational property (7.36) of φ^0 yields the following extremal property of $\mathcal{M}^0_{AC}$:

$$\mathcal{M}^0_{Ac}(u_N, u_N) = \max_{\varphi \in \mathcal{C}^*} \left\{ -\int_{\Omega_F} \rho_F |\nabla\varphi|^2 \, dx + 2\int_\Sigma \rho_F \varphi u_N \, d\sigma \right\} \tag{7.51}$$

where $\mathcal{C}^*$ is defined by (7.29). In fact, the calculation process for $\boldsymbol{M}^0_{Ac}$ may be interpreted as an application of this property in discretized form.

Modal decomposition of $\mathcal{M}^\omega_{AC}$

On substituting (7.43) into (7.46) we find:

$$\boxed{\mathcal{M}^\omega_{Ac}(u_N, \delta u_N) = \mathcal{M}^0_{Ac}(u_N, \delta u_N) + \sum_{\alpha \geq 1} \frac{\omega^2}{-\omega^2 + \omega_\alpha^2} \mathcal{M}_\alpha(u_N, \delta u_N)} \tag{7.52}$$

where $\mathcal{M}_\alpha$ is the *modal acoustic mass* defined by:

$$\boxed{\mathcal{M}_\alpha(u_N, \delta u_N) = \frac{1}{\mu_\alpha} \left(\int_\Sigma \rho_F \varphi_\alpha u_N \, d\sigma \right) \left(\int_\Sigma \rho_F \varphi_\alpha \delta u_N \, d\sigma \right)} \tag{7.53}$$

Summation rule for modal acoustic masses

We show that the series for acoustic masses $\mathcal{M}_\alpha$ is convergent, i.e.:

$$\boxed{\mathcal{M}^0_{Ac}(u_N, u_N) = \sum_{\alpha \geq 1} \mathcal{M}_\alpha(u_N, u_N)} \tag{7.54}$$

1. First, we establish the modal expansion of φ^0. Putting $\omega = 0$ in (7.35), we obtain:

$$\boxed{\varphi^0 = \sum_{\alpha \geq 1} \frac{\int_\Sigma \rho_F \varphi_\alpha u_N \, d\sigma}{\mu_\alpha} \varphi_\alpha} \tag{7.55}$$

   We deduce from this that:

$$\lim_{n \to \infty} \int_{\Omega_F} \left| \nabla \left(\varphi^0 - \sum_{\alpha=1}^{n} \frac{\int_\Sigma \rho_F \varphi_\alpha u_N \, d\sigma}{\mu_\alpha} \varphi_\alpha \right) \right|^2 dx = 0 \tag{7.56}$$

2. Expanding (7.56), and using (7.38), we obtain the stated result (7.54).

3. Relation (7.54), truncated to the order n, yields an approximate expression of the mass matrix $\boldsymbol{M}^0_{Ac}$.

7.6 Case of a compressible fluid with a free surface

Here, we use the above results to study the effects of compressibility on the vibrations of a fluid with a free surface Γ, (effects due to gravity are neglected) (Fig. 7.2). In this case, the surface condition is written:

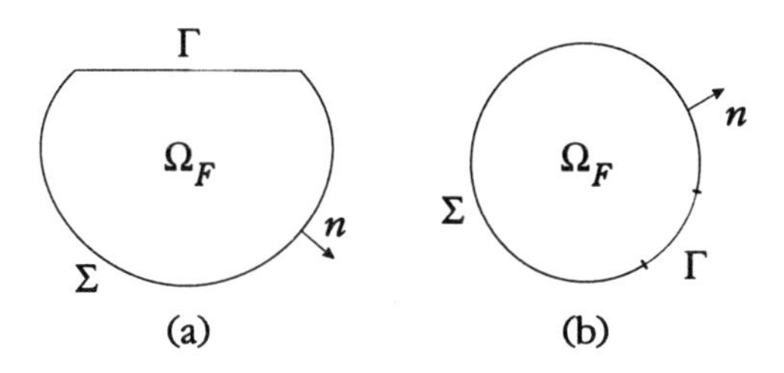

Figure 7.2: Fluid with a free surface

$$\boxed{p = 0 \quad \text{on } \Gamma} \tag{7.57}$$

which results from the nullity of the fluctuation of pressure on Γ — the lagrangian and eulerian fluctuations being indistinguishable in the absence of gravity for a homogeneous fluid — (*cf.* (2.20) and (2.8)).
We note that this boundary condition covers the two following distinct situations:

1. The case of a compressible liquid contained in a reservoir having a horizontal free surface Γ (Fig. 7.2(a)).

2. The case of an open cavity containing a gas communicating with an external medium assumed to be at constant pressure along a surface Γ of any form (Fig. 7.2(b)).

Boundary value problem in φ to prescribed displacements

Here we use the first definition (2.24) of the potential $\varphi = p/(\rho_F\omega^2)$, which now yields a well-posed problem for $\omega = 0$, as we shall see below.
In the harmonic regime, equations (7.1) in terms of p, φ are written:

$$\boxed{\begin{array}{rcllr} p &=& \rho_F\omega^2\varphi & \text{in } \Omega_F & (a) \\ p &=& -\rho_F c^2 \Delta\varphi & \text{in } \Omega_F & (b) \\ \dfrac{\partial\varphi}{\partial n} &=& u_N & \text{on } \Sigma & (c) \\ p &=& 0 & \text{on } \Gamma & (d) \end{array}} \tag{7.58}$$

Replacing p by its expression (7.58a) in the other equations yields (after division of ω^2 by (7.58d)):

$$\boxed{\begin{array}{rcllr} \Delta\varphi + \dfrac{\omega^2}{c^2}\varphi & = & 0 & \text{in } \Omega_F & (a) \\ \dfrac{\partial\varphi}{\partial n} & = & u_N & \text{on } \Sigma & (b) \\ \varphi & = & 0 & \text{on } \Gamma & (c) \end{array}} \quad (7.59)$$

Solution φ^0 for $\omega = 0$ — For $\omega = 0$, (7.58) is written:

$$\boxed{\begin{array}{rcllr} \Delta\varphi^0 & = & 0 & \text{in } \Omega_F & (a) \\ \dfrac{\partial\varphi^0}{\partial n} & = & u_N & \text{on } \Sigma & (b) \\ \varphi^0 & = & 0 & \text{on } \Gamma & (c) \end{array}} \quad (7.60)$$

which is a well-posed Neumann-Dirichlet problem.

Remarks on φ^0 — We point out that φ^0 *coincides* with the potential of the displacements introduced in chapter 5 in the study of vibrations of an incompressible weightless liquid with a free surface Γ contained in an elastic reservoir. We also note that this potential coincides with the potential φ^∞ introduced in (3.81) in the study of sloshing vibrations.

Alternative formulations

- It can be shown that, by choosing $l(\varphi) = \int_\Gamma \varphi\, d\sigma$ in the second definition (2.26) of φ, the potential thus defined coincides with the potential φ introduced in the present formulation.

- On the other hand, the formulation in p, which is deduced from (7.12) by addition of the boundary condition $p = 0$ on Γ, yields an ill-posed problem for $\omega = 0$.

Associated spectral problem

The solutions of the spectral problem deduced from (7.59) for $u_N = 0$ (with $\lambda = \frac{\omega^2}{c^2}$) describing the acoustic modes of a fluid occupying a domain Ω_F bounded by a fixed wall Σ and a free surface Γ, verify the boundary value problem:

$$\boxed{\begin{array}{rcllr} \Delta\varphi + \lambda\varphi & = & 0 & \text{in } \Omega_F & (a) \\ \dfrac{\partial\varphi}{\partial n} & = & 0 & \text{on } \Sigma & (b) \\ \varphi & = & 0 & \text{on } \Gamma & (c) \end{array}} \quad (7.61)$$

of which the variational formulation is stated:
find $\lambda \in \mathbb{R}, \varphi \in \mathcal{C}^*$, such that $\forall \delta\varphi \in \mathcal{C}^*$:

$$\boxed{\int_{\Omega_F} \nabla\varphi \cdot \nabla\delta\varphi \, dx - \lambda \int_{\Omega_F} \varphi\delta\varphi \, dx = 0} \tag{7.62}$$

$$\text{with:} \quad \boxed{\mathcal{C}^* = \left\{\varphi \mid \varphi = 0 \text{ on } \Gamma\right\}} \tag{7.63}$$

Relations (7.23), (7.24), (7.25) and (7.26) are applied with the current definition of φ_α. The corresponding matrix eigenvalue problem is deduced from (7.28) by canceling the rows and columns corresponding to the nodal values of φ on Γ.

Modal analysis of the vibratory response

The variational formulation (7.59) is written with the space $\mathcal{C}^*$ defined by (7.63): for given ω and u_N, find $\varphi \in \mathcal{C}^*$ such that:

$$\boxed{\int_{\Omega_F} \nabla\varphi \cdot \nabla\delta\varphi \, dx - \frac{\omega^2}{c^2} \int_{\Omega_F} \varphi\delta\varphi \, dx = \int_\Sigma u_N \delta\varphi \, d\sigma \quad \forall\delta\varphi \in \mathcal{C}^*} \tag{7.64}$$

Formulas (7.35) and (7.43) which give the modal decomposition of φ apply in the present case of a free surface. The definitions of $\mathcal{M}^\omega_{Ac}$, $\mathcal{M}^0_{Ac}$, and of $\mathcal{M}_\alpha$, together with the summation rules (7.52) and (7.54) also apply.

Comparison of $\mathcal{M}^0_{Ac}$ and $\mathcal{M}_A$ — We have noted that the solution φ^0 to (7.60) coincides with the "hydroelastic" potential introduced in chapter 5 (solution to equations (5.21)). Consequently, referring to the definition (5.25) of $\mathcal{M}_A$, we have the property:

$$\boxed{\mathcal{M}^0_{Ac} = \mathcal{M}_A} \tag{7.65}$$

Summation rule (7.54) is then stated:

$$\boxed{\mathcal{M}_A = \sum_{\alpha \geq 1} \mathcal{M}_\alpha} \tag{7.66}$$

where $\mathcal{M}_\alpha$ are the modal acoustic mass operators corresponding to the case of a free surface. Relation (7.66) may thus be interpreted as a decomposition of the incompressible hydroelastic added mass $\mathcal{M}_A$ into a sum of acoustic modal masses.

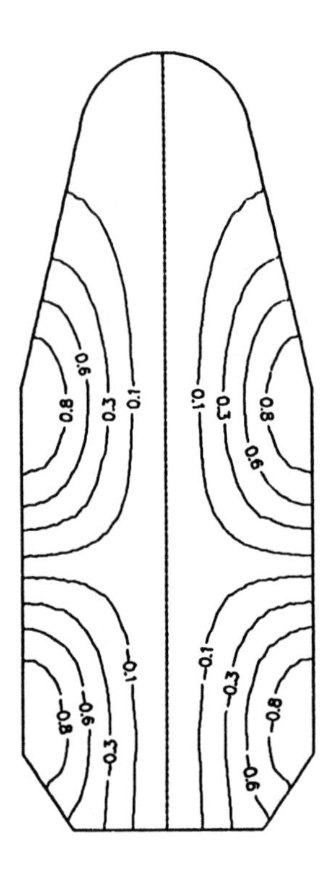

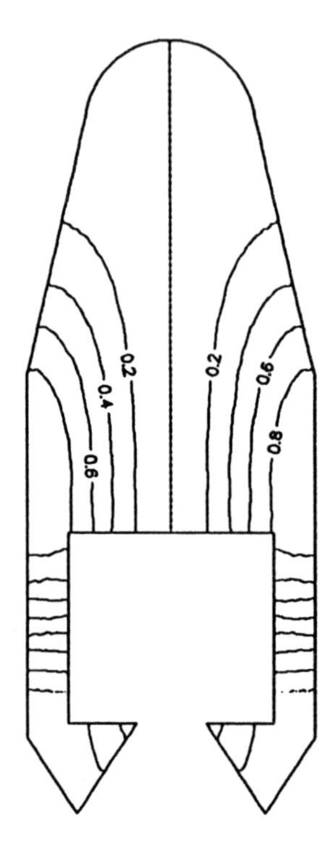

Influence of a satellite on the pressure distribution coresponding to the acoustic eigenmodes in the payload of the launcher Ariane 4.
The figures represent, in a meridian section, the iso-pressures of the second mode n=2, for the two configurations (without and with a satellite) carried out by finite element computations (*cf.* (7.28)).

CHAPTER 8

Structural-acoustic vibrations

8.1 Introduction

We examine here the *interior problem* of harmonic vibrations of an elastic structure [1] containing a weightless compressible fluid, with or without a free surface. Examples of applications include:

- problems of the vibroacoustic environment concerning not only launcher payloads but also aircraft and automobiles.

- study of the effect of the compressibility of liquids on hydroelastic vibrations (e.g. in launcher liquid hydrogen reservoirs).

The aim of this chapter is to establish various *symmetric* variational formulations in view of a direct treatment by finite elements; this will involve *first order* operators, in the general case of elastic and fluid three-dimensional mediums.
A natural way of tackling this problem consists of describing the fluid by the pressure field p (the structure being described by the displacement field u).
We shall begin by establishing the standard variational formulation of this problem, for which the numerical treatment by finite elements has a linear eigenvalue problem in $\lambda = \omega^2$ involving *non-symmetric* matrices.
We shall next establish, through introduction of the additional unknown field — viz, the displacement potentials φ of the fluid — two variational formulations of the spectral problem, respectively "with mass coupling" and "with stiffness coupling", which when discretized by finite elements yields a linear boundary value problem in $\lambda = \omega^2$ involving *symmetric matrices*. We shall see that this problem can, by means of an elimination procedure, be reduced to an eigenvalue problem respectively in (u, p) and (u, φ), which introduces the concepts of *added mass* and *added stiffness* operators, in continuum and discretized cases.
Finally, we shall present various alternative formulations.

[1] In view of coupling analysis with the external environment.

8.2 Unsymmetric (u, p) variational formulation

In this chapter we neglect the gravity effects. Under these conditions, the pressure and density of the the fluid at equilibrium are *constant*. [2] Furthermore, the lagrangian and eulerian fluctuations in pressure and density coincide.
We begin by recalling a standard formulation of the problem of the response of an elasto-acoustic coupled system to forces applied on the structure.
The corresponding boundary value problem is written, with $\lambda = \omega^2$:

$$\begin{array}{|rcllr|}
\sigma_{ij,j}(u) + \omega^2 \rho_S u_i & = & 0 & \text{in } \Omega_S & (a) \\
\sigma_{ij}(u) n_j^S & = & F_i^d & \text{on } \partial\Omega_S \setminus \Sigma & (b) \\
\sigma_{ij}(u) n_j^S & = & p n_i & \text{on } \Sigma & (c) \\
\dfrac{\partial p}{\partial n} & = & \omega^2 \rho_F u \cdot n & \text{on } \Sigma & (d) \\
\Delta p + \dfrac{\omega^2}{c^2} p & = & 0 & \text{in } \Omega_F & (e)
\end{array} \quad (8.1)$$

- (8.1a) is the elastodynamic equation (*cf.* (1.24a)). [3]
- (8.1b) is the boundary condition resulting from the surface density of the forces F_i^d applied on the structure.
- (8.1c) results from the action of pressure forces exerted by the fluid on the structure $\sigma_{ij} n_j^S = -p n_i^S = p n_i$, n denoting the normal external to the fluid.
- (8.1d) and (8.1e) are respectively the contact condition for the fluid on Σ, and the Helmholtz equation (*cf.* (7.12)).

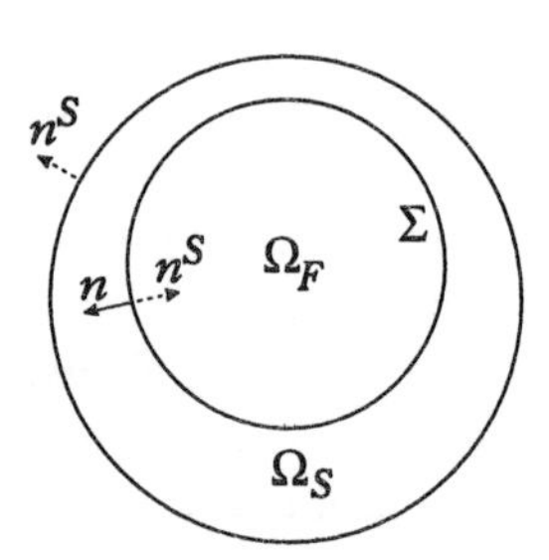

Figure 8.1: Elastic structure containing a gas

[2] The fluid medium is assumed homogeneous.
[3] We neglect here the prestress terms due to pressurization (see chapter 9).

We apply the test-functions method. We proceed in two steps, treating successively the equations relating to the structure ("subject to fluid pressure actions"), then the equations relating to the fluid ("subject to a wall displacement Σ") (Fig. 8.1).
First we introduce the space $\mathcal{C}_u$ of the smooth functions u defined in Ω_S. Multiplying equation (8.1a) by any function $\delta u \in \mathcal{C}_u$, then applying Green's formula (1.8), and finally, taking into account (8.1bc), we arrive, for $u \in \mathcal{C}_u$ and $\forall \delta u \in \mathcal{C}_u$, at:

$$\boxed{\int_{\Omega_S} \sigma_{ij}(u)\epsilon_{ij}(\delta u)\, dx \; - \; \omega^2 \int_{\Omega_S} \rho_S\, u \cdot \delta u\, dx - \int_\Sigma p\, n \cdot \delta u\, d\sigma = \int_{\partial\Omega_S \setminus \Sigma} F^d \cdot \delta u\, d\sigma} \qquad (8.2)$$

which is interpreted as the variational formulation of the problem of harmonic response of the structure, to pressure actions of the fluid.
In the second step, we consider the space $\mathcal{C}_p$ of the smooth p in Ω_F. We multiply equation (8.1e) by $\delta p \in \mathcal{C}_p$, then we integrate over Ω_F; on applying Green's formula (3.16), and finally, taking into account (8.1d), we arrive (after division by ρ_F)) at the following property of $p \in \mathcal{C}_p$ verified by $\forall \delta p \in \mathcal{C}_p$:

$$\boxed{\frac{1}{\rho_F} \int_{\Omega_F} \nabla p \cdot \nabla \delta p\, dx \; - \; \frac{\omega^2}{\rho_F c^2} \int_{\Omega_F} p\, \delta p\, dx \; - \; \omega^2 \int_\Sigma u \cdot n^F \delta p\, d\sigma \; = \; 0} \qquad (8.3)$$

and which is interpreted as the variational formulation of the harmonic response problem of the fluid to a movement of the wall. The variational formulation of the problem consists, for given ω and F^d, of finding, $(u, p) \in \mathcal{C}_u \times \mathcal{C}_p$ verifying (8.2) and (8.3).

Finite elements discretization

The matrices corresponding to the various bilinear or linear forms involved in the variational formulation (8.2, 8.3) are defined by:

$$\boxed{\begin{array}{rcll}
\displaystyle\int_{\Omega_S} \sigma_{ij}(u)\epsilon_{ij}(\delta u)\, dx & \Longrightarrow & \delta U^T K U & (a) \\
\displaystyle\int_{\Omega_S} \rho_S\, u \cdot \delta u\, dx & \Longrightarrow & \delta U^T M U & (b) \\
\displaystyle\frac{1}{\rho_F} \int_{\Omega_F} \nabla p \cdot \nabla \delta p\, dx & \Longrightarrow & \delta p^T F p & (c) \\
\displaystyle\frac{1}{\rho_F c^2} \int_{\Omega_F} p\, \delta p\, dx & \Longrightarrow & \delta p^T K_p p & (d) \\
\displaystyle\int_\Sigma p\, \delta u \cdot n\, d\sigma & \Longrightarrow & \delta U^T C p & (e) \\
\displaystyle\int_{\partial\Omega_S \setminus \Sigma} F^d \cdot \delta u\, d\sigma & \Longrightarrow & \delta U^T F^d & (f)
\end{array}} \qquad (8.4)$$

The variational equations (8.2) and (8.3) are written in discretized form:

$$\boxed{\begin{bmatrix} K & -C \\ 0 & F \end{bmatrix} \begin{bmatrix} U \\ p \end{bmatrix} - \omega^2 \begin{bmatrix} M & 0 \\ C^T & K_p \end{bmatrix} \begin{bmatrix} U \\ p \end{bmatrix} = \begin{bmatrix} F^d \\ 0 \end{bmatrix}} \qquad (8.5)$$

Bibliography

From computational point of view, this formulation has the advantage of only introducing one unknown per node to describe the fluid. On the other hand, it has the disadvantage of yielding *non-symmetric matrices*.
This formulation, the treatment of which by finite elements is described in Zienkiewicz & Newton [215], has been the subject of many investigations (*cf.* Gladwell [86], Craggs [46], Craggs & Stead [47], Zienkiewicz & Bettess [216], Petyt & Lim [182], Zienkiewicz & Taylor [217] vol. 2, Dowell, Gorman & Smith [60], Ramakrishnan & Koval [189])). And in the field of the automobile industry, Nefske, Wolf & Howell [159].
Various procedures for symmetrization at the matrix level have been proposed — e.g. by multiplication on the left of (8.5) by appropriate matrices, or by change of variables — (*cf.* Irons [102], Coppolino [43], Daniel [52], [53], Müller [155], Felippa [69], [70], Felippa & Ohayon [71], [174]).

Remark

The boundary value problem (8.1) — and therefore the variational formulation (8.2),(8.3) — yields an ill-posed problem for $\omega = 0$, since p is indeterminate according to (8.1de), (*cf.* (7.12)). This indeterminacy is accompanied by the existence of the non physical unwanted eigenvalue $\omega^2 = \lambda = 0$, corresponding to the solution ($u = 0, p =$ constant), of the acoustic spectral problem in p (*cf.* § 7.12).

"Non standard" formulations

Symmetric formulation in (u, u^F)

One way of arriving at a symmetric variational formulation is to describe the fluid by a field of displacement u^F (*cf.* Hamdi, Ousset & Verchery [92], Chen & Taylor [31], Bermudez & Rodriguez [19]). In this case it is easy to establish a symmetric variational formulation of this problem.
We have simply to consider the fluid as a special elastic medium, with the constitutive law $p = -\rho_F c^2 div u^F$ (*cf.* (2.15)).
We then arrive at the following formulation of the problem:
For a given ω^2, find $(u, u^F) \in \mathcal{V}$, such that for all $(\delta u, \delta u^F) \in \mathcal{V}$, we have:

$$\begin{aligned}\int_{\Omega_S} \sigma_{ij}(u)\epsilon_{ij}(\delta u)dx + \rho_F c^2 \int_{\Omega_F} div u^F div \delta u^F dx \\ -\omega^2 \left(\int_{\Omega_S} \rho_S u \cdot \delta u dx + \rho_F \int_{\Omega_F} u^F \cdot \delta u^F dx \right) = \int_{\partial\Omega_S \setminus \Sigma} F^d \cdot \delta u \, d\sigma\end{aligned} \tag{8.6}$$

where the admissible space $\mathcal{V}$ is defined by:

$$\mathcal{V} = \left\{ u, u^F \text{smooth} \mid (u - u^F) \cdot n|_\Sigma = 0, \; curl\, u^F = 0|_{\Omega_F} \right\} \tag{8.7}$$

This symmetric formulation has the disadvantage of requiring the discretization of the constraint $curl\, u^F = 0$, which is a complex numerical problem (*cf.* Nédélec [158]).
On the other hand, this displacements formulation has been adapted and applied, in one-dimensional case, to ducts (without flow, we refer to Ohayon [171] and with flow *cf.* Piet-Lahanier & Ohayon [184]).
If in addition the fluid is incompressible, it is necessary to take into account the constraint $div u^F = 0$ (*cf.* Girault & Raviart [85], Pironneau [187]).
One alternative consists of satisfying this constraint by replacing u^F by $\nabla\varphi$ in formulation (8.6), (8.7): we thus arrive at a symmetric formulation in which the fluid is described by a scalar field, involving a second order operator in φ (*cf.* Sanchez-Hubert & Sanchez-Palencia [195]).

Formulation using the velocity potential

This formulation consists of reformulating the boundary value problem (8.1), by substituting for p the *velocity potential* $\psi = j\omega\varphi$ defined by $p = -j\omega\rho_F\psi$ (*cf.* Everstine [64]).
The variational formulation of the problem (which is deduced from (8.2), (8.3) on replacing p by $-j\omega\rho_F\psi$), yields the following matrix equations [4] (valid for $\omega \neq 0$):

$$\left\{\begin{bmatrix} \boldsymbol{K} & 0 \\ 0 & \boldsymbol{F} \end{bmatrix} + j\omega \begin{bmatrix} 0 & \boldsymbol{C} \\ -\boldsymbol{C}^T & 0 \end{bmatrix} - \omega^2 \begin{bmatrix} \boldsymbol{M} & 0 \\ 0 & \boldsymbol{K}_\psi \end{bmatrix}\right\} \begin{bmatrix} \boldsymbol{U} \\ \boldsymbol{\psi} \end{bmatrix} = \begin{bmatrix} \boldsymbol{F}^d \\ 0 \end{bmatrix} \tag{8.8}$$

where $\boldsymbol{K}$ and $\boldsymbol{M}$ are the mass and stiffness matrices for the structure, where $\boldsymbol{F}$ here discretizes $\int_{\Omega_F} \rho_F \nabla\psi \cdot \nabla\delta\psi\, dx$, $\boldsymbol{K}_\psi$ discretizes $\frac{\rho_F}{c^2} \int_{\Omega_F} \psi\delta\psi\, dx$ and where $\boldsymbol{C}$ discretizes $\int_\Sigma \rho_F \psi \delta u \cdot n\, d\sigma$ (and $\boldsymbol{C}^T$ discretizes $\int_\Sigma \rho_F u \cdot n \delta\psi\, d\sigma$).
On changing the sign of the second equation of (8.8), we arrive at a system involving three symmetric matrix equations.
We note that formulation (8.8) is not valid for $\omega = 0$ and that the associated spectral problem has a non physical unwanted solution for $\omega = 0$.
The solution of the latter problem requires, in addition, non standard algorithms.
This formulation has been adapted to apply to the case $\omega = 0$, by Olson & Vandini [176].

Remark

By multiplying the first equation of (8.5) by ω^2, we can arrive at a quadratic matrix system in ω^2, involving three symmetric matrices.

---

[4] sometimes known as "gyroscopic type".

8.3 Basic (u, p, φ) equations

The symmetric formulations that we consider below are based on the following equations in (u, p, φ) of the spectral acoustic problem whose solutions are the elasto-acoustic modes: [5]

$$
\begin{array}{|ll|ll|}
\hline
 & & \nabla p - \rho_F \lambda \nabla \varphi = 0 \ |_{\Omega_F} & (d) \\
\sigma_{ij,j}(u) + \rho_S \lambda u_i = 0 \ |_{\Omega_S} & (a) & p = -\rho_F c^2 \Delta \varphi \ |_{\Omega_F} & (e) \\
\sigma_{ij}(u) n_j^S = 0 \ |_{\partial\Omega_S \setminus \Sigma} & (b) & \dfrac{\partial \varphi}{\partial n} = u \cdot n \ |_{\Sigma} & (f) \\
\sigma_{ij}(u) n_j^S = p n_i \ _{\Sigma} & (c) & l(\varphi) = 0, \ l(1) \neq 0 & (g) \\
\hline
\end{array}
\qquad (8.9)
$$

Solution for $\lambda = 0$ — For $\lambda = 0$, (8.9d) is written $\nabla p = 0$ which leads to $p = \text{constant} = p^0$, p^0 being given explicitly — after integrating (8.9e) in Ω_F and using (8.9f) — by:

$$
\boxed{p^0 = -\frac{\rho_F c^2}{Vol\,(\Omega_F)} \int_\Sigma u \cdot n \, d\sigma} \qquad (8.10)
$$

The displacement u then satisfies the following equations:

$$
\begin{aligned}
\sigma_{ij,j}(u) &= 0 \quad \text{in } \Omega_S && (a) \\
\sigma_{ij}(u) n_j^S &= 0 \quad \text{on } \partial\Omega_S \setminus \Sigma && (b) \\
\sigma_{ij}(u) n_j^S &= \left(-\frac{\rho_F c^2}{Vol\,(\Omega_F)} \int_\Sigma u \cdot n \, d\sigma\right) n_i \quad \text{on } \Sigma && (c)
\end{aligned}
\qquad (8.11)
$$

We show that the solution of (8.11) corresponds to the rigid body modes $u = u^{\mathcal{R}}$ of the structure.

It is sufficient to multiply (8.11a) by u, integrate in Ω_S, then apply Green's formula (1.8), taking account of (8.11b) and (8.11c). We obtain the expression $\int_{\Omega_S} \sigma_{ij}(u) \epsilon_{ij}(u) \, dx + \frac{\rho_F c^2}{Vol(\Omega_F)} \left(\int_\Sigma u \cdot n \, d\sigma\right)^2 = 0$, which is in the form of two positive terms.

Cancellation of the first term yields $u = u^{\mathcal{R}}$, and cancellation of the second to $p^0 = 0$ (*cf.* (8.10)), which result from the conservation of volume of the cavity Ω_F in rigid body displacements.

Finally, for $\lambda = 0$, the solution φ — denoted $\varphi^{\mathcal{R}}$ — verifies $\Delta \varphi^{\mathcal{R}} = 0$ in Ω_F, $\partial \varphi^{\mathcal{R}} / \partial n = u^{\mathcal{R}} \cdot n$ on Σ, with unicity condition (8.9g).

In conclusion, for $\lambda = 0$, and for a free structure, the solution is given by $u = u^{\mathcal{R}}$, $p = 0$, $\varphi = \varphi^{\mathcal{R}}$.

[5] Equations (8.9def) are the basic equations (2.27, 2.30) and (2.31) of chapter 2 with $\lambda = \omega^2$.

8.4 Symmetric formulation in $(\boldsymbol{u}, \boldsymbol{p}, \boldsymbol{\varphi})$ with mass coupling

In what follows, we successively establish:

1. A symmetric formulation with mass coupling using the potential $\varphi = p/(\rho_F \omega^2)$ (*cf.* definition (2.24) of φ). We discuss the non physical unwanted solutions present in this formulation for $\lambda = 0$, related to definition (2.24) of φ. We show that introducing the potential φ defined by $p = \rho_F \omega^2 \varphi + \pi$ (*cf.* 2.26), enables derivation from this formulation, of a formulation which has no unwanted zero eigenvalue and which is called "regularized".

2. A *symmetric* alternative formulation with *stiffness coupling*.

Boundary value problem in $(\boldsymbol{u}, \boldsymbol{p}, \boldsymbol{\varphi})$

We use here the displacement potential φ defined by $p = \rho_F \lambda \varphi$ (first definition (2.24) of φ).
The boundary value problem (8.9) can then be written:

$$
\begin{array}{rcllr}
\sigma_{ij,j}(u) + \rho_S \lambda u_i & = & 0 & \text{in } \Omega_S & (a) \\
\sigma_{ij}(u) n_j^S & = & 0 & \text{on } \partial\Omega_S \setminus \Sigma & (b) \\
\sigma_{ij}(u) n_j^S & = & \rho_F \lambda \varphi n_i & \text{on } \Sigma & (c) \\
\dfrac{\partial \varphi}{\partial n} & = & u \cdot n & \text{on } \Sigma & (d) \\
\rho_F \Delta \varphi + \dfrac{p}{c^2} & = & 0 & \text{in } \Omega_F & (e) \\
\dfrac{p}{\rho_F c^2} & = & \dfrac{\lambda}{c^2} \varphi & \text{in } \Omega_F & (f)
\end{array}
\qquad (8.12)
$$

where (8.12f) replaces (8.9d) and (8.9g), and where (8.12c) results from the replacement of p by $\rho_F \lambda \varphi$ in (8.9c).

$(\boldsymbol{u}, \boldsymbol{p}, \boldsymbol{\varphi})$ variational formulation

We shall establish the variational formulation of the solutions $(\lambda \neq 0, u, p, \varphi)$ to (8.12).
We proceed in three steps, by the test-functions method:

1. We introduce the space $\mathcal{C}_u$ of the smooth functions u defined in Ω_S. On multiplying (8.12a) by δu, applying Green's formula (1.8), and, finally, taking

into acccount (8.12b) and (8.12c), we arrive, for $u \in \mathcal{C}_u$ and $\forall \delta u \in \mathcal{C}_u$, at:

$$\boxed{\int_{\Omega_S} \sigma_{ij}(u)\epsilon_{ij}(\delta u)\,dx - \lambda \int_{\Omega_S} \rho_S u \cdot \delta u\,dx - \lambda \int_\Sigma \rho_F \varphi n \cdot \delta u\,d\sigma = 0} \quad (8.13)$$

2. In the second step, we consider the space $\mathcal{C}_\varphi$ of the smooth functions φ defined in Ω_F. Multiplying equation (8.12e) by $\delta\varphi$, integrating in Ω_F, then applying Green's formula (3.16), and finally, taking account of (8.12d), we arrive, for $\varphi \in \mathcal{C}_\varphi$ and $\forall \delta\varphi \in \mathcal{C}_\varphi$, at:

$$\boxed{-\int_{\Omega_F} \rho_F \nabla\varphi \cdot \nabla\delta\varphi\,dx + \int_\Sigma \rho_F u \cdot n \delta\varphi\,d\sigma + \int_{\Omega_F} \frac{p}{c^2}\,\delta\,\varphi\,dx = 0} \quad (8.14)$$

3. We consider the space $\mathcal{C}_p$ of the functions p defined in Ω_F. Multiplying equation (8.12f) by δp and integrating in Ω_F, we obtain, for $p \in \mathcal{C}_p$ and $\forall \delta p \in \mathcal{C}_p$:

$$\boxed{\int_{\Omega_F} \frac{p}{\rho_F c^2} \delta p\,dx - \lambda \int_{\Omega_F} \frac{\varphi}{c^2} \delta p\,dx = 0} \quad (8.15)$$

The variational formulation of the spectral problem consists therefore of finding $\lambda \in \mathbb{R}$ and $(u, p, \varphi) \in V = \mathcal{C}_u \times \mathcal{C}_p \times \mathcal{C}_\varphi$ verifying (8.13), (8.14), (8.15).
Conversely, by going formally through the calculations, we can verify that the variational properties (8.13), (8.15) and (8.14) enable equations (8.12) to be retrieved.
In the case of *slender structures* (beams, plates, shells), it is sufficient to replace the bilinear form $\int_\Omega \sigma_{ij}(u)\epsilon_{ij}(v)\,dx$ by the corresponding variational expressions.

Symmetric variational formulation in (u, p, φ) — On adding equations (8.13), (8.15) and equation (8.14) (after multiplying by $(-\lambda)$), we obtain the following symmetric variational formulation, verified for $(u, p, \varphi) \in V$ and $\forall (\delta u, \delta p, \delta\varphi) \in V$:

$$\boxed{\int_{\Omega_S} \sigma_{ij}(u)\epsilon_{ij}(\delta u)\,dx + \int_{\Omega_F} \frac{p}{\rho_F c^2} \delta p\,dx - \lambda \int_{\Omega_S} \rho_S u \cdot \delta u\,dx - \lambda \widehat{m}(\cdot,\cdot) = 0} \quad (8.16)$$

where $\widehat{m}(\cdot,\cdot)$ is the *symmetric bilinear* form on $V \times V$ defined by:

$$\boxed{\begin{aligned} \widehat{m} &= -\int_{\Omega_F} \rho_F \nabla\varphi \cdot \nabla\delta\varphi\,dx + \cdots \\ &+ \left(\int_\Sigma \rho_F u \cdot n \delta\varphi\,d\sigma + \int_\Sigma \rho_F \varphi n \cdot \delta u\,d\sigma\right) + \left(\int_{\Omega_F} \frac{p}{c^2}\,\delta\varphi\,dx + \int_{\Omega_F} \frac{\varphi}{c^2} \delta p\,dx\right) \end{aligned}} \quad (8.17)$$

Remark — It can be shown that $\mathcal{C}_u = (H^1(\Omega_S))^3$, $\mathcal{C}_p = L^2(\Omega_F)$, and $\mathcal{C}_\varphi = H^1(\Omega_F)$ (*cf.* Boujot [24]).

Matrix equations

Discretization by finite elements introduces, in addition to the matrices $\boldsymbol{K}$ and $\boldsymbol{M}$ defined respectively by (8.4ab) and (8.4d), the symmetric matrix $\boldsymbol{F}$ (of which the present definition differs from the preceding definition (8.4c) by a constant multiplying factor) and the rectangular coupling matrices $\boldsymbol{A}$ and $\boldsymbol{B}$ defined as follows:

$$\begin{array}{rcll} \int_{\Omega_F} \rho_F \nabla\varphi\cdot\nabla\delta\varphi\, dx & \Longrightarrow & \delta\boldsymbol{\Phi}^T\boldsymbol{F}\boldsymbol{\Phi} & (a) \\ \int_\Sigma \rho_F\varphi\,\delta u\cdot n\, d\sigma + \int_\Sigma \rho_F\delta\varphi\, u\cdot n\, d\sigma & \Longrightarrow & \delta\boldsymbol{U}^T\boldsymbol{A}\boldsymbol{\Phi} + \delta\boldsymbol{\Phi}^T\boldsymbol{A}^T\boldsymbol{U} & (b) \\ \int_{\Omega_F} \frac{1}{c^2}\varphi\,\delta p\, dx + \int_{\Omega_F}\frac{1}{c^2}\delta\varphi\, p\, dx & \Longrightarrow & \delta\boldsymbol{p}^T\boldsymbol{B}\boldsymbol{\Phi} + \delta\boldsymbol{\Phi}^T\boldsymbol{B}^T\boldsymbol{p} & (c) \end{array} \tag{8.18}$$

In discretized form, (8.16) then yields the following *symmetric* system:

$$\begin{bmatrix} \boldsymbol{K} & 0 & 0 \\ 0 & \boldsymbol{K}_p & 0 \\ 0 & 0 & 0 \end{bmatrix} \begin{bmatrix} \boldsymbol{U} \\ \boldsymbol{p} \\ \boldsymbol{\Phi} \end{bmatrix} = \lambda \begin{bmatrix} \boldsymbol{M} & 0 & \boldsymbol{A} \\ 0 & 0 & \boldsymbol{B} \\ \boldsymbol{A}^T & \boldsymbol{B}^T & -\boldsymbol{F} \end{bmatrix} \begin{bmatrix} \boldsymbol{U} \\ \boldsymbol{p} \\ \boldsymbol{\Phi} \end{bmatrix} \tag{8.19}$$

Elimination of Φ — The third matrix equation of (8.19) is written:

$$\boldsymbol{A}^T\boldsymbol{U} + \boldsymbol{B}^T\boldsymbol{p} - \boldsymbol{F}\boldsymbol{\Phi} = 0 \tag{8.20}$$

In the general case, it does not enable $\boldsymbol{\Phi}$ to be eliminated, since the matrix $\boldsymbol{F}$ is of rank $N_\varphi - 1$; $\int_{\Omega_F} \rho_F \nabla\varphi\cdot\nabla\delta\varphi\, dx = 0$ if and only if $\varphi = \text{constant}$. Consequently, $\boldsymbol{F}\boldsymbol{\Phi} = 0 \Leftrightarrow \boldsymbol{\Phi} = \text{constant}\boldsymbol{1}_{N_\varphi}$, denoting by $\boldsymbol{1}_{N_\varphi}$ the vector of $\mathbb{R}^{N_\varphi}$ of components equal to 1.

We now define a partitioning of nodal values $\boldsymbol{\Phi}$ by letting $\boldsymbol{\Phi} = \begin{bmatrix} \phi_1 \\ \boldsymbol{\Phi}_2 \end{bmatrix}$ where ϕ_1 denotes a particular component of $\boldsymbol{\Phi}$ (here, the first), which induces a partitioning of matrices $\boldsymbol{A}$, $\boldsymbol{B}$ and $\boldsymbol{F}$.

With this notation, (8.19) is then written explicitly:

$$\left\{ \begin{array}{ll} \boldsymbol{K}\boldsymbol{U} = \lambda(\boldsymbol{M}\boldsymbol{U} + A_1\phi_1 + \boldsymbol{A}_2\boldsymbol{\Phi}_2) & (a) \\ \boldsymbol{K}_p\boldsymbol{p} = \lambda B_1\phi_1 + \boldsymbol{B}_2\boldsymbol{\Phi}_2 & (b) \\ {A_1}^T\boldsymbol{U} + {B_1}^T\boldsymbol{p} - F_{11}\phi_1 = \boldsymbol{F}_{12}\boldsymbol{\Phi}_2 & (c) \\ {\boldsymbol{A}_2}^T\boldsymbol{U} + {\boldsymbol{B}_2}^T\boldsymbol{p} - {\boldsymbol{F}_{12}}^T\phi_1 = \boldsymbol{F}_{22}\boldsymbol{\Phi}_2 & (d) \end{array} \right. \tag{8.21}$$

$\boldsymbol{F}_{22}$ being nonsingular (*cf.* § 3.4), (8.21d) enables $\boldsymbol{\Phi}_2$ to be eliminated from equations (8.21abc).

We proceed by substitution, replacing $\boldsymbol{\Phi_2}$ by its expression in terms of $\boldsymbol{U}, \boldsymbol{p}, \phi_1$ taken from (8.21d), which is written:

$$\boldsymbol{\Phi}_2 = \boldsymbol{F}_{22}^{-1}{\boldsymbol{A}_2}^T\boldsymbol{U} + \boldsymbol{F}_{22}^{-1}{\boldsymbol{B}_2}^T\boldsymbol{p} - \boldsymbol{F}_{22}^{-1}{\boldsymbol{F}_{12}}^T\phi_1 \tag{8.22}$$

(8.21c) is then written:

$$(A_1{}^T - \boldsymbol{F}_{12}\boldsymbol{F}_{22}^{-1}\boldsymbol{A}_2^T)\boldsymbol{U} + (B_1{}^T - \boldsymbol{F}_{12}\boldsymbol{F}_{22}^{-1}\boldsymbol{B}_2^T)\boldsymbol{p} = (F_{11} - \boldsymbol{F}_{12}\boldsymbol{F}_{22}^{-1}\boldsymbol{F}_{12}^T)\phi_1 \quad (8.23)$$

This latter relation simplifies as follows:

- We show that $F_{11} - \boldsymbol{F}_{12}\boldsymbol{F}_{22}^{-1}\boldsymbol{F}_{12}^T = 0$.

  Thus, writing equation $\boldsymbol{F}\mathbf{1}_{N_\varphi} = 0$ in partitioned form, we obtain:

$$\begin{bmatrix} F_{11} & \boldsymbol{F}_{12} \\ \boldsymbol{F}_{12}^T & \boldsymbol{F}_{22} \end{bmatrix} \begin{bmatrix} 1 \\ \mathbf{1}_{N_\varphi - 1} \end{bmatrix} = \begin{bmatrix} 0 \\ \mathbf{0} \end{bmatrix} \quad (8.24)$$

  As $\boldsymbol{F}_{22}$ is nonsingular, the second equation of (8.24) yields:

$$\mathbf{1}_{N_\varphi - 1} = -\boldsymbol{F}_{22}^{-1}\boldsymbol{F}_{12}^T \quad (8.25)$$

  Using this result in the first equation of (8.24), we find that $F_{11} - \boldsymbol{F}_{12}\boldsymbol{F}_{22}^{-1}\boldsymbol{F}_{12}^T = 0$.

- The coefficient of $\boldsymbol{U}$ in (8.23) also simplifies on using (8.25):

$$(A_1{}^T - \boldsymbol{F}_{12}\boldsymbol{F}_{22}^{-1}\boldsymbol{A}_2^T) = \begin{bmatrix} 1 & \mathbf{1}_{N_\varphi - 1}{}^T \end{bmatrix} \begin{bmatrix} A_1{}^T \\ \boldsymbol{A}_2^T \end{bmatrix} = \begin{bmatrix} \mathbf{1}_{N_\varphi}{}^T \end{bmatrix} \begin{bmatrix} \boldsymbol{A}^T \end{bmatrix} \quad (8.26)$$

  Similarly, the coefficient of $\boldsymbol{p}$ in (8.23) simplifies to:

$$(B_1{}^T - \boldsymbol{F}_{12}\boldsymbol{F}_{22}^{-1}\boldsymbol{B}_2^T) = \begin{bmatrix} 1 & \mathbf{1}_{N_\varphi - 1}{}^T \end{bmatrix} \begin{bmatrix} B_1{}^T \\ \boldsymbol{B}_2^T \end{bmatrix} = \begin{bmatrix} \mathbf{1}_{N_\varphi}{}^T \end{bmatrix} \begin{bmatrix} \boldsymbol{B}^T \end{bmatrix} \quad (8.27)$$

Letting:

$$\boldsymbol{a}^T = \begin{bmatrix} \mathbf{1}_{N_\varphi}^T \end{bmatrix} \begin{bmatrix} \boldsymbol{A}^T \end{bmatrix}, \quad \boldsymbol{b}^T = \begin{bmatrix} \mathbf{1}_{N_\varphi}^T \end{bmatrix} \begin{bmatrix} \boldsymbol{B}^T \end{bmatrix} \quad (8.28)$$

relation (8.23) is finally written:

$$\boxed{\boldsymbol{a}^T \boldsymbol{U} + \boldsymbol{b}^T \boldsymbol{p} = 0} \quad (8.29)$$

Finally, we replace $\boldsymbol{\Phi}_2$ by its expression (8.22) in (8.21ab) (using relations (8.26) and (8.27) in transposed form). We thus arrive at the following matrix system, symmetrical in $(\boldsymbol{U}, \boldsymbol{p}, \phi_1)$:

$$\begin{bmatrix} \boldsymbol{K} & 0 & 0 \\ 0 & \boldsymbol{K}_p & 0 \\ 0 & 0 & 0 \end{bmatrix} \begin{bmatrix} \boldsymbol{U} \\ \boldsymbol{p} \\ \phi_1 \end{bmatrix} = \lambda \begin{bmatrix} \boldsymbol{M} + \boldsymbol{A}_2\boldsymbol{F}_{22}^{-1}\boldsymbol{A}_2^T & \boldsymbol{A}_2\boldsymbol{F}_{22}^{-1}\boldsymbol{B}_2^T & \boldsymbol{a} \\ \boldsymbol{B}_2\boldsymbol{F}_{22}^{-1}\boldsymbol{A}_2^T & \boldsymbol{B}_2\boldsymbol{F}_{22}^{-1}\boldsymbol{B}_2^T & \boldsymbol{b} \\ \boldsymbol{a}^T & \boldsymbol{b}^T & 0 \end{bmatrix} \begin{bmatrix} \boldsymbol{U} \\ \boldsymbol{p} \\ \phi_1 \end{bmatrix} \quad (8.30)$$

which exhibits the matrix $\boldsymbol{M}_{(u,p,\phi_1)}^A$, called the *added mass matrix*, which is expressed:

$$\boldsymbol{M}_{(u,p,\phi_1)}^A = \begin{bmatrix} \boldsymbol{A}_2\boldsymbol{F}_{22}^{-1}\boldsymbol{A}_2^T & \boldsymbol{A}_2\boldsymbol{F}_{22}^{-1}\boldsymbol{B}_2^T & \boldsymbol{a} \\ \boldsymbol{B}_2\boldsymbol{F}_{22}^{-1}\boldsymbol{A}_2^T & \boldsymbol{B}_2\boldsymbol{F}_{22}^{-1}\boldsymbol{B}_2^T & \boldsymbol{b} \\ \boldsymbol{a}^T & \boldsymbol{b}^T & 0 \end{bmatrix} \quad (8.31)$$

Interpretation of the relation between u and p — We note that the relation $\boldsymbol{a}^T \boldsymbol{U} + \boldsymbol{b}^T \boldsymbol{p} = 0$ (equation 8.29), is deduced by multiplication on the left of (8.20) by $\mathbf{1}^T$.
It corresponds therefore to the discretization of the variational property (8.14) (which gives (8.20)), for $\delta\varphi = \text{constant} = 1$:

$$\boxed{\int_\Sigma \rho_F u \cdot n \, d\sigma + \int_{\Omega_F} \frac{p}{c^2} dx = 0} \Longrightarrow \boxed{\boldsymbol{a}^T \boldsymbol{U} + \boldsymbol{b}^T \boldsymbol{p} = 0} \tag{8.32}$$

This latter equation thus results, in discretized form, from the conservation of total mass of fluid contained in Ω_F.

Bibliography — For formulation (8.16) and its treatment by finite elements (8.19), (8.30) *cf.* Morand & Ohayon [138], [142], (*cf.* also Ohayon & Valid [168], Ohayon, Meidinger & Berger [172], in the nuclear field, Gibert [84], for ductings, *cf.* Axisa *et al* [7], Everstine & Yang [65], in biomechanics *cf.* Coquart, Depeursinge, Curnier & Ohayon [44]). Variants may be found in Tabarrok [205] and Liu & Uras [121].

Practical construction of the added mass matrix

The "mass" matrix denoted $\boldsymbol{M}^A_{(u,p,\phi_1)}$ generalizes the concept of added mass, established for a incompressible fluid in chapter 5, to the case of an compressible fluid.

1. First, we discretize the bilinear form $\widehat{m}$ defined by (8.17) which, from (8.18), yields:

$$\begin{bmatrix} 0 & 0 & \boldsymbol{A} \\ 0 & 0 & \boldsymbol{B} \\ \boldsymbol{A}^T & \boldsymbol{B}^T & -\boldsymbol{F} \end{bmatrix} \tag{8.33}$$

   Construction of the matrix $\begin{bmatrix} 0 & \boldsymbol{A} \\ \boldsymbol{A}^T & 0 \end{bmatrix}$ and $\begin{bmatrix} 0 & \boldsymbol{B} \\ \boldsymbol{B}^T & 0 \end{bmatrix}$ results from the simultaneous assemblage of elementary contributions corresponding to the discretization of bilinear forms $(\int_\Sigma \rho_F u \cdot n \delta\varphi \, d\sigma + \int_\Sigma \rho_F \varphi n \cdot \delta u \, d\sigma)$ and $\frac{1}{c^2}(\int_{\Omega_F} p \, \delta\varphi \, dx + \int_{\Omega_F} \varphi \delta p \, dx)$.

2. Secondly, in (8.33), we distinguish the value ϕ_1 from the other nodal values $\boldsymbol{\Phi}_2$ of φ. Eliminating $\boldsymbol{\Phi}_2$ by means of the elimination algorithm (1.64), we obtain the matrix $\boldsymbol{M}^A_{(u,p,\phi_1)}$.

Remark — Elimination techniques include the Gauss *frontal elimination*, which enables the nodal values of φ to by eliminated one by one during the assemblage.

Axisymmetric cavity

We work in cylindrical coordinates: z is the axis of revolution, u_r, u_z, u_θ are the components of u in the local coordinates (r, z, θ)(*cf.* chapter 1). We may decompose u, p, φ into a Fourier series (*cf.* § 1.7 in the case of a vector field u and § 3.5 in the case of scalar fields p and φ).

1. The method of calculating the axisymmetric modes ($n = 0$) results from the preceding discussion ($\boldsymbol{F}$ is a singular matrix as in the three-dimensional case).

2. The method of calculating the non-axisymmetric modes ($n \geq 1$) is simplified, since $\boldsymbol{F}$ is nonsingular (we continue to denote by $\boldsymbol{F}$ the matrix discretizing the restriction of $\int_{\Omega_F} \rho_F \nabla\varphi \cdot \nabla\delta\varphi \, dx$ to functions of the type $\varphi_n(r, z) \cos n\theta$).

   Consequently, $\boldsymbol{\Phi}$ can be eliminated from (8.19) by means of (8.20), yielding the following matrix problem:

$$\begin{bmatrix} \boldsymbol{K} & 0 \\ 0 & \boldsymbol{K}_p \end{bmatrix} \begin{bmatrix} \boldsymbol{U} \\ \boldsymbol{p} \end{bmatrix} = \lambda \begin{bmatrix} \boldsymbol{M} + \boldsymbol{A}\boldsymbol{F}^{-1}\boldsymbol{A}^T & \boldsymbol{A}\boldsymbol{F}^{-1}\boldsymbol{B}^T \\ \boldsymbol{B}\boldsymbol{F}^{-1}\boldsymbol{A}^T & \boldsymbol{B}\boldsymbol{F}^{-1}\boldsymbol{B}^T \end{bmatrix} \begin{bmatrix} \boldsymbol{U} \\ \boldsymbol{p} \end{bmatrix} \tag{8.34}$$

   which exhibits the added mass matrix:

$$\boldsymbol{M}^A_{(u,p)} = \begin{bmatrix} \boldsymbol{A}\boldsymbol{F}^{-1}\boldsymbol{A}^T & \boldsymbol{A}\boldsymbol{F}^{-1}\boldsymbol{B}^T \\ \boldsymbol{B}\boldsymbol{F}^{-1}\boldsymbol{A}^T & \boldsymbol{B}\boldsymbol{F}^{-1}\boldsymbol{B}^T \end{bmatrix} \tag{8.35}$$

Case of a free surface

We consider here the transposition of the above results to the case of a fluid having a free surface Γ (*cf.* Morand & Ohayon [142]). We emphasize that this boundary condition covers the following two physical properties:

1. The case of a compressible liquid contained in a reservoir and having a free surface Γ, and neglecting gravity effects (Fig. 8.2(a)).

2. The case of an *open cavity* containing a gas in communication with an external medium assumed at constant pressure along a surface Γ of any shape (Fig. 8.2(b)).

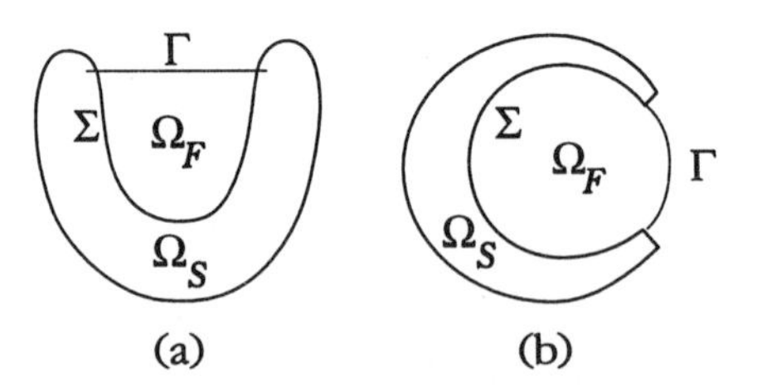

Figure 8.2: Case of a free surface

- The boundary value problem (8.12) is modified, in this case, by addition of the boundary condition $\varphi = 0$ on Γ (*cf.* § 7.6 and Fig. 7.2).

- The variational formulation (8.16) is modified by replacing $\mathcal{C}_\varphi$ with:

$$\boxed{\mathcal{C}_\varphi^* = \{\varphi \in \mathcal{C}_\varphi \mid \varphi = 0 \quad \text{on } \Gamma\}} \tag{8.36}$$

  which results in the corresponding rows and columns being eliminated in (8.19).

- In this case, the matrix $\boldsymbol{F}^*$ thus obtained is nonsingular, which makes it possible completely to eliminate the remaining nodal values $\boldsymbol{\Phi}^*$ — the final equations being similar to (8.34).

Elastic cavity completely filled with an incompressible liquid

We give here a few indications on the variational formulation of this problem, and the matrix structure of the discretized problem. [6]
The variational formulation in (u, φ) of this problem is deduced by putting $c = \infty$ in equations (8.13)(8.14)(8.15).
The formulation obtained is made up of (8.13), and equation (8.14) deprived of its third term. Its discretization yields a system of matrix equations in $(\boldsymbol{U}, \boldsymbol{\Phi})$, which is none other than system (8.19) stripped of the rows and columns corresponding to $\boldsymbol{p}$. In this case, the elimination of $\boldsymbol{\Phi}$ by means of the relation $\boldsymbol{A}^T\boldsymbol{U} - \boldsymbol{F}\boldsymbol{\Phi} = 0$ (i.e. (8.20) without the term in $\boldsymbol{p}$), results from the discussion of the elimination of $\boldsymbol{\Phi}$ (*cf.* § 8.4). We obtain a matrix system in $\boldsymbol{U}$ and ϕ_1 (which coincides with (8.30) stripped of the rows and columns corresponding to $\boldsymbol{p}$).

Elimination of the unwanted zero eigenvalue

We shall show that (8.19) and (8.30) have unwanted zero eigenvalues which can be eliminated by a suitable modification of these formulas (the three-dimensional case or axisymmetric modes of a closed cavity, with no free surface).

Unwanted zero eigenvalues of the non-reduced matrix problem

For $\lambda = 0$, (8.19) reduces to $\boldsymbol{K}\boldsymbol{U} = 0, \boldsymbol{K}_p\boldsymbol{p} = 0, [\,0\,]\ \boldsymbol{\Phi} = 0$. We deduce from this that:
- $\boldsymbol{U} = \boldsymbol{U}^{\mathcal{R}}$, where $\boldsymbol{U}^{\mathcal{R}}$ are the rigid body modes of the structure (*cf.* (1.33)),
- $\boldsymbol{p} = 0$, which results from the non-singularity of the matrix $\boldsymbol{K}_p$ due to the positive definite character of the quadratic form $\int_\Omega p^2\,dx$ (*cf.* (8.4d)),
- and that $\boldsymbol{\Phi}$ is arbitrary, which signifies the existence — in addition to the zero eigenvalues corresponding to the rigid body modes of the structure — of N_φ zero unwanted eigenvalues (where N_φ is the dimension of $\boldsymbol{\Phi}$). We see that the elimination of nodal values of Φ diminishes with the number of these non physical unwanted eigenvalues.

[6] For a reservoir partially filled with an incompressible fluid, we find equations (5.14), (5.15) by cancelling in (8.19) the rows and columns relating to the nodal values of p — which is formally equivalent to putting $c = \infty$ in the equations.

Zero eigenvalues of the reduced problem

For the case where $\boldsymbol{F}$ is singular, after reduction, from (8.30) there remains a non physical unwanted zero eigenvalue corresponding to an arbitrary ϕ_1.
We shall show that it is easily possible to modify (8.30) so as to eliminate the arbitrary unwanted solution ϕ_1 through a change of variable.
We begin by writing (8.30) in abbreviated form, introducing the vector $\boldsymbol{X}$ of components $\boldsymbol{U}$ and $\boldsymbol{p}$, the symmetric positive matrices $\boldsymbol{K}'$, $\boldsymbol{M}'$, and the coupling vector $\boldsymbol{c}$ as follows:

$$\boxed{\begin{bmatrix} \boldsymbol{K}' & 0 \\ 0 & 0 \end{bmatrix} \begin{bmatrix} \boldsymbol{X} \\ \phi_1 \end{bmatrix} = \lambda \begin{bmatrix} \boldsymbol{M}' & \boldsymbol{c} \\ \boldsymbol{c}^T & 0 \end{bmatrix} \begin{bmatrix} \boldsymbol{X} \\ \phi_1 \end{bmatrix}} \tag{8.37}$$

which is written explicitly:

$$\begin{aligned} \boldsymbol{K}'\boldsymbol{X} &= \lambda \boldsymbol{M}'\boldsymbol{X} + \lambda \boldsymbol{c}\phi_1 & (a) \\ \lambda \boldsymbol{c}^T\boldsymbol{X} &= 0 & (b) \end{aligned} \tag{8.38}$$

Carrying out the change of variables:

$$\nu_1 = \lambda \phi_1 \tag{8.39}$$

equation (8.38a) and equation (8.38b) — divided by λ — finally yield:

$$\boxed{\begin{bmatrix} \boldsymbol{K}' & -\boldsymbol{c} \\ -\boldsymbol{c}^T & 0 \end{bmatrix} \begin{bmatrix} \boldsymbol{X} \\ \nu_1 \end{bmatrix} = \lambda \begin{bmatrix} \boldsymbol{M}' & 0 \\ 0 & 0 \end{bmatrix} \begin{bmatrix} \boldsymbol{X} \\ \nu_1 \end{bmatrix}} \tag{8.40}$$

Discussion

1. $\lambda \neq 0$: (8.37) and (8.40) are trivially equivalent (same eigenvalues and same eigenvectors $\boldsymbol{X}$).

2. $\lambda = 0$: (8.40) is written explicitly:

$$\begin{aligned} \boldsymbol{K}'\boldsymbol{X} - \boldsymbol{c}\nu_1 &= 0 & (a) \\ \boldsymbol{c}^T\boldsymbol{X} &= 0 & (b) \end{aligned} \tag{8.41}$$

We show that (8.41) is equivalent to $\boldsymbol{K}'\boldsymbol{X} = 0$ and to $\nu_1 = 0$.
Multiplying (8.41a) on the left by $\boldsymbol{X}^T$, and (8.41b) by ν_1, we obtain:

$$\begin{aligned} \boldsymbol{X}^T\boldsymbol{K}'\boldsymbol{X} - \boldsymbol{X}^T\boldsymbol{c}\nu_1 &= 0 & (a) \\ \nu_1\boldsymbol{c}^T\boldsymbol{X} &= 0 & (b) \end{aligned} \tag{8.42}$$

Noting that $\nu_1\boldsymbol{c}^T\boldsymbol{X} = \boldsymbol{X}^T\boldsymbol{c}\nu_1$, we find $\boldsymbol{X}^T\boldsymbol{K}'\boldsymbol{X} = 0$, hence $\boldsymbol{K}'\boldsymbol{X} = 0$. Referring now to (8.41a), we have $\boldsymbol{c}\nu_1 = 0$, which requires $\nu_1 = 0$ (since $\boldsymbol{c}$ has a value other than zero). The matrix system (8.40) therefore possesses no unwanted zero eigenvalue.
Equation (8.40) may therefore be called a "regularized" matrix formulation of (8.37).

Number of physical eigenvalues

Equations (8.40) may be interpreted as the solution to a variational problem *subject to a constraint* — the scalar ν_1 being interpreted as the *associated Lagrange multiplier* — which is written:
find λ and $\boldsymbol{X}$ verifiying $\boldsymbol{c}^T\boldsymbol{X} = 0$, such that, for all $\delta\boldsymbol{X}$ verifying $\boldsymbol{c}^T\delta\boldsymbol{X} = 0$, we have:

$$\delta\boldsymbol{X}^T\boldsymbol{K}'\boldsymbol{X} = \lambda\,\delta\boldsymbol{X}^T\boldsymbol{M}'\boldsymbol{X} \tag{8.43}$$

- If N is the dimension of $\boldsymbol{K}'$, (8.43) has $N-1$ eigenvalues, including the multiple physical zero eigenvalue corresponding to the rigid body modes. The equivalent matrix system (8.40), of dimension $N+1$, has the same $N-1$ eigenvalues as (8.43) (the difference in dimension resulting from the presence in (8.40) of a "double infinite eigenvalue").

  On the other hand, (8.37) has, in addition to the $N-1$ physical eigenvalues, the unwanted zero eigenvalue (corresponding to the solution $\phi_1 = 1$), together with an "infinite eigenvalue".

- From (8.38b), we thus see that, for $\lambda = 0$, formulation (8.37) violates the constraint $\boldsymbol{c}^T\boldsymbol{X} = 0$ which corresponds, from (8.29) and (8.32), to the mass conservation law.

  Formulation (8.40) corrects this anomaly by restoring the mass conservation law (*cf.* Géradin, Roberts & Huck [81]).

  This observation is the basis of the detailed variational analysis discussed later, which yields a continuum-based formulation in (u, p, φ) having no unwanted zero eigenvalue.

Matrix equations

With the initial notations, the "regularized" form of (8.30) is thus written:

$$\begin{bmatrix} \boldsymbol{K} & 0 & -\boldsymbol{a} \\ 0 & \boldsymbol{K}_p & -\boldsymbol{b} \\ -\boldsymbol{a}^T & -\boldsymbol{b}^T & 0 \end{bmatrix} \begin{bmatrix} \boldsymbol{U} \\ \boldsymbol{p} \\ \nu_1 \end{bmatrix} = \lambda \begin{bmatrix} \boldsymbol{M} + \boldsymbol{A}_2\boldsymbol{F}_{22}^{-1}\boldsymbol{A}_2^T & \boldsymbol{A}_2\boldsymbol{F}_{22}^{-1}\boldsymbol{B}_2^T & 0 \\ \boldsymbol{B}_2\boldsymbol{F}_{22}^{-1}\boldsymbol{A}_2^T & \boldsymbol{B}_2\boldsymbol{F}_{22}^{-1}\boldsymbol{B}_2^T & 0 \\ 0 & 0 & 0 \end{bmatrix} \begin{bmatrix} \boldsymbol{U} \\ \boldsymbol{p} \\ \nu_1 \end{bmatrix} \tag{8.44}$$

For the practical application of the latter formulation, we proceed by static condensation of (8.33), which is equivalent to eliminating $N_\varphi - 1$ nodal values in terms of $\boldsymbol{U}, \boldsymbol{p}, \phi_1$.
It is then a matter of extracting from this condensed matrix the row and column corresponding to ϕ_1, to construct the stiffness matrix.
In what follows, we give a continuum-based interpretation of the preceding matrix equations (8.44).

8.5 Mass coupling formulation without unwanted zero eigenvalue

We shall show that use of the displacement potential defined by $p = \rho_F \omega^2 \varphi + \pi$ (*cf.* (2.26)), yields a variational formulation having no unwanted zero eigenvalue.

Preliminary remark — The constant π is identically null in the following cases (which have no unwanted zero eigenvalue):

1. calculation of the non-axisymmetric elasto-acoustic modes for structures of revolution (*cf.* the expansion into a Fourier series (4.53)).

2. calculation of the elasto-acoustic modes when the (weightless) fluid has a free surface. In this case, the free surface condition $p = 0$ is replaced, from (8.36), by $\varphi = 0$ on Γ, which yields $\pi = 0$.

Boundary value problem in (u, p, φ)

Equations (8.9) lead to the following boundary value problem:

$$\begin{array}{lll|lll}
\sigma_{ij,j}(u) + \rho_S \lambda u_i = 0 & |_{\Omega_S} & (a) & p = \rho_F \lambda \varphi + \pi & |_{\Omega_F} & (d) \\
\sigma_{ij}(u) n_j^S = 0 & |_{\partial\Omega_S \backslash \Sigma} & (b) & \rho_F \Delta\varphi + \dfrac{p}{c^2} = 0 & |_{\Omega_F} & (e) \\
\sigma_{ij}(u) n_j^S = (\rho_F \lambda \varphi + \pi) n_i & |_{\Sigma} & (c) & \dfrac{\partial \varphi}{\partial n} = u \cdot n & |_{\Sigma} & (f) \\
 & & & l(\varphi) = 0,\ l(1) \neq 0 & & (g)
\end{array} \tag{8.45}$$

where (8.45d) replaces (8.9d), and where (8.45c) is the result of replacing p by $\rho_F \lambda \varphi + \pi$ in (8.9c).

Solution for $\lambda = 0$ — For $\lambda = 0$, we can verify that (8.45), like (8.9), has the solution $u = u^{\mathcal{R}}$, $p = \pi = 0$, and $\varphi = \varphi^{\mathcal{R}}$, showing that the boundary value problem has no unwanted mode for $\lambda = 0$.

Variational formulation

We proceed in three steps by the test-functions method:

1. We introduce the space $\mathcal{C}_u$ of the smooth functions δu defined in Ω_S. Multiplying (8.45a) by δu, applying Green's formula (1.8), and finally taking account of (8.45bc), we arrive, for $u \in \mathcal{C}_u$ and $\forall \delta u \in \mathcal{C}_u$, at:

$$\boxed{\int_{\Omega_S} \sigma_{ij}(u) \epsilon_{ij}(\delta u) dx - \pi \int_{\Sigma} \delta u \cdot n \, d\sigma - \lambda \int_{\Omega_S} \rho_S u \cdot \delta u dx - \lambda \int_{\Sigma} \rho_F \varphi n \cdot \delta u \, d\sigma = 0} \tag{8.46}$$

2. In the second step, we introduce the admissible class $\mathcal{C}_p$ of test-functions δp defined in Ω_F.

   Multiplying (8.45d) by $\delta p \in \mathcal{C}_p$ and integrating in Ω_F we obtain $\forall \delta p \in \mathcal{C}_p$ (after division by $\rho_F c^2$):

$$\int_{\Omega_F} \frac{p}{\rho_F c^2} \delta p dx - \pi \int_{\Omega_F} \frac{1}{\rho_F c^2} \delta p \, dx - \lambda \int_{\Omega_F} \frac{\varphi}{c^2} \delta p \, dx = 0 \tag{8.47}$$

3. In the third step, we introduce the admissible space $\mathcal{C}_\varphi$ of the functions $\delta\varphi$ smooth in Ω_F. Multiplying (8.45e) by $\delta\varphi \in \mathcal{C}_\varphi$, integrating in Ω_F, then applying Green's formula (3.16), and finally taking account of (8.45f), we obtain $\forall \delta\varphi \in \mathcal{C}_\varphi$:

$$\int_{\Omega_F} \rho_F \nabla\varphi \cdot \nabla\delta\varphi \, dx - \int_\Sigma \rho_F u \cdot n \delta\varphi \, d\sigma - \int_{\Omega_F} \frac{p}{c^2} \delta\varphi \, dx = 0 \tag{8.48}$$

Conversely, going formally through the calculations, we can verify that variational properties (8.46),(8.47) and (8.48), completed by (8.45g), characterize the solutions u, p, φ of the spectral problem (8.45).

Remark — We note that in (8.48), the test-functions $\delta\varphi$ are not subject to condition (8.45g) which must satisfy φ.

In order to establish a variational formulation where φ and $\delta\varphi$ belong to the same admissible space, we have to introduce (*cf.* § 4.5), the space $\mathcal{C}_\varphi^l$ defined by:

$$\mathcal{C}_\varphi^l = \{\varphi \in \mathcal{C}_\varphi \mid l(\varphi) = 0 \, , l(1) \neq 0\} \tag{8.49}$$

We note that $\varphi \in \mathcal{C}_\varphi$ can be written uniquely, in the form:

$$\varphi = \varphi^l + \text{constant} \quad \text{with } \varphi \in \mathcal{C}_\varphi \text{ and } \varphi^l \in \mathcal{C}_\varphi^l \tag{8.50}$$

On applying l to (8.50), and using $l(\varphi^l) = 0$, we find the value of the constant: constant $= l(\varphi)/l(1)$. Conversely, (8.50) is verified by $\varphi^l = \varphi - l(\varphi)/l(1)$. Consequently, $\mathcal{C}_\varphi$ expands according to the direct sum:

$$\mathcal{C}_\varphi = \mathcal{C}_\varphi^l \oplus \mathbb{R} \tag{8.51}$$

Considering this expansion, (8.48) may be replaced by the equivalent set of two variational equations, obtained by restricting (8.48) successively to the class of test-functions $\delta\varphi \in \mathcal{C}_\varphi^l$, and to the class of constant test-functions $\delta\pi \in \mathbb{R}$, which yields, $\forall \delta\varphi \in \mathcal{C}_\varphi^l$, and $\forall \delta\pi \in \mathbb{R}$:

$$\begin{aligned} &\int_{\Omega_F} \rho_F \nabla\varphi \cdot \nabla\delta\varphi \, dx - \int_\Sigma \rho_F u \cdot n \delta\varphi \, d\sigma - \int_{\Omega_F} \frac{p}{c^2} \delta\varphi \, dx = 0 \quad (a) \\ &\delta\pi \int_{\Omega_F} \frac{p}{c^2} \, dx + \delta\pi \int_\Sigma \rho_F u \cdot n \, d\sigma = 0 \quad (b) \end{aligned} \tag{8.52}$$

We note that in (8.52b) we find the equation for conservation of mass (8.32).

Conclusion — The variational formulation (8.46), (8.47), and (8.48) of equations (8.45) can be stated in the following equivalent form:
find $\lambda \in \mathbb{R}^+$ and $(u, p, \varphi, \pi) \in \mathcal{C}_u \times \mathcal{C}_p \times \mathcal{C}_\varphi^l \times \mathbb{R}$ such that (8.46), (8.47), (8.52ab) are satisfied, $\forall (\delta u, \delta p, \delta\varphi, \delta\pi) \in \mathcal{C}_u \times \mathcal{C}_p \times \mathcal{C}_\varphi^l \times \mathbb{R}$.

Remark — From a mathematical point of view, we can show that the admissible spaces introduced are $\mathcal{C}_u = (H^1(\Omega_S))^3$, $\mathcal{C}_p = L^2(\Omega_F)$, $\mathcal{C}_\varphi = H^1(\Omega_F)$.
We note that the discretization of p does not imply the continuity of p.

Matrix equations

The discretization of (8.46), (8.47), (8.52ab) by finite elements involves the same matrices as those involved in (8.19), together with the matrices $\boldsymbol{a}$ and $\boldsymbol{b}$ defined in (8.28), which enable the respective forms $\delta\pi \int_\Sigma \rho_F u \cdot n \, d\sigma$, and $\delta\pi \int_\Sigma \frac{p}{c^2} \, dx$ to be discretized (*cf.* (8.32)).

Practical choice of the constraint imposed on φ — For numerical application, it is convenient to choose for l, a cancellation condition for a nodal value of φ, e.g. the first: $\phi_1 = 0$. We denote by $\boldsymbol{\Phi}_2$ the truncated vector of the $N_\varphi - 1$ remaining nodal values.
The matrix equations are then written:

$$
\begin{aligned}
\boldsymbol{K}\boldsymbol{U} - \frac{\boldsymbol{a}}{\rho_F}\pi - \lambda \boldsymbol{M}\boldsymbol{U} - \lambda \boldsymbol{A}_2 \boldsymbol{\Phi}_2 &= 0 \quad (a) \\
\boldsymbol{K}_p \boldsymbol{p} - \frac{\boldsymbol{b}}{\rho_F}\pi - \lambda \boldsymbol{B}_2 \boldsymbol{\Phi}_2 &= 0 \quad (b) \\
\boldsymbol{F}_{22}\boldsymbol{\Phi}_2 - \boldsymbol{A}_2^T \boldsymbol{U} - \boldsymbol{B}_2^T \boldsymbol{p} &= 0 \quad (c) \\
\boldsymbol{a}^T \boldsymbol{U} + \boldsymbol{b}^T \boldsymbol{p} &= 0 \quad (d)
\end{aligned}
\tag{8.53}
$$

where the matrices $\boldsymbol{A}_2$, $\boldsymbol{B}_2$ and $\boldsymbol{F}_{22}$ are those introduced previously by partitioning in § 8.4.

Elimination of Φ_2 — Since $\boldsymbol{F}_{22}$ is nonsingular, equation (8.53c) enables $\boldsymbol{\Phi}_2$ to be eliminated in terms of $\boldsymbol{U}$ and $\boldsymbol{p}$, which after substitution into (8.53ab) and putting $\boldsymbol{a}' = \boldsymbol{a}/\rho_F$ and $\boldsymbol{b}' = \boldsymbol{b}/\rho_F$, yields the *following symmetric* matrix equations:

$$
\begin{bmatrix} \boldsymbol{K} & 0 & -\boldsymbol{a}' \\ 0 & \boldsymbol{K}_p & -\boldsymbol{b}' \\ -\boldsymbol{a}'^T & -\boldsymbol{b}'^T & 0 \end{bmatrix}
\begin{bmatrix} \boldsymbol{U} \\ \boldsymbol{p} \\ \pi \end{bmatrix}
= \lambda
\begin{bmatrix} \boldsymbol{M} + \boldsymbol{A}_2 \boldsymbol{F}_{22}^{-1} \boldsymbol{A}_2^T & \boldsymbol{A}_2 \boldsymbol{F}_{22}^{-1} \boldsymbol{B}_2^T & 0 \\ \boldsymbol{B}_2 \boldsymbol{F}_{22}^{-1} \boldsymbol{A}_2^T & \boldsymbol{B}_2 \boldsymbol{F}_{22}^{-1} \boldsymbol{B}_2^T & 0 \\ 0 & 0 & 0 \end{bmatrix}
\begin{bmatrix} \boldsymbol{U} \\ \boldsymbol{p} \\ \pi \end{bmatrix}
\tag{8.54}
$$

Putting $\nu_1 = \pi/\rho_F$ in (8.54), we find exactly the regularized matrix equations (8.44).

Remark — In conformity with (8.43), the above matrix equations (8.54) may be written in the following form:

$$\begin{bmatrix} \boldsymbol{K} & 0 \\ 0 & \boldsymbol{K}_p \end{bmatrix} \begin{bmatrix} \boldsymbol{U} \\ \boldsymbol{p} \end{bmatrix} = \lambda \begin{bmatrix} \boldsymbol{M} + \boldsymbol{A}_2 \boldsymbol{F}_{22}^{-1} \boldsymbol{A}_2^T & \boldsymbol{A}_2 \boldsymbol{F}_{22}^{-1} \boldsymbol{B}_2^T \\ \boldsymbol{B}_2 \boldsymbol{F}_{22}^{-1} \boldsymbol{A}_2^T & \boldsymbol{B}_2 \boldsymbol{F}_{22}^{-1} \boldsymbol{B}_2^T \end{bmatrix} \begin{bmatrix} \boldsymbol{U} \\ \boldsymbol{p} \end{bmatrix} \tag{8.55}$$

$$\text{with the constraint} \quad \boldsymbol{a}^T \boldsymbol{U} + \boldsymbol{b}^T \boldsymbol{p} = 0$$

The fluid is involved through a stiffness matrix $\boldsymbol{K}_p$ and also by the *added mass matrix* $\boldsymbol{M}_A$ coupling $\boldsymbol{U}$ and $\boldsymbol{p}$, expressed:

$$\boldsymbol{M}_A = \begin{bmatrix} \boldsymbol{A}_2 \boldsymbol{F}_{22}^{-1} \boldsymbol{A}_2^T & \boldsymbol{A}_2 \boldsymbol{F}_{22}^{-1} \boldsymbol{B}_2^T \\ \boldsymbol{B}_2 \boldsymbol{F}_{22}^{-1} \boldsymbol{A}_2^T & \boldsymbol{B}_2 \boldsymbol{F}_{22}^{-1} \boldsymbol{B}_2^T \end{bmatrix} \tag{8.56}$$

8.6 Condensed formulation in (u,p) and continuum-based added mass operator

We propose here to give a continuum-based interpretation of the reduced matrix equations in (u,p) in the general case of a three dimensional cavity filled with fluid.
The variational equations (8.46), (8.47), (8.52ab) may be interpreted by considering (8.52b) as a constraint and π as the associated Lagrange multiplier.
By forcing u and p to satisfy this constraint, we can eliminate π from these equations. We thus have to consider the admissible class $\mathcal{C}_{u,p}$ of the u,p, defined by:

$$\mathcal{C}_{u,p} = \left\{ u \in \mathcal{C}_u, p \in \mathcal{C}_p \mid \int_\Sigma \rho_F u \cdot n \, d\sigma + \int_{\Omega_F} \frac{p}{c^2} \, dx = 0 \right\} \tag{8.57}$$

On adding (8.46) and (8.47), and keeping (8.52a), we obtain an equivalent formulation of the equations which can be stated: find $\lambda \in \mathbb{R}^+$, $(u,p) \in \mathcal{C}_{u,p}$ and $\varphi \in \mathcal{C}_\varphi^l$ such that, $\forall (\delta u, \delta p) \in \mathcal{C}_{u,p}$ and $\forall \delta\varphi \in \mathcal{C}_\varphi^l$, we have:

$$\begin{aligned} \int_{\Omega_S} \sigma_{ij}(u) \epsilon_{ij}(\delta u) \, dx + \int_{\Omega_F} \frac{p}{\rho_F c^2} \delta p \, dx - \lambda \int_{\Omega_S} \rho_S u \cdot \delta u \, dx + \cdots \\ -\lambda \left[\int_\Sigma \rho_F \varphi n \cdot \delta u \, d\sigma + \int_{\Omega_F} \frac{\varphi}{c^2} \delta p \, dx \right] = 0 \end{aligned} \tag{8.58}$$

$$\int_{\Omega_F} \rho_F \nabla\varphi \cdot \nabla\delta\varphi \, dx - \int_\Sigma \rho_F u \cdot n \delta\varphi \, d\sigma - \int_{\Omega_F} \frac{p}{c^2} \, \delta\varphi \, dx = 0 \tag{8.59}$$

Elimination of φ and added mass operator

The variational equation (8.59) considered in isolation for given (u,p), characterizes the unique solution φ to the Neumann problem (8.45e), (8.45fg), whose condition for existence is simply the constraint imposed on the admissible couple $(u,p) \in \mathcal{C}_{u,p}$. It is easily verified that this solution — denoted $\varphi_{u,p}$ — is linearly dependent on u and p.

The elimination consists of putting $\varphi_{u,p}$ into (8.58), which yields the following *variational formulation in* u, p, whose symmetry is established in what follows:

$$\begin{aligned}\int_{\Omega_S} \sigma_{ij}(u)\epsilon_{ij}(\delta u)\, dx + \int_{\Omega_F} \frac{p\,\delta p}{\rho_F c^2}\, dx - \lambda \int_{\Omega_S} \rho_S u\cdot\delta u\, dx + \cdots \\ -\lambda \left[\int_\Sigma \rho_F \varphi_{u,p} n\cdot\delta u\, d\sigma + \int_{\Omega_F} \frac{\varphi_{u,p}\,\delta p}{c^2}\, dx \right] = 0\end{aligned} \tag{8.60}$$

We introduce the following bilinear form $\mathcal{M}_A(u,p|\delta u, \delta p)$ — called *elasto-acoustic added mass operator* — defined by (*cf.* Morand & Ohayon [138]):

$$\mathcal{M}_A = \int_\Sigma \rho_F \varphi_{u,p} n\cdot\delta u\, d\sigma + \int_{\Omega_F} \frac{\varphi_{u,p}\,\delta p}{c^2}\, dx \tag{8.61}$$

We show that *this operator is symmetric.*
Consider the solution $\varphi_{\delta u,\delta p}$ of (8.59) for given $(\delta u, \delta p)$, taking as special test-function $\delta\varphi = \varphi_{u,p}$; φ thus verifies:

$$\int_\Sigma \rho_F \varphi_{u,p} \delta u\cdot n\, d\sigma + \int_{\Omega_F} \frac{\varphi_{u,p}\,\delta p}{c^2}\, dx = \int_{\Omega_F} \rho_F \nabla\varphi_{\delta u,\delta p}\cdot\nabla\varphi_{u,p}\, dx \tag{8.62}$$

which enables $\mathcal{M}_A$ to be written in the following symmetric form:

$$\boxed{\mathcal{M}_A = \int_{\Omega_F} \rho_F \nabla\varphi_{u,p}\cdot\nabla\varphi_{\delta u,\delta p}\, dx} \tag{8.63}$$

In conclusion, the matrix equations (8.55) and (8.56) are interpreted, taking account of (8.61) and (8.63), as a discretization of (8.60).

Extremal properties

- **Rayleigh quotient** —The variational property (8.60) results from the stationarity of the Rayleigh quotient $R(u, p)$ defined for $(u, p) \in \mathcal{C}_{u,p}$ by:

$$\boxed{R(u,p) = \frac{\displaystyle\int_{\Omega_S} \sigma_{ij}(u)\epsilon_{ij}(u)\, dx + \int_{\Omega_F} \frac{p^2}{\rho_F c^2}\, dx}{\displaystyle\int_\Omega \rho_S |u|^2\, dx + \mathcal{M}_A(u,p|u,p)}} \tag{8.64}$$

  The numerator is the sum of the potential energies of deformation of the structure and of the fluid. The denominator is the sum of two terms. The first corresponds classically to the inertia of the structure and the second corresponds to the inertia of the fluid: since $u^F = \nabla\varphi$, from (8.63) we can write: $\mathcal{M}_A(u,p|u,p) = \int_{\Omega_F} \rho_F |u^F|^2 dx$.

- **Extremal property of $\mathcal{M}_A$** — We shall show that $\mathcal{M}_A(u,p|u,p)$ verify the property:

$$\mathcal{M}_A(\cdot|\cdot) = \max_{\varphi \in \mathcal{C}^l_\varphi} \left\{ -\int_{\Omega_F} \rho_F |\nabla\varphi|^2\, dx + 2\int_\Sigma \rho_F u\cdot n\varphi\, d\sigma + 2\int_{\Omega_F} \frac{p\varphi}{c^2}\, dx \right\} \tag{8.65}$$

The demonstration is based on the following classical property of the solution u of the problem $a(u,v) = < f,v >$ for any v with $a(\cdot,\cdot) > 0$. If $J(v) = -a(v,v) + 2 < f,v >$, u verifies $J(u) = \max_v J(v)$. Using the variational equation satisfied by u, we deduce that $J(u) = \max_v J(v)$ $= a(u,u) = 2 < f,u >$. The stated property is obtained on applying the above result for $a(\varphi,\delta\varphi) = \int_{\Omega_F} \rho_F \nabla\varphi \cdot \nabla\delta\varphi \, dx$ and $< f,\delta\varphi >=$ $\int_\Sigma \rho_F u \cdot n \delta\varphi \, d\sigma + \int_{\Omega_F} \frac{p\delta\varphi}{c^2} dx$.

- **Discretization** — We can show, from (8.65), that the eigenvalues of the discretized problem are higher than those of the same rank of the continuum problem (to within one domain approximation).

Special cases

- **Symmetry of revolution** — In the calculation of non-axisymmetric modes, we note that the constraint appearing in (8.57) is trivially verified through the variation in $\cos n\theta$ of p and of $u \cdot n$. Moreover, $\mathcal{C}_\varphi^l$ must be replaced by $\mathcal{C}_\varphi$ (no constraint on φ) in the variational formulation (8.58), (8.59), in the condensed formulation (8.60) and in the expressions (8.64) and (8.65).

- **Case of a free surface** — In equations (8.58) and (8.59), and in (8.64), (8.65): $\mathcal{C}_{u,p}$ is replaced by $\mathcal{C}_u \times \mathcal{C}_p$, and $\mathcal{C}_\varphi^l$ by $\mathcal{C}_\varphi^*$ (*cf.* (8.36)).

Remarks

- Summarizing, we have established a symmetric variational formulation in (u,p) having no unwanted zero eigenvalue (*cf.* Morand & Ohayon [149]).

- We recall that the "direct" variational formulation of the spectral problem in (u,p) (deduced from (8.2) and (8.3) for $F = 0$) yields non-symmetric matrix equations and has an unwanted eigenvalue. We can verify that this formulation can be regularized for $\lambda = 0$ by imposing the constraint (8.32). [7]

- **Fluid-fluid coupling** — The above results can readily be adapted to the case of two fluids (inviscid, immiscible) in contact along a surface Γ — one of the fluids may be incompressible - taking into account the coupling conditions on Γ (*cf.* (2.34)).

- **Basis of structural-acoustic modes** — We can show that the "elasto-acoustic" eigenvectors (u,p), solutions to (8.60), form a basis of $\mathcal{C}_{u,p}$. The corresponding modal projection of the response of the internal structure-fluid coupled system to forces applied on the structure, yields equations similar to equations (1.68). [8]

[7] which is verified naturally for $\lambda \neq 0$.

[8] using the orthogonality of the elasto-acoustic modes which can be deduced from the Rayleigh quotient (8.64).

8.7 (u, p, φ) symmetric formulation with stiffness coupling

We shall establish an alternative variational formulation of the basic boundary value problem (8.9).
We proceed in three steps by the test-functions method:

1. We introduce the space $\mathcal{C}_u$ of the smooth functions u defined in Ω_S. Multiplying (8.9a) by δu, applying Green's formula (1.8), and finally, taking into account (8.9bc), then for $u \in \mathcal{C}_u$ and $\forall \delta u \in \mathcal{C}_u$, we arrive at:

$$\boxed{\int_{\Omega_S} \sigma_{ij}(u)\epsilon_{ij}(\delta u)\, dx \;-\; \lambda \int_{\Omega_S} \rho_S u \cdot \delta u\, dx - \int_{\Sigma} p n \cdot \delta u\, d\sigma \;=\; 0} \tag{8.66}$$

2. Secondly, we introduce the space $\mathcal{C}_p$ of smooth p in Ω_F.

   Multiplying (8.9e) by δp, integrating in Ω_F, then applying Green's formula (3.16), and finally, taking account of (8.9f), then for $p \in \mathcal{C}_p$ and $\forall \delta p \in \mathcal{C}_p$, we arrive at:

$$\boxed{\int_{\Omega_F} \nabla\varphi \cdot \nabla\delta p\, dx - \int_{\Omega_F} \frac{p\,\delta p}{\rho_F c^2}\, dx - \int_{\Sigma} u \cdot n \delta p\, d\sigma = 0} \tag{8.67}$$

3. Finally, we consider the space $\mathcal{C}_\varphi$ of the smooth functions $\delta\varphi$ defined in Ω_F.

   We show that (8.9d) can be written in variational form, $\forall \delta\varphi \in \mathcal{C}_\varphi$:

$$\boxed{\int_{\Omega_F} \nabla p \cdot \nabla\delta\varphi\, dx - \lambda \int_{\Omega_F} \rho_F \nabla\varphi \cdot \nabla\delta\varphi\, dx = 0} \tag{8.68}$$

The demonstration is based on the following remark: any constant field α in Ω_F simply-connected (and thus verifying $\nabla\alpha = 0$) satisfies the variational property:
$\int_{\Omega_F} \nabla\alpha \cdot \nabla\delta\varphi dx = 0 \ \forall\delta\varphi$ — which constitutes the variational property of the Neumann problem $\Delta\alpha = 0$ in Ω_F, $\partial\alpha/\partial n = 0$ on $\partial\Omega_F$, satisfied by the constant functions α. Applying this result to $\alpha = p - \rho_F\lambda\varphi$ which is constant from (8.9d), we then find (8.68).
We denote by $\mathcal{C}_\varphi^l$ the space of $\varphi \in \mathcal{C}_\varphi$ verifying $l(\varphi) = 0$ (*cf.* (8.49)). The variational formulation in u, p, φ then consists of finding $\lambda \in \mathbb{R}$, $(u, p, \varphi) \in \mathcal{C}_u \times \mathcal{C}_p \times \mathcal{C}_\varphi^l$ which verifies (8.66), (8.67) and (8.68), $\forall(\delta u, \delta p, \delta\varphi) \in \mathcal{C}_u \times \mathcal{C}_p \times \mathcal{C}_\varphi^l$.
Conversely, by going formally through the calculations, we can verify that this property enables (8.9) to be retrieved.

- From a mathematical point of view, it can be shown that $\mathcal{C}_u = (H^1(\Omega_S))^3, \mathcal{C}_p = H^1(\Omega_F)$, and $\mathcal{C}_\varphi = H^1(\Omega_F)$. We note that p and φ belong to the same space, and that their discretization requires similiar interpolations (unlike the case of the formulation with mass coupling).

- **Solution for $\lambda = 0$** — As the boundary value problem (8.9) has no unwanted zero eigenvalue, the same is true for the variational formulation (8.66), (8.67), (8.68).

Symmetric matrix equations

The discretization of this formulation by finite elements involves the matrices $\boldsymbol{K}$, $\boldsymbol{M}$ and $\boldsymbol{F}$ defined in (8.4a), (8.4ab), as well as the coupling term $\boldsymbol{C}$ defined in (8.4e). We introduce also the matrix $\boldsymbol{D}$ defined by:

$$\int_{\Omega_F} \nabla\varphi\cdot\nabla\delta p\, dx + \int_{\Omega_F} \nabla\delta\varphi\cdot\nabla p\, dx \Longrightarrow \delta\boldsymbol{p}^T\boldsymbol{D}^T\boldsymbol{\Phi} + \delta\boldsymbol{\Phi}^T\boldsymbol{D}\boldsymbol{p} \tag{8.69}$$

With these notations, the variational formulation (8.66),(8.68), (8.67) yields the following *symmetric* matrix equations:

$$\boxed{\begin{bmatrix} \boldsymbol{K} & 0 & -\boldsymbol{C} \\ 0 & 0 & \boldsymbol{D} \\ -\boldsymbol{C}^T & \boldsymbol{D}^T & -\boldsymbol{K}_p \end{bmatrix}\begin{bmatrix} \boldsymbol{U} \\ \boldsymbol{\Phi} \\ \boldsymbol{p} \end{bmatrix} = \lambda\begin{bmatrix} \boldsymbol{M} & 0 & 0 \\ 0 & \boldsymbol{F} & 0 \\ 0 & 0 & 0 \end{bmatrix}\begin{bmatrix} \boldsymbol{U} \\ \boldsymbol{\Phi} \\ \boldsymbol{p} \end{bmatrix}} \tag{8.70}$$

As the constraint $l(\varphi)$ is arbitrary, it is sufficient in practice to cancel any nodal value of φ, i.e. a component of $\boldsymbol{\Phi}$, which is equivalent finally to cancelling the corresponding row and column in (8.70).
For a bibliography on this formulation, *cf.* Ohayon [170], [173], [175], *cf.* also Antoniadis & Kanarachos [107], [4], Sandberg & Göransson [196], and for inclusion of wall damping (impedance) terms, *cf.* Kehr-Candille & Ohayon [108].

Condensed matrix equations in $(\boldsymbol{u}, \varphi)$

The third equation of (8.70) enables $\boldsymbol{p}$ to be expressed in terms of $\boldsymbol{U}$ and $\boldsymbol{\Phi}$ as follows:

$$\boldsymbol{p} = -\boldsymbol{K}_p^{-1}\boldsymbol{C}^T\boldsymbol{U} + \boldsymbol{K}_p^{-1}\boldsymbol{D}^T\boldsymbol{\Phi} \tag{8.71}$$

The matrix $\boldsymbol{K}_p$ defined by (8.4d) is indeed non-singular, since it is associated with the positive definite quadratic form $\int_{\Omega_F} \frac{p^2}{\rho_F c^2} dx$.
On replacing $\boldsymbol{p}$ by this expression in the first two equations of (8.70), we arrive at the following condensed matrix equations:

$$\boxed{\begin{bmatrix} \boldsymbol{K} + \boldsymbol{C}\boldsymbol{K}_p^{-1}\boldsymbol{C}^T & -\boldsymbol{C}\boldsymbol{K}_p^{-1}\boldsymbol{D}^T \\ -\boldsymbol{D}\boldsymbol{K}_p^{-1}\boldsymbol{C}^T & \boldsymbol{D}\boldsymbol{K}_p^{-1}\boldsymbol{D}^T \end{bmatrix}\begin{bmatrix} \boldsymbol{U} \\ \boldsymbol{\Phi} \end{bmatrix} = \lambda\begin{bmatrix} \boldsymbol{M} & 0 \\ 0 & \boldsymbol{F} \end{bmatrix}\begin{bmatrix} \boldsymbol{U} \\ \boldsymbol{\Phi} \end{bmatrix}} \tag{8.72}$$

- The constraint on φ (cancellation of a component of $\boldsymbol{\Phi}$), can either be carried out through (8.70) (before eliminating $\boldsymbol{p}$), or through (8.72) (i.e. after elimination).

- The presence of the fluid results in a **symmetric added stiffness** matrix $\boldsymbol{K}_A$, coupling $\boldsymbol{U}$ and $\boldsymbol{\Phi}$, which is expressed:

$$\boldsymbol{K}_A = \begin{bmatrix} \boldsymbol{C}\boldsymbol{K}_p^{-1}\boldsymbol{C}^T & -\boldsymbol{C}\boldsymbol{K}_p^{-1}\boldsymbol{D}^T \\ -\boldsymbol{D}\boldsymbol{K}_p^{-1}\boldsymbol{C}^T & \boldsymbol{D}\boldsymbol{K}_p^{-1}\boldsymbol{D}^T \end{bmatrix} \tag{8.73}$$

Special cases

- **Symmetry of revolution** — In calculations of non-axisymmetric modes, it is not necessary to apply a constraint on φ.
- **Case of a free surface** — In equations (8.66), (8.67), (8.68), $\mathcal{C}_p$ has to be replaced by $\mathcal{C}_p^* = \{p \in \mathcal{C}_p | p = 0 \text{ on } \Gamma\}$. For discretization, cancellation of the nodal values of p on Γ must be carried out before eliminating $\boldsymbol{p}$ in (8.70).

8.8 Symmetric formulation in (u, γ, p)

The derivation of the above symmetric variational formulations is based on a description of the fluid by two scalar fields which differ essentially in the multiplying factor ω^2 — the structure being described by u.
An alternative method is to describe the fluid by a scalar field — e.g. p — and, for the structure, to introduce the additional variable γ defined by $\gamma = -\omega^2 u$ (i.e. the acceleration or the dynamic reaction forces if necessary). The boundary value problem (8.1) can then be reformulated in the spectral case in the following form:

$$\begin{array}{rcllr} \sigma_{ij,j}(u) - \rho_S \gamma_i & = & 0 & \text{in } \Omega_S & (a) \\ \sigma_{ij}(u) n_j^S & = & 0 & \text{on } \partial\Omega_S \setminus \Sigma & (b) \\ \sigma_{ij}(u) n_j^S & = & p n_i & \text{on } \Sigma & (c) \\ \gamma_i + \lambda u_i & = & 0 & \text{in } \Omega_S & (d) \\ \dfrac{\partial p}{\partial n} & = & \rho_F \lambda u \cdot n & \text{on } \Sigma & (e) \\ \Delta p + \dfrac{\omega^2}{c^2} p & = & 0 & \text{in } \Omega_F & (f) \end{array} \tag{8.74}$$

We apply the method of test-functions:
— multiply (8.74a) by $\delta u \in \mathcal{C}_u$, integrate in Ω_S, then take into account (8.74bc) after applying Green's formula,
— multiply (8.74d) par $\delta\gamma \in \mathcal{C}_\gamma$, and integrate in Ω_S, and finally,
— multiply (8.74f) by $\delta p \in \mathcal{C}_p$, integrate in Ω_F, then take into account (8.74e) after applying Green's formula.
We arrive at the symmetric variational formulation in (u, γ, p) with *mass coupling* [9], which may be characterized by the stationarity of the following general-

[9] We can also establish a symmetric formulation in (u, γ, p) with *stiffness coupling* (*cf.*

ized Rayleigh quotient $R(u, \gamma, p)$ (*cf.* Ohayon [169], Deneuvy [56]):

$$R = \frac{\int_{\Omega_S} \rho_S |\gamma|^2 \, dx + \frac{1}{\rho_F} \int_{\Omega_F} |\nabla p|^2 \, dx}{\frac{1}{\rho_F c^2} \int_{\Omega_F} p^2 \, dx - \int_{\Omega_S} \sigma_{ij}(u) \epsilon_{ij}(u) dx - \int_{\Omega_S} \rho_S \gamma \cdot u \, dx + \int_\Sigma p u \cdot n \, d\sigma} \quad (8.75)$$

From a mathematical point of view, the spaces introduced by this formulation are as follows: $\mathcal{C}_u = (H^1(\Omega_S))^3$, $\mathcal{C}_\gamma = (L^2(\Omega_S))^3$ and $\mathcal{C}_p = H^1(\Omega_F)$.
We can also verify that in the absence of rigid body movements (partially clamped structure), u can be eliminated in terms of γ and p.

8.9 Conclusion and open problems

Problems under development include the formalization of the various techniques used — both continuum or discrete — to establish the symmetric formulations. [10] Open problems include a mathematical study of the convergence of the various mass-coupling or stiffness-coupling formulations after finite element discretization procedures.

Ohayon [169], [173], Deneuvy [56], *cf.* also Antoniadis & Kanarachos [4], [107]). Similar situations have been analysed for structure-structure coupling (*cf.* Ohayon [165], Ohayon & Valid [167]).

[10] A mathematical study of symmetrization techniques may be found in Bluman & Kumei [21].

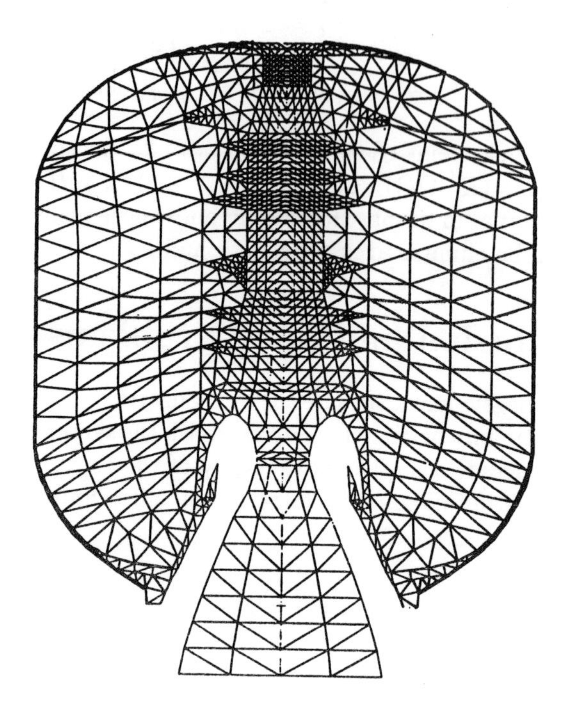

Mesh of a solid propergol booster for structural-acoustic response (structure of revolution).

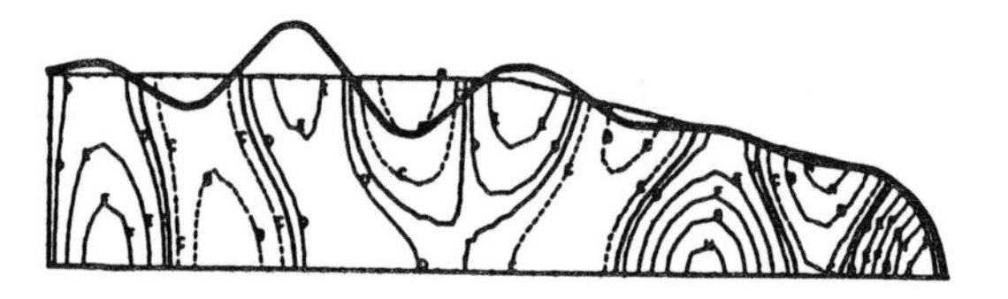

Structural-acoustic mode of the payload of the launcher Ariane 4 (structural deformation and iso-pressures).

CHAPTER 9

Modal reduction in fluid-structure interaction

9.1 Introduction

In chapter 1, using modal projection of the variational formulation in displacements u, we derived a superelement involving the displacements u_Σ of a surface of interaction Σ, together with the generalized coordinates q_α relating to the modes u_α of the structure fixed on Σ.
This method of condensation yields matrices $\widehat{\boldsymbol{K}}$ and $\widehat{\boldsymbol{M}}$ which have simply to be assembled – according to the degrees of freedom of the coupling surface Σ — with the mass and stiffness matrices of the adjacent structure. The aim is to generalize this approach to the following two cases:

- *hydroelastic vibration in the presence of gravity*
- *structural-acoustic vibrations*

We saw in chapters 6 and 8 that a description of the fluid by a variable of the pressure type p or displacement potential type φ leads naturally to a non-symmetric formulation of the fluid-structure interaction problem. This is why we introduced various *symmetric* variational formulations involving auxiliary variables, for the purpose of direct finite element treatment by means of standard algorithms.
By applying the techniques of modal projection to these latter formulations we can establish "reduced" *symmetric* matrix models.
In the present case we shall show that the "reduced" symmetric matrix models may be obtained easily, starting from non-symmetric two-field formulations.
We shall be concerned mainly with developing modal projection techniques for the field describing the fluid, keeping the degrees of freedom $\boldsymbol{U}$ of the structure discretized by finite elements.
We thus arrive at two *symmetric* formulations, where coupling of the degrees of freedom $\boldsymbol{U}$ of the structure and of generalized coordinates relating to the fluid is carried out either by the mass matrix or by the stiffness matrix of the system.

Following the same procedure as in the first chapter, extension of the modal projection procedure to structure variables will finally yield various dynamic substructuring schemes for fluid-structure interaction problems.

9.2 Hydroelastic vibrations in the presence of gravity

We develop successively two methods for the approximate solution of the dynamic substructuring type based on application of the Ritz-Galerkin method to the variational formulation in (u, φ) of the problem of hydroelastic vibrations in the presence of gravity (*cf.* chapter 6):

1. A method of modal reduction "of the fluid" consisting of representing the coupled fluid-structure system by the physical coordinates $\boldsymbol{U}$ relating to a discretization of the structure by finite elements, and by generalized coordinates associated with the sloshing modes of the fluid.

2. A method of modal reduction "of the structure" and "of the fluid", consisting of representing the coupled system, on the one hand by means of generalized coordinates relating to the sloshing modes of the fluid, and on the other hand by means of various systems of generalized coordinates associated with the eigenmodes of the coupled fluid-structure system corresponding to a non-resonant behaviour of the liquid defined above.

Non-resonant behaviour of the liquid

We recall that the study in chapter 3 of the modal analysis of the harmonic response φ of a liquid, with gravity effects, to a harmonic displacement of the wall of amplitude u_N and circular frequency ω, led to the introduction of two displacement potentials $\varphi^0_{u_N}$ and $\varphi^\infty_{u_N}$, solutions respectively to problem (3.10) and (3.81).
The mappings $u_N \longrightarrow \varphi^0$ and $u_N \longrightarrow \varphi^\infty$ — which are independent of ω — therefore define two *non-resonant* models of the liquid, which can considered, respectively, as the limit of the harmonic response φ for $\omega \longrightarrow 0$ and $\omega \longrightarrow \infty$.
In the coupled problem, the description of the liquid by means of $\varphi^0_{u_N}$ or $\varphi^\infty_{u_N}$ defines a problem which only involves the variables u relating to the structure, and to the introduction of the added mass operators M_A (*cf.* (5.25)) and M^0_B (*cf.* (3.96)).
In what follows, we shall be using the fluid-structure coupled model taking into account the non-resonant behaviour of the liquid, by means respectively of φ^0 and φ^∞ (this latter case corresponds to the problem of "hydroelastic vibrations" developed in chapter 5).

Review of (u, φ) variational formulation

The notation is that of chapter 6: we consider a structure Ω_S containing a liquid (incompressible) Ω_F. We denote by Γ the free surface of the liquid at rest and by Σ the fluid-structure contact surface.
We denote by $\widetilde{F}$ the applied external forces on $\Sigma^d = \partial\Omega_S \setminus \Sigma$. We refer to chapter 6, section 6.3. The structure is described by its displacement u and and the liquid by the *canonical* displacement potential φ (satisfying the condition $\int_\Gamma \varphi\, d\sigma = 0$). We denote by $\mathcal{C}_u$ the space of admissible u and by $\mathcal{C}^*_\varphi$ the space of admissible φ, $\mathcal{C}^*_\varphi = \{\varphi \mid \int_\Gamma \varphi\, d\sigma = 0\}$.
The variational formulation (6.33) of the coupled problem is stated:
For given ω and $\widetilde{F}$, find $(u,\varphi) \in \mathcal{C}_u \times \mathcal{C}^*_\varphi$ such that $\forall(\delta u, \delta\varphi) \in \mathcal{C}_u \times \mathcal{C}^*_\varphi$, we have:

$$\boxed{\begin{array}{ll} k^0(u,\delta u) - \omega^2 \int_{\Omega_S} \rho_S u \cdot \delta u\, dx - \omega^2 \int_\Sigma \rho_F \varphi \delta u_N\, d\sigma = \int_{\Sigma_d} \widetilde{F} \cdot \delta u\, d\sigma & (a) \\ \int_{\Omega_F} \rho_F \nabla\varphi \cdot \nabla\delta\varphi\, dx - \omega^2 \int_\Gamma \frac{\rho_F}{g} \varphi \delta\varphi\, d\sigma \; - \; \int_\Sigma \rho_F u_N \delta\varphi\, d\sigma = 0 & (b) \end{array}} \tag{9.1}$$

where k^0 is the operator defined by (6.34), which we shall here consider as the sum of the "elastic" term k_E — taking into account if necessary, for a closed cavity, the geometric rigidity term due to internal pressurization — and a term "due to gravity" denoted $k^{(g)}$ as follows (*cf.* (6.32), (6.34) and (6.35)):

$$\boxed{k^0 \; = \; k_E + k^{(g)} \quad \text{with } k^{(g)} = k_G + k_\Sigma + \mathcal{K}^0_B} \tag{9.2}$$

Substructuring in (u, κ_α) with mass coupling

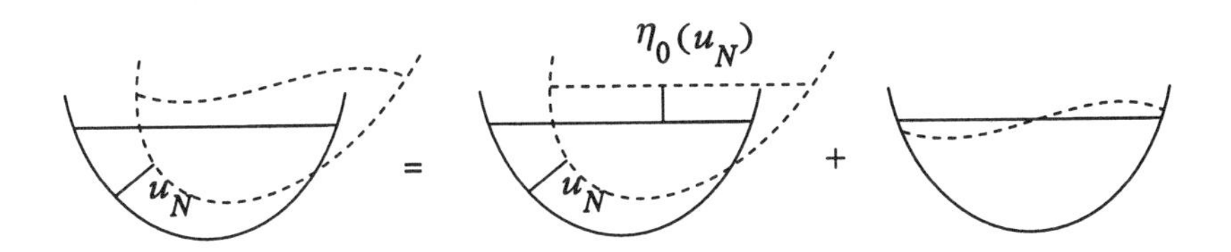

Figure 9.1: Illustration of modal decomposition

Our modal reduction procedure consists of seeking an approximate solution to (9.1), by approximating $\mathcal{C}_u$ with finite elements, and $\mathcal{C}^*_\varphi$ with the subspace spanned by the n first eigenfunctions φ_α, seeking φ in the form (3.79) (*cf.* § 3.8 and Fig. 9.1):

$$\boxed{\varphi = \varphi^0_{u_N} \; + \sum_{\alpha=1}^{n} \kappa_\alpha \varphi_\alpha} \tag{9.3}$$

where $\varphi^0_{u_N}$ is the solution to (3.10).

We replace φ by its expression (9.3) in (9.1), and we put successively $\delta\varphi = \varphi_1, \varphi_2, \ldots, \varphi_n$.
Taking into account the orthogonality relations of φ_α (3.34cd), together with the conjugate relation of (3.76) between $\varphi^0_{u_N}$ and φ_α, we obtain $\forall \delta u \in \mathcal{C}_u$ and for $1 \leq \alpha \leq n$:

$$\begin{aligned} & k^0(u, \delta u) - \omega^2 \int_{\Omega^S} \rho_S u \cdot \delta u \; dx \\ & \quad = \omega^2 \int_\Sigma \rho_F \varphi^0_{u_N} \delta u_N + \omega^2 \sum_\alpha^n \kappa_\alpha \int_\Sigma \rho_F \varphi_\alpha \delta u_N \, d\sigma + \int_{\Sigma_d} \widetilde{F} \cdot \delta u \, d\sigma \qquad (a) \\ & \qquad (-\omega^2 + \omega_\alpha^2) \mu_\alpha \kappa_\alpha = \omega^2 \int_\Sigma \rho_F u_N \, \varphi_\alpha \, d\sigma \qquad (b) \end{aligned} \tag{9.4}$$

From (3.96), we se that $\int_\Sigma \rho_F \varphi^0_{u_N} \delta u_N \, d\sigma$ is none other than the "hydrostatic" mass operator $\mathcal{M}^0_B(u_N, \delta u_N)$.

Symmetric matrix equations in (u, κ_α)

We introduce the matrices which correspond to the various bilinear (or linear) forms involved in (9.4) as follows:

$$\begin{aligned} \int_{\Omega_S} a_{ijkh} \epsilon_{kh}(u) \epsilon_{ij}(\delta u) \, dx &\Longrightarrow \delta \boldsymbol{U}^T \boldsymbol{K}_E \boldsymbol{U} & (a) \\ \int_{\Omega_S} \rho_S u \cdot \delta u \, dx &\Longrightarrow \delta \boldsymbol{U}^T \boldsymbol{M} \boldsymbol{U} & (b) \\ k^{(g)}(u, \delta u) &\Longrightarrow \delta \boldsymbol{U}^T \boldsymbol{K}^{(g)} \boldsymbol{U} & (c) \\ \mathcal{M}^0_B(u_N, \delta u_N) &\Longrightarrow \delta \boldsymbol{U}^T \boldsymbol{M}^0_B \boldsymbol{U} & (d) \\ \int_\Sigma \rho_F \varphi_\alpha u_N \, d\sigma &\Longrightarrow \boldsymbol{C}^T_\alpha \boldsymbol{U} & (e) \\ \int_{\Sigma_d} \widetilde{F} \cdot \delta u \, d\sigma &\Longrightarrow \delta \boldsymbol{U}^T \widetilde{\boldsymbol{F}} & (f) \end{aligned} \tag{9.5}$$

Equations (9.4) may then be written in discretized form, for $1 \leq \alpha \leq n$:

$$\begin{aligned} \left(\boldsymbol{K}_E + \boldsymbol{K}^{(g)} \right) \boldsymbol{U} - \omega^2 (\boldsymbol{M} + \boldsymbol{M}^0_B) \boldsymbol{U} &= \omega^2 \sum_{\alpha=1}^n \boldsymbol{C}_\alpha \kappa_\alpha + \widetilde{\boldsymbol{F}} & (a) \\ (-\omega^2 + \omega_\alpha^2) \mu_\alpha \kappa_\alpha &= \omega^2 {\boldsymbol{C}_\alpha}^T \boldsymbol{U} & (b) \end{aligned} \tag{9.6}$$

which is written in the following *symmetric* matrix form:

$$\left[\begin{array}{c|ccc} \boldsymbol{K}^0 & \cdots & 0 & \cdots \\ \hline \vdots & \ddots & & \\ 0 & & \omega_\alpha^2 \mu_\alpha & \\ \vdots & & & \ddots \end{array} \right] \begin{bmatrix} \boldsymbol{U} \\ \vdots \\ \kappa_\alpha \\ \vdots \end{bmatrix} - \omega^2 \left[\begin{array}{c|ccc} \boldsymbol{M}^0 & \cdots & \boldsymbol{C}_\alpha & \cdots \\ \hline \vdots & \ddots & & \\ {\boldsymbol{C}_\alpha}^T & & \mu_\alpha & \\ \vdots & & & \ddots \end{array} \right] \begin{bmatrix} \boldsymbol{U} \\ \vdots \\ \kappa_\alpha \\ \vdots \end{bmatrix} = \begin{bmatrix} \widetilde{\boldsymbol{F}} \\ \vdots \\ 0 \\ \vdots \end{bmatrix} \tag{9.7}$$

where we have let:

$$\begin{aligned} \boldsymbol{K}^0 &= \boldsymbol{K}_E + \boldsymbol{K}^{(g)} & (a) \\ \boldsymbol{M}^0 &= \boldsymbol{M} + \boldsymbol{M}^0_B & (b) \end{aligned} \tag{9.8}$$

Calculation of $\boldsymbol{M}_B^0$

- The direct calculation of $\boldsymbol{M}_B^0$ has already been discussed (*cf.* (3.101)).
- $\boldsymbol{M}_B^0$ may also be calculated by means of the following summation rule (*cf.* (3.107), (9.5e)):

$$\boldsymbol{M}_B^0 = \boldsymbol{M}_A + \sum_{\alpha=1}^{\infty} \boldsymbol{M}_\alpha \quad \text{with } \boldsymbol{M}_\alpha = \frac{\boldsymbol{C}_\alpha \boldsymbol{C}_\alpha{}^T}{\mu_\alpha} \tag{9.9}$$

which involves the discretized form $\boldsymbol{M}_\alpha$ of the modal sloshing operators M_α (*cf.* (3.105)).

The matrix serie $\sum_{\alpha=1}^{\infty} \boldsymbol{M}_\alpha$ involves the *infinite* serie of sloshing modes defined in the *approximate domain* Ω_F resulting from the discretization of the structure.

In practice, the sloshing modes are calculated by finite elements, solving (3.36).

In this case, the modal development of $\boldsymbol{M}_B^0$ is written $\boldsymbol{M}_A + \sum_{\alpha=1}^{N_B} \boldsymbol{M}_\alpha$, where N_B denotes the total number of eigenvalues of the discretized problem. Now, we have seen that the serie of $\boldsymbol{M}_\alpha$ converges *rapidly* (*cf.* (3.107)), which in practice allows its approximate calculation by severely limiting the number of terms $n \ll N_B$.

We may thus write $\boldsymbol{M}_B^0 \sim \boldsymbol{M}_A + \sum_{\alpha=1}^{n} \boldsymbol{M}_\alpha$.

Fluid superelement

— The matrix equations (9.7) may be interpreted as an assemblage, according to the degrees of freedom of $\boldsymbol{U}_\Sigma$:

1. of the "hydroelastic" matrix model $(\boldsymbol{K}_E, \boldsymbol{M} + \boldsymbol{M}_A)$,
2. and of the "fluid superelement" in $(\boldsymbol{U}_\Sigma, \kappa_\alpha)$ made up of the following stiffness and mass matrices (with $1 \leq \alpha \leq n$):

$$\left\{ \left[\begin{array}{c|ccc} \boldsymbol{K}^{(g)} & \cdots & 0 & \cdots \\ \hline \vdots & \ddots & & \\ 0 & & \omega_\alpha^2 \mu_\alpha & \\ \vdots & & & \ddots \end{array} \right], \left[\begin{array}{c|ccc} \sum_{\alpha=1}^{n} \boldsymbol{M}_\alpha & \cdots & \boldsymbol{C}_\alpha & \cdots \\ \hline \vdots & \ddots & & \\ \boldsymbol{C}_\alpha{}^T & & \mu_\alpha & \\ \vdots & & & \ddots \end{array} \right] \right\} \tag{9.10}$$

Remark — It should be noted that the coupling method presented here, which only involves the values of φ_α on Σ, enables the calculated sloshing modes to be used, for example, through methods of the boundary integral equations type.

Energy interpretation

The *symmetric* matrix equations (9.7) (transposed to time by substituting $\ddot{\boldsymbol{U}}(t)$ and $\ddot{\kappa}_\alpha(t)$ for $-\omega^2\boldsymbol{U}$ and $-\omega^2\kappa_\alpha$), are the Lagrange equations of the coupled fluid-structure system described by the generalized system of coordinates $\{\boldsymbol{U}, \kappa_1, \ldots, \kappa_n\}$, whose potential energy E_P and kinetic energy E_C are expressed: [1]

$$\begin{aligned} E_P &= \frac{1}{2}\boldsymbol{U}^T(\boldsymbol{K}_E + \boldsymbol{K}^{(g)})\boldsymbol{U} + \frac{1}{2}\sum_{\alpha=1}^{n} \omega_\alpha^2 \mu_\alpha \kappa_\alpha^2 & (a) \\ E_C &= \frac{1}{2}\dot{\boldsymbol{U}}^T(\boldsymbol{M} + \boldsymbol{M}_B^0)\dot{\boldsymbol{U}} + \frac{1}{2}\sum_{\alpha=1}^{n} \dot{\kappa}_\alpha^2 \mu_\alpha + \sum_{\alpha=1}^{n} \dot{\kappa}_\alpha {\boldsymbol{C}_\alpha}^T \dot{\boldsymbol{U}} & (b) \end{aligned} \tag{9.11}$$

Equivalent $(\boldsymbol{u}, \boldsymbol{\tau}_\alpha)$ symmetric formulation with stiffness coupling

If in (9.11) we perform the following generalized coordinate change:

$$\tau_\alpha = \kappa_\alpha + \frac{{\boldsymbol{C}_\alpha}^T \boldsymbol{U}}{\mu_\alpha} \tag{9.12}$$

we obtain (using the modal decomposition (9.9) of $\boldsymbol{M}_B^0$):

$$\begin{aligned} E_P &= \frac{1}{2}\boldsymbol{U}^T\left(\boldsymbol{K}_E + \boldsymbol{K}^{(g)} + \sum_{\alpha=1}^{n} \boldsymbol{K}_\alpha\right)\boldsymbol{U} - \sum_{\alpha=1}^{n} \omega_\alpha^2 \tau_\alpha {\boldsymbol{C}_\alpha}^T \boldsymbol{U} + \frac{1}{2}\sum_{\alpha=1}^{n} \omega_\alpha^2 \tau_\alpha^2 \mu_\alpha & (a) \\ E_C &= \frac{1}{2}\dot{\boldsymbol{U}}^T\left(\boldsymbol{M} + \boldsymbol{M}_A + \sum_{n+1}^{\infty} \boldsymbol{M}_\alpha\right)\dot{\boldsymbol{U}} + \frac{1}{2}\sum_{\alpha=1}^{n} \dot{\tau}_\alpha^2 \mu_\alpha & (b) \end{aligned} \tag{9.13}$$

where $\boldsymbol{K}_\alpha$ denotes the "sloshing stiffness" matrix of the αth mode (the definition of $\boldsymbol{M}_\alpha$ is recalled in (9.9)):

$$\boldsymbol{K}_\alpha = \omega_\alpha^2 \frac{\boldsymbol{C}_\alpha {\boldsymbol{C}_\alpha}^T}{\mu_\alpha} = \omega_\alpha^2 \boldsymbol{M}_\alpha \tag{9.14}$$

The Lagrange equations deduced in (9.13) then yield, in harmonic regime, to the following *symmetric* matrix equations:

$$\begin{aligned} \left(\boldsymbol{K}_E + \boldsymbol{K}^{(g)} + \sum_{\alpha=1}^{n} \boldsymbol{K}_\alpha\right)\boldsymbol{U} - \omega^2\left(\boldsymbol{M} + \boldsymbol{M}_A + \sum_{n+1}^{\infty} \boldsymbol{M}_\alpha\right)\boldsymbol{U} &= \sum_{\alpha=1}^{n} \omega_\alpha^2 \boldsymbol{C}_\alpha \tau_\alpha + \widetilde{\boldsymbol{F}} & (a) \\ (-\omega^2 + \omega_\alpha^2)\mu_\alpha \tau_\alpha &= \omega_\alpha^2 {\boldsymbol{C}_\alpha}^T \boldsymbol{U} & (b) \end{aligned} \tag{9.15}$$

which can immediately be written in *symmetric* matrix form with "stiffness coupling".

[1] $\widetilde{\boldsymbol{F}}$ being considered as generalized forces.

Fluid superelement with stiffness coupling

If in (9.15), we replace M_B^0 by its modal expansion $M_A + \sum_{\alpha=1}^{\infty} M_\alpha$, these equations may be interpreted as an assemblage, according to the degrees of freedom U_Σ:

1. of **the hydroelastic matrix model** $(K_E, M + M_A)$,
2. and of the **"superelement of fluid"** in (U_Σ, τ_α) made up of the following stiffness (showing the coupling between U_σ and τ_α) and mass matrices:

$$\left\{ \left[\begin{array}{c|c} K^{(g)} + \sum_{\alpha=1}^{n} K_\alpha & \cdots \; -\omega_\alpha^2 C_\alpha \; \cdots \\ \hline \vdots & \ddots \\ -\omega_\alpha^2 {C_\alpha}^T & \omega_\alpha^2 \mu_\alpha \\ \vdots & \quad \ddots \end{array} \right], \left[\begin{array}{c|c} \sum_{n+1}^{\infty} M_\alpha & \cdots \; 0 \; \cdots \\ \hline \vdots & \ddots \\ 0 & \mu_\alpha \\ \vdots & \quad \ddots \end{array} \right] \right\} \tag{9.16}$$

Direct derivation of (u, τ_α) stiffness coupling formulation

We seek an approximate solution to (9.1), using the following alternative modal decomposition of φ (*cf.* 3.91 and Fig. 9.2):

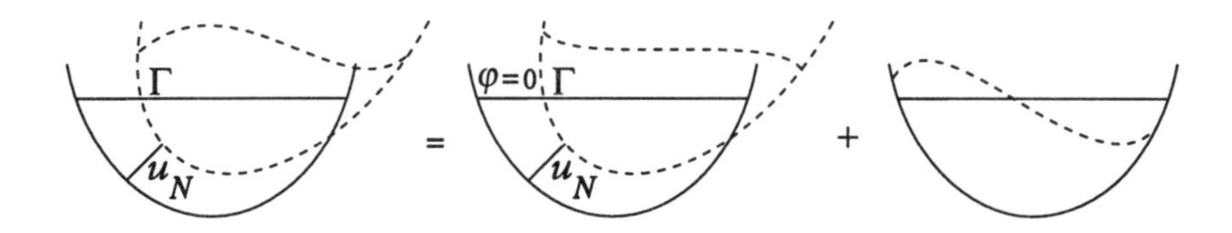

Figure 9.2: Illustration of modal decomposition

$$\boxed{\varphi = \varphi_{u_N}^{\infty} + \sum_{\alpha=1}^{n} \tau_\alpha \varphi_\alpha} \tag{9.17}$$

where $\varphi_{u_N}^{\infty}$ is the solution to the *static problem for a prescribed wall displacement* u_N (3.81).
The direct derivation consists of applying the Ritz-Galerkin method, replacing φ by its expression (9.17) in the variational formulation (9.1).
Putting successively $\delta\varphi = \varphi_1, \varphi_2, \ldots$, and taking into account the orthogonality relations (3.34cd), together with the conjugate relation (3.88) between $\varphi_{u_N}^{\infty}$ and φ_α, we obtain $\forall \delta u \in \mathcal{C}_u$ and for $1 \leq \alpha \leq n$:

$$\begin{aligned} & k^0(u, \delta u) - \omega^2 \int_{\Omega^S} \rho_S u \cdot \delta u \; dx - \omega^2 \int_\Sigma \rho_F \varphi_{u_N}^{\infty} \delta u_N \, d\sigma \\ & \quad = \omega^2 \sum_\alpha^n \tau_\alpha \int_\Sigma \rho_F \varphi_\alpha \delta u_N \, d\sigma + \int_{\Sigma_d} \widetilde{F} \cdot \delta u \, d\sigma & (a) \\ & \qquad (-\omega^2 + \omega_\alpha^2) \mu_\alpha \tau_\alpha = \omega_\alpha^2 \int_\Sigma \rho_F u_N \varphi_\alpha \, d\sigma & (b) \end{aligned} \tag{9.18}$$

We recall that $\int_\Sigma \rho_F \varphi^\infty_{u_N} \delta u_N \, d\sigma = \mathcal{M}^\infty_B(u_N, \delta u_N)$ (*cf.* (3.103)), and that $\varphi^\infty_{u_N}$ is none other than the potential φ introduced in chapter 5 — equations (5.1abc) and (3.81) being identical. Consequently, $\mathcal{M}^\infty_B = \mathcal{M}_A$.

Symmetric matrix equations in u, τ_α

From (9.5) and (9.8a), equations (9.18) are written in discretized form, for $1 \leq \alpha \leq n$:

$$\begin{aligned} \left(\boldsymbol{K}_E + \boldsymbol{K}^{(g)}\right)\boldsymbol{U} - \omega^2(\boldsymbol{M} + \boldsymbol{M}_A)\boldsymbol{U} &= \omega^2 \sum_{\alpha=1}^{n} \boldsymbol{C}_\alpha \tau_\alpha + \widetilde{\boldsymbol{F}} \quad (a) \\ (-\omega^2 + \omega_\alpha^2)\mu_\alpha \tau_\alpha &= \omega_\alpha^2 \boldsymbol{C}_\alpha{}^T \boldsymbol{U} \quad (b) \end{aligned} \tag{9.19}$$

Equations (9.19) may be written in *symmetric* matrix form by proceding as follows: in the second term of (9.19a), $\omega^2 \tau_\alpha$ is replaced by its expression $\omega_\alpha^2 \tau_\alpha - (\omega_\alpha^2/\mu_\alpha)\boldsymbol{C}_\alpha{}^T \boldsymbol{U}$ from (9.19b). Equations (9.19) are then written, using definition (9.14) of $\boldsymbol{K}_\alpha$, in the following equivalent form, with $1 \leq \alpha \leq n$:

$$\begin{aligned} \left(\boldsymbol{K}_E + \boldsymbol{K}^{(g)} + \sum_{\alpha=1}^{n} \boldsymbol{K}_\alpha\right)\boldsymbol{U} - \omega^2(\boldsymbol{M} + \boldsymbol{M}_A)\boldsymbol{U} &= \sum_{\alpha=1}^{n} \omega_\alpha^2 \boldsymbol{C}_\alpha \tau_\alpha + \widetilde{\boldsymbol{F}} \quad (a) \\ (-\omega^2 + \omega_\alpha^2)\mu_\alpha \tau_\alpha &= \omega_\alpha^2 \boldsymbol{C}_\alpha{}^T \boldsymbol{U} \quad (b) \end{aligned} \tag{9.20}$$

which can immediately be written in *symmetric* matrix form.

These last equations may be interpreted as the assemblage, according to the degrees of freedom $\boldsymbol{U}_\Sigma$, of the *hydroelastic matrix model* $(\boldsymbol{K}_E, \boldsymbol{M} + \boldsymbol{M}_A)$, and of a *superelement of fluid* whose stiffness matrix can be verified to coincide with that of (9.16), and whose mass matrix differs from that of (9.16) in the replacement of $\sum_{\alpha=n+1}^{\infty} \boldsymbol{M}_\alpha$ by 0.

Truncation effects

We know that in the limit cases $\omega = 0$ and $\omega = \infty$, the "exact" behaviour of the liquid is described respectively by the potentials $\varphi^0_{u_N}$ and $\varphi^\infty_{u_N}$ introduced in chapter 3.

In what follows, we analyse the effects of modal truncation, introduced in the above formulations, on the representation of the dynamic behaviour of the liquid for $\omega \longrightarrow 0$ and for $\omega \longrightarrow \infty$.

To begin, we recall the modal expansion (3.109) of $\varphi^0_{u_N} - \varphi^\infty_{u_N}$, which for discretized u, is written:

$$\varphi^0_{u_N} = \varphi^\infty_{u_N} + \sum_{\alpha=1}^{\infty} \frac{\boldsymbol{C}_\alpha{}^T \boldsymbol{U}}{\mu_\alpha} \varphi_\alpha \tag{9.21}$$

Symmetric matrix formulation with mass coupling

- For $\omega \to 0$, from (9.6b), κ_α tends to 0, and from (9.3), φ tends to the exact value $\varphi^0_{u_N}$.

- For $\omega \to \infty$, $\kappa_\alpha \to -\frac{\boldsymbol{C}_\alpha{}^T\boldsymbol{U}}{\mu_\alpha}$. From (9.3), $\varphi \to \varphi^0_{u_N} - \sum_{\alpha=1}^{n} \frac{\boldsymbol{C}_\alpha{}^T\boldsymbol{U}}{\mu_\alpha}\varphi_\alpha$. Referring to (9.21), we see that $\varphi \to \varphi^\infty_{u_N} + \sum_{n+1}^{\infty} \frac{\boldsymbol{C}_\alpha{}^T\boldsymbol{U}}{\mu_\alpha}\varphi_\alpha$.

Symmetric matrix formulations with stiffness coupling

We begin by noting that the above conclusions apply to the symmetric matrix formulation with stiffness coupling (9.15), since is is equivalent, in its construction, to the formulation with mass coupling (9.6).
Formulation (9.20), however obtained by direct modal analysis of (9.1) by means of (9.17), has the following particularities:

- For $\omega \to 0$, from (9.20b) $\tau_\alpha \to \frac{\boldsymbol{C}_\alpha{}^T\boldsymbol{U}}{\mu_\alpha}$. Consequently, from (9.17) and (9.21), $\varphi \to \varphi^0_{u_N} - \sum_{n+1}^{\infty} \frac{\boldsymbol{C}_\alpha{}^T\boldsymbol{U}}{\mu_\alpha}\varphi_\alpha$.

- For $\omega \to \infty$, from (9.20b), τ_α tends to 0 and therefore φ tends to the exact value of $\varphi^\infty_{u_N}$.

In conclusion, if the limit behaviour for $\omega \to 0$ is to be restored correctly, we have a choice between symmetric formulations with mass coupling (9.6) or the equivalent formulation with stiffness coupling (9.15).

Complete dynamic substructuring

In the above, we have developed a technique for dynamic "partial" substructuring based on a hybrid representation of the problem of fluid-structure interaction, in terms of "physical coordinates" $\boldsymbol{U}$, and "generalized coordinates" κ_α or τ_α relating to the sloshing modes.
We now develop a technique for "total" dynamic substructuring of the coupled system, which consists, while retaining the description in $\boldsymbol{\kappa} = \{\kappa_\alpha\}$ or $\boldsymbol{\tau} = \{\tau_\alpha\}$ of the fluid, of substituting into the variables $\boldsymbol{U}$, n_S generalized coordinates q_σ as follows:

$$\boxed{\boldsymbol{U} = \sum_{\sigma=1}^{n_S} q_\sigma \boldsymbol{U}_\sigma} \tag{9.22}$$

where $\{\boldsymbol{U}_\sigma\}$ is a projection basis chosen so as to diagonalize the matrix relating to the degrees of freedom $\boldsymbol{U}$ involved in equations (9.6) and (9.15).
The projection technique consists of applying rule (1.20) to the matrix equations in $(\boldsymbol{U}, \boldsymbol{\kappa})$ or $(\boldsymbol{U}, \boldsymbol{\tau})$, with the projection matrix $\boldsymbol{H}$ defined by:

$$\boldsymbol{H} = \left[\begin{array}{c|c} \boldsymbol{U}_1 \cdots \boldsymbol{U}_{n_S} & 0 \\ \hline 0 & \boldsymbol{I} \end{array}\right] \tag{9.23}$$

where $\boldsymbol{I}$ is the unity matrix of rank n.

Diagonalization results in taking for $\boldsymbol{U}_\sigma$, the eigenvectors which are solutions of two distinct eigenvalue problems which we shall analyses successively.

Coupling by the mass matrix — Diagonalization of the matrix relating to $\boldsymbol{U}$ in (9.6), leads to seeking the eigenvectors $\{\boldsymbol{U}_\sigma\}$ which are solutions to:

$$\left(\boldsymbol{K}_E + \boldsymbol{K}^{(g)}\right)\boldsymbol{U} = \lambda(\boldsymbol{M}+\boldsymbol{M}_B^0)\boldsymbol{U} \tag{9.24}$$

We denote by $\boldsymbol{U}_\sigma$ the eigenvectors of (9.24) assumed to be normalized to 1 in generalized mass, i.e. verifying the orthogonality relations:

$$\boldsymbol{U}_\sigma{}^T(\boldsymbol{M} + \boldsymbol{M}_B^0)\boldsymbol{U}_{\sigma'} = \delta_{\sigma\sigma'} \quad , \quad \boldsymbol{U}_\sigma{}^T(\boldsymbol{K}_E + \boldsymbol{K}^{(g)})\boldsymbol{U}_{\sigma'} = \omega_\sigma^2\delta_{\sigma\sigma'} \tag{9.25}$$

This problem may have negative eigenvalues:

1. $\boldsymbol{K}_E$ is a positive matrix and $\boldsymbol{K}^{(g)}$ a symmetric matrix, not necessarily positive. (We see from (9.1a) that the contributions from the terms k_G and k_Σ (*cf.* (6.23)) are not necessarily positive; the positivity of $\boldsymbol{K}_E + \boldsymbol{K}^{(g)}$ is equivalent to the condition for stability of the equilibrium position as was discussed at the end of chapter 6).

2. $\boldsymbol{M}$ and $\boldsymbol{M}_B^0$ are positive symmetric matrices.

Taking for $\boldsymbol{U}_\sigma$ in (9.23), the eigenvectors of (9.24) with conditions (9.25), projection of equations (9.6) yields:

$$\boxed{\begin{bmatrix} \textbf{Diag}\,\omega_\sigma^2 & 0 \\ 0 & \textbf{Diag}\,\omega_\alpha^2 \end{bmatrix}\begin{bmatrix} \boldsymbol{q} \\ \boldsymbol{\kappa} \end{bmatrix} - \omega^2\begin{bmatrix} \boldsymbol{I}_{n_S} & [C_{\sigma\alpha}] \\ [C_{\alpha\sigma}] & \boldsymbol{I}_n \end{bmatrix}\begin{bmatrix} \boldsymbol{q} \\ \boldsymbol{\kappa} \end{bmatrix} = \begin{bmatrix} \widetilde{\boldsymbol{f}} \\ 0 \end{bmatrix}} \tag{9.26}$$

where $\boldsymbol{I}_{n_S}$ and $\boldsymbol{I}_n$ are the unit matrices of respective dimensions n_S and n, where $\widetilde{\boldsymbol{f}}$ denotes the vector of components $\widetilde{f}_\sigma = \boldsymbol{U}_\sigma{}^T\widetilde{\boldsymbol{F}}$, and where we have let:

$$C_{\sigma\alpha} = \boldsymbol{C}_\alpha{}^T\boldsymbol{U}_\sigma \tag{9.27}$$

- We note that the problem in $\boldsymbol{U}$ (*cf.* 9.24) takes into account the non-resonant behaviour of the fluid through $\boldsymbol{K}^{(g)}$ and $\boldsymbol{M}_B^0$, *independently of the modal truncation* introduced for the description of the fluid.

- However, from a practical point of view this formulation has the disadvantage of depending on a modal basis $\{\boldsymbol{U}_\sigma\}$, which cannot be experimentally demonstrated.

  The physical interpretation of (9.24) is as follows: solutions $\{\boldsymbol{U}_\sigma\}$ may be considered as solutions to (9.6) within the approximation $\kappa_\alpha = 0$, which, from (9.6b), is only valid for $\omega \ll \omega_\alpha$, $\forall\alpha$.

  For example, in the case of a launcher containing liquids, the frequency of the first sloshing modes is of the order of a Hertz, which is lower than or of the same magnitude as that of the first bending modes of the launcher.

Coupling by stiffness matrix — Diagonalization of the matrix relating to $\boldsymbol{U}$ in (9.15), involves seeking for the eigenvectors $\{\boldsymbol{U}_\sigma\}$ which are solutions of:

$$\left(\boldsymbol{K}_E + \boldsymbol{K}^{(g)} + \sum_{\alpha=1}^{n} \boldsymbol{K}_\alpha\right)\boldsymbol{U} = \lambda\left(\boldsymbol{M}+\boldsymbol{M}_A+\sum_{n+1}^{\infty} \boldsymbol{M}_\alpha\right)\boldsymbol{U} \tag{9.28}$$

We denote by $\boldsymbol{U}_\sigma$ the eigenvectors of (9.28), which we assume to be normalized to 1 in generalized mass, i.e. verifying the orthogonality relations:

$$\begin{aligned} \boldsymbol{U}_\sigma{}^T\left(\boldsymbol{M}+\boldsymbol{M}_A+\textstyle\sum_{n+1}^{\infty}\boldsymbol{M}_\alpha\right)\boldsymbol{U}_{\sigma'} &= \delta_{\sigma\sigma'} \quad &(a)\\ \boldsymbol{U}_\sigma{}^T\left(\boldsymbol{K}_E + \boldsymbol{K}^{(g)} + \textstyle\sum_{\alpha=1}^{n}\boldsymbol{K}_\alpha\right)\boldsymbol{U}_{\sigma'} &= \omega_\sigma^2\delta_{\sigma\sigma'} \quad &(b)\end{aligned} \tag{9.29}$$

This problem can have negative eigenvalues for the same reasons as for (9.24). Taking for $\boldsymbol{U}_\sigma$ in (9.23) the eigenvectors of (9.28) with conditions (9.29), projection of equations (9.15) yields:

$$\boxed{\begin{bmatrix} \textbf{Diag}\,\omega_\sigma^2 & -[\omega_\alpha^2 C_{\sigma\alpha}] \\ -[C_{\alpha\sigma}\omega_\alpha^2] & \textbf{Diag}\,\omega_\alpha^2 \end{bmatrix}\begin{bmatrix}\boldsymbol{q}\\ \boldsymbol{\tau}\end{bmatrix} - \omega^2\begin{bmatrix}\boldsymbol{I}_{ns} & 0\\ 0 & \boldsymbol{I}_n\end{bmatrix}\begin{bmatrix}\boldsymbol{q}\\ \boldsymbol{\tau}\end{bmatrix} = \begin{bmatrix}\tilde{\boldsymbol{f}}\\ 0\end{bmatrix}} \tag{9.30}$$

with the previous definitions of $\boldsymbol{I}_{n_S}, \boldsymbol{I}_n, \tilde{\boldsymbol{f}}$ and $C_{\sigma\alpha}$ (*cf.* (9.27)).

Discussion — The problem in $\boldsymbol{U}$ (*cf.* 9.28) contains two terms which are dependent on the truncation of n: $\sum_{\alpha=1}^{n}\boldsymbol{K}_\alpha$ and the residual term $\sum_{n+1}^{\infty}\boldsymbol{M}_\alpha$. Given the rapid convergence of the serie of the $\boldsymbol{M}_\alpha$, this latter term may be neglected for a sufficiently high n.

Example of launchers — Studies of launcher attitude control require modeling of the interaction between the elastic structure and the liquid, taking into account the effects of the apparent gravity $(\vec{g} - \vec{\gamma})$. This model is the result of coupling between sloshing modes and hydroelastic modes, generally calculated in neglecting the additional stiffness terms due to gravity.
Consequently, we take for $\{\boldsymbol{U}_\sigma\}$ the hydroelastic modes which are solutions of $\boldsymbol{K}_E\boldsymbol{U} = \lambda(\boldsymbol{M} + \boldsymbol{M}_A)\boldsymbol{U}$, and which verify the orthogonality relations $\boldsymbol{U}_\sigma{}^T(\boldsymbol{M} + \boldsymbol{M}_A)\boldsymbol{U}_{\sigma'} = \delta_{\sigma\sigma'}$ and $\boldsymbol{U}_\sigma{}^T\boldsymbol{K}_E\boldsymbol{U}_{\sigma'} = \omega_\sigma^2\delta_{\sigma\sigma'}$.
Using this basis in (9.22), and distinguishing the $n_{\mathcal{R}}$ rigid body modes $\boldsymbol{U}_i^{\mathcal{R}}$ $(\lambda = 0)$, from the $n'_S = n_S - n_{\mathcal{R}}$ elastic modes $(\omega'_\sigma > 0, \boldsymbol{U}'_\sigma)$, leads us to seek $\boldsymbol{U}$ in the form:

$$\boldsymbol{U} = \sum_{i=1}^{n_{\mathcal{R}}} q_i^{\mathcal{R}}\boldsymbol{U}_i^{\mathcal{R}} + \sum_{\sigma=1}^{n'_S} q'_\sigma\boldsymbol{U}'_\sigma \tag{9.31}$$

Applying the technique of modal projection, we arrive, after simplification, at the following matrix equations:

$$\begin{bmatrix} [\widetilde{K}_{ij}] & 0 & -[\omega_\alpha^2 C_{i\alpha}^{\mathcal{R}}] \\ 0 & \textbf{Diag}\,\omega_\sigma'^2 & -[\omega_\alpha^2 C'_{\sigma\alpha}] \\ -[C_{\alpha i}^{\mathcal{R}}\omega_\alpha^2] & -[C'_{\alpha\sigma}\omega_\alpha^2] & \textbf{Diag}\,\omega_\alpha^2 \end{bmatrix}\begin{bmatrix}\boldsymbol{q}^{\mathcal{R}}\\ \boldsymbol{q}'\\ \boldsymbol{\tau}\end{bmatrix} - \omega^2\begin{bmatrix}\boldsymbol{I}_{n_{\mathcal{R}}} & 0 & 0\\ 0 & \boldsymbol{I}_{n_S} & 0\\ 0 & 0 & \boldsymbol{I}_n\end{bmatrix}\begin{bmatrix}\boldsymbol{q}^{\mathcal{R}}\\ \boldsymbol{q}'\\ \boldsymbol{\tau}\end{bmatrix} = \begin{bmatrix}\widetilde{\boldsymbol{f}^{\mathcal{R}}}\\ \widetilde{\boldsymbol{f}'}\\ 0\end{bmatrix} \tag{9.32}$$

where we have let $\widetilde{K}_{ij} = U_i^{\mathcal{R}^T} \widetilde{K} U_j^{\mathcal{R}}$ with $\widetilde{K} = K^{(g)} + \sum_{\alpha=1}^{n} K_\alpha$.
The following approximations (legitimate for launchers) are made:

1. Calculation shows that a very small number of sloshing modes is sufficient for the residual term $\sum_{n+1}^{\infty}$ appearing in (9.15) to be negligible, hence the diagonal nature of the mass matrix (9.32).

2. It can be verified that the coupling terms $U_i^{\mathcal{R}^T} \widetilde{K} U'_\sigma$ together with $U'^T_\sigma \widetilde{K} U'_{\sigma'}$ may be neglected in comparison with ω'^2_σ.

   This approximation is equivalent to taking into account the additional stiffness terms $\widetilde{K} = K^{(g)} + \sum_{\alpha=1}^{n} K_\alpha$ in the rigid body modes, and neglecting their influence in comparison with that of the elastic term K_E, for the elastic modes.

Pendulum approximation — In the special case of a rigid reservoir of axisymmetric shape, (with the apparent weight vector directed along the axis of revolution), it can be shown that equations (9.32) (cancelling q') are equivalent to the equations which describe a system of pendulums suspended at certain points on the axis of revolution. This type of pendulum model is applicable to launchers whose bending behaviour is assimilated to a refined beam model - which is equivalent to neglecting local deformations in reservoirs.

9.3 Structural-acoustic vibrations

Starting from a variational formulation in (u, φ) of the problem of structural-acoustic vibrations, we develop two methods of dynamic substructuring, proceeding in a way analogous to that used for hydroelastic vibrations under gravity: [2]

1. A method of modal reduction of the "fluid" which consists of representing the coupled fluid-structure system by physical coordinates U relative to a discretization of the structure by finite elements, and by generalized coordinates associated with the acoustic modes of the fluid.

2. A method of modal reduction "of the structure" and "of the fluid", which consists of representing the coupled system, on the one hand, by generalized coordinates relative to the acoustic modes of the fluid, and on the other hand by two generalized systems of coordinates associated respectively with the eigenmodes of the coupled elastic cavity-fluid system corresponding to a non-resonant behaviour of the fluid which will be specified below, and the eigenmodes of the elastic structure *in vacuo*.

These methods allow a simplification of the internal elasto-acoustic problem with a view to coupling with an external environment model. [3]

[2] Structural-acoustic coupling has also been studied by asymptotic methods, *cf.* Sanchez-Hubert & Sanchez-Palencia [195].

[3] Among the works on acoustic radiation of structure are Junger & Feit [104], Fahy [67], Lesueur [117]. For general methods, see Dautray & Lions [54] vol. 6. For fluid-structure

Non-resonant behaviour of the fluid— We recall that the study in chapter 7 of the response of a compressible fluid to a wall displacement led to the introduction of a displacement potential $\varphi^0_{u_N}$ which is a solution to the problem (7.10).
The mapping $u_N \longrightarrow \varphi^0$ — which is independent of ω — therefore defines a *non-resonant* behaviour model of the fluid.
Modeling of the fluid by a potential $\varphi^0_{u_N}$ in the coupled problem leads us, as we shall see, to a problem involving only the variables u relative to the structure, by introducing the added mass operator M^0_{Ac} defined by (7.47).

(u, φ) variational formulation

The notation is that of chapter 8: we denote by Ω_S the domain occupied by the structure, by Ω_F the domain occupied by the fluid, by $\Sigma = \partial\Omega_F$ the fluid-structure contact surface, and by n the normal external to the fluid. We denote by F the external forces applied on $\Sigma^d = \partial\Omega_S \setminus \Sigma$. This formulation results from the following two variational properties:

1. The variational formulation (7.30) of the boundary value problem in terms of φ [4] (7.8), corresponding to the harmonic response of the fluid to a normal displacement $u_N = u{\cdot}n$ of the wall, established in chapter 7, with definition (7.29) of the admissible space $\mathcal{C}^*$ *here denoted* $\mathcal{C}^*_\varphi$. The pressure p is related to φ and u_N by (7.9):

$$\boxed{p = \rho_F\omega^2\varphi - \frac{\rho_F c^2}{Vol\,(\Omega_F)}\int_\Sigma u_N\, d\sigma \quad \text{with} \quad \int_{\Omega_F}\varphi\, dx = 0} \tag{9.33}$$

2. The variational property in u (*cf.* 1.22) of the harmonic response of the structure to forces F applied on Σ^d and to the pressure p of the liquid on Σ, then replacing p by its expression (9.33) in terms of φ and u_N.

   We denote by $\mathcal{C}_u$ the space of the admissible u.

The (non symmetric) variational formulation of the coupled problem is then written:
For given ω and F, find $(u, \varphi) \in \mathcal{C}_u \times \mathcal{C}^*_\varphi$ such that, $\forall(\delta u, \delta\varphi) \in \mathcal{C}_u \times \mathcal{C}^*_\varphi$, we have:

$$\boxed{\begin{array}{ll} k^0(u,\delta u) - \omega^2\displaystyle\int_{\Omega_S}\rho_S u\cdot\delta u\, dx - \omega^2\int_\Sigma \rho_F\varphi\delta u_N\, d\sigma = \int_{\Sigma^d} F\cdot\delta u\, d\sigma & (a) \\ \displaystyle\int_{\Omega_F}\rho_F\nabla\varphi\cdot\nabla\delta\varphi\, dx - \omega^2\int_{\Omega_F}\frac{\rho_F}{c^2}\varphi\delta\varphi\, dx - \int_\Sigma \rho_F u_N\delta\varphi\, d\sigma = 0 & (b)\end{array}} \tag{9.34}$$

---

interactions, see Belytschko & Geers [12], Geers *et al* [78], [79], Pinsky & Abboud [186], Soize, Desanti & David [203], Amini *et al* [3], Harari & Hughes [93].

[4] This is the *canonical* displacement potential, i.e. it verifies the condition $\int_{\Omega_F}\varphi\, dx = 0$.

where we have let: [5]

$$k^0 = \int_{\Omega_S} \sigma_{ij}(u)\epsilon_{ij}(\delta u)\, dx + \frac{\rho_F c^2}{Vol\,(\Omega_F)} \left(\int_\Sigma u_N \, d\sigma\right)\left(\int_\Sigma \delta u_N \, d\sigma\right) \tag{9.35}$$

and where $\mathcal{C}^*_\varphi = \{\varphi \mid \int_{\Omega_F} \varphi\, dx = 0\}$.

Prestress effects — In what follows, the prestress effects from pressurization of the reservoirs (and any weight of the structure) have been neglected. Otherwise, the following contribution $k^{(p)}$ must be added to k^0:

$$k^{(p)} = \int_{\Omega_S} \sigma^0_{hj} u_{i,h} \delta u_{i,j} dx + \int_\Sigma P^0 n_1(u)\cdot \delta u \, d\sigma \tag{9.36}$$

which we have already met (cf. 6.18) in the linearization of the equations of a structure subject to a pressure field[6] and to gravity (*cf.* chapter 6 and Morand & Ohayon [137]).

Substructuring in $(\boldsymbol{u}, \boldsymbol{\kappa_\alpha})$ with mass coupling

The modal reduction procedure we consider consists of seeking an approximate solution to (9.34) by approximating $\mathcal{C}_u$ by finite elements and $\mathcal{C}^*_\varphi$ by the subspace spanned by the n first eigenfunctions φ_α, seeking φ in the form (*cf.* (7.39,7.41)):

$$\boxed{\varphi = \varphi^0_{u_N} + \sum_{\alpha=1}^{n} \kappa_\alpha \varphi_\alpha} \tag{9.37}$$

where $\varphi^0_{u_N}$ is the solution to (7.10).
We replace φ by its expression (9.37) in (9.34), and we successively put $\delta\varphi = \varphi_1, \varphi_2, \ldots$.
Taking into account the orthogonality relations of φ_α (7.26cd), together with the conjugate relation (7.37) between $\varphi^0_{u_N}$ and φ_α, we arrive, $\forall \delta u \in \mathcal{C}_u$ and for $1 \le \alpha \le n$ at:

$$\begin{aligned} & k^0(u, \delta u) - \omega^2 \int_{\Omega^S} \rho_S u\cdot\delta u \; dx \\ & = \omega^2 \int_\Sigma \rho_F \varphi^0_{u_N} \delta u_N + \omega^2 \sum_\alpha^n \kappa_\alpha \int_\Sigma \rho_F \varphi_\alpha \delta u_N \, d\sigma + \int_{\Sigma_d} F\cdot\delta u \, d\sigma & (a) \\ & \qquad (-\omega^2 + \omega^2_\alpha)\mu_\alpha \kappa_\alpha = \omega^2 \int_\Sigma \rho_F u_N \, \varphi_\alpha \, d\sigma & (b) \end{aligned} \tag{9.38}$$

From(7.47), we see that $\int_\Sigma \rho_F \varphi^0_{u_N} \delta u_N \, d\sigma$ is none other than the mass operator $\mathcal{M}^0_{Ac}(u_N, \delta u_N)$ (positif).

[5]The positive operator k^0 is made up of the sum of the elastic stiffness operator k_E of the structure, and an additional term $k^{(c)}$, which, for a closed axisymmetric cavity, is null for a Fourier circumferential index other than 0.

[6]In the special case of a thin shell, P^0 is the difference between static pressures exerted on the internal and external surfaces.

Symmetric matrix equations in (u, κ_α)

We introduce the matrices corresponding to the various bilinear (or linear) forms involved in (9.38) defined by:

$$
\boxed{
\begin{array}{rcll}
\displaystyle\int_{\Omega_S} \sigma_{ij}(u)\epsilon_{ij}(\delta u)\, dx & \Longrightarrow & \delta \boldsymbol{U}^T \boldsymbol{K}_E \boldsymbol{U} & (a) \\
\displaystyle\int_{\Omega_S} \rho_S u \cdot \delta u \, dx & \Longrightarrow & \delta \boldsymbol{U}^T \boldsymbol{M} \boldsymbol{U} & (b) \\
\displaystyle\frac{\rho_F c^2}{Vol\,(\Omega_F)} \left(\int_\Sigma u_N \, d\sigma\right)\left(\int_\Sigma \delta u_N \, d\sigma\right) & \Longrightarrow & \delta \boldsymbol{U}^T \boldsymbol{K}^{(c)} \boldsymbol{U} & (c) \\
M^0_{Ac}(u_N, \delta u_N) & \Longrightarrow & \delta \boldsymbol{U}^T \boldsymbol{M}^0_{Ac} \boldsymbol{U} & (d) \\
\displaystyle\int_\Sigma \rho_F \varphi_\alpha u_N \, d\sigma & \Longrightarrow & \boldsymbol{C}^T_\alpha \boldsymbol{U} & (e) \\
\displaystyle\int_{\Sigma_d} F \cdot \delta u \, d\sigma & \Longrightarrow & \delta \boldsymbol{U}^T \boldsymbol{F} & (f)
\end{array}
}
\tag{9.39}
$$

Equations (9.38) are then written in discretized form, for $1 \leq \alpha \leq n$:

$$
\begin{array}{ll}
\left(\boldsymbol{K}_E + \boldsymbol{K}^{(c)}\right) \boldsymbol{U} - \omega^2 (\boldsymbol{M} + \boldsymbol{M}^0_{Ac}) \boldsymbol{U} = \omega^2 \displaystyle\sum_{\alpha=1}^{n} \boldsymbol{C}_\alpha \kappa_\alpha + \boldsymbol{F} & (a) \\
(-\omega^2 + \omega_\alpha^2)\mu_\alpha \kappa_\alpha = \omega^2 {\boldsymbol{C}_\alpha}^T \boldsymbol{U} & (b)
\end{array}
\tag{9.40}
$$

which is written in the following *symmetric* matrix form:

$$
\left[\begin{array}{c|ccc}
\boldsymbol{K}^0 & \cdots & 0 & \cdots \\
\hline
\vdots & \ddots & & \\
0 & & \omega_\alpha^2 \mu_\alpha & \\
\vdots & & & \ddots
\end{array}\right]
\begin{bmatrix} \boldsymbol{U} \\ \vdots \\ \kappa_\alpha \\ \vdots \end{bmatrix}
- \omega^2
\left[\begin{array}{c|ccc}
\boldsymbol{M}^0 & \cdots & \boldsymbol{C}_\alpha & \cdots \\
\hline
\vdots & \ddots & & \\
{\boldsymbol{C}_\alpha}^T & & \mu_\alpha & \\
\vdots & & & \ddots
\end{array}\right]
\begin{bmatrix} \boldsymbol{U} \\ \vdots \\ \kappa_\alpha \\ \vdots \end{bmatrix}
=
\begin{bmatrix} \boldsymbol{F} \\ \vdots \\ 0 \\ \vdots \end{bmatrix}
\tag{9.41}
$$

where we have let:

$$
\begin{array}{rcll}
\boldsymbol{K}^0 & = & \boldsymbol{K}_E + \boldsymbol{K}^{(c)} & (a) \\
\boldsymbol{M}^0 & = & \boldsymbol{M} + \boldsymbol{M}^0_{Ac} & (b)
\end{array}
\tag{9.42}
$$

Superelement of fluid

Matrix equations (9.41) may be interpreted as an assemblage, according to the degrees of freedom $\boldsymbol{U}_\Sigma$:

1. of the matrix model of the structure in vacuo $(\boldsymbol{K}_E, \boldsymbol{M})$,
2. and of the "superelement of fluid" in $(\boldsymbol{U}_\Sigma, \kappa_\alpha)$ made up of the following stiffness and mass matrices (with $1 \leq \alpha \leq n$):

$$\left\{ \left[\begin{array}{c|ccc} \boldsymbol{K}^{(c)} & \cdots & 0 & \cdots \\ \hline \vdots & \ddots & & \\ 0 & & \omega_\alpha^2 \mu_\alpha & \\ \vdots & & & \ddots \end{array}\right] , \left[\begin{array}{c|ccc} \boldsymbol{M}^0_{Ac} & \cdots & \boldsymbol{C}_\alpha & \cdots \\ \hline \vdots & \ddots & & \\ \boldsymbol{C}_\alpha{}^T & & \mu_\alpha & \\ \vdots & & & \ddots \end{array}\right] \right\} \tag{9.43}$$

Energy interpretation — The *symmetric* matrix equations 9.40) (transposed in time by substituting $\ddot{\boldsymbol{U}}(t)$ and $\ddot{\kappa}_\alpha(t)$ for $-\omega^2 \boldsymbol{U}$ and $-\omega^2\kappa_\alpha$), are the Lagrange equations of a coupled fluid-structure system described by the generalized system of coordinates $\{\boldsymbol{U}, \kappa_1, \ldots, \kappa_n\}$, whose potential energy E_P and kinetic energy E_C are expressed: [7]

$$\boxed{\begin{aligned} E_P &= \frac{1}{2}\boldsymbol{U}^T(\boldsymbol{K}_E + \boldsymbol{K}^{(c)})\boldsymbol{U} + \frac{1}{2}\sum_{\alpha=1}^{n} \omega_\alpha^2 \mu_\alpha \kappa_\alpha^2 && (a) \\ E_C &= \frac{1}{2}\dot{\boldsymbol{U}}^T(\boldsymbol{M} + \boldsymbol{M}^0_{Ac})\dot{\boldsymbol{U}} + \frac{1}{2}\sum_{\alpha=1}^{n} \dot{\kappa}_\alpha^2 \mu_\alpha + \sum_{\alpha=1}^{n} \dot{\kappa}_\alpha \boldsymbol{C}_\alpha{}^T \dot{\boldsymbol{U}} && (b) \end{aligned}} \tag{9.44}$$

Equivalent symmetric formulation in $(\boldsymbol{u}, \boldsymbol{\tau}_\alpha)$ with stiffness coupling

If in (9.44) we carry out the following change of generalized coordinates:

$$\boxed{\tau_\alpha = \kappa_\alpha + \frac{\boldsymbol{C}_\alpha{}^T \boldsymbol{U}}{\mu_\alpha}} \tag{9.45}$$

we obtain (using the modal expansion (7.54) of $\boldsymbol{M}^0_{Ac}$):

$$\begin{aligned} E_P &= \frac{1}{2}\boldsymbol{U}^T\left(\boldsymbol{K}_E + \boldsymbol{K}^{(c)} + \sum_{\alpha=1}^{n}\boldsymbol{K}_\alpha\right)\boldsymbol{U} - \sum_{\alpha=1}^{n}\omega_\alpha^2 \tau_\alpha \boldsymbol{C}_\alpha{}^T\boldsymbol{U} + \frac{1}{2}\sum_{\alpha=1}^{n}\omega_\alpha^2\tau_\alpha^2\mu_\alpha && (a) \\ E_C &= \frac{1}{2}\dot{\boldsymbol{U}}^T\left(\boldsymbol{M} + \sum_{n+1}^{\infty}\boldsymbol{M}_\alpha\right)\dot{\boldsymbol{U}} + \frac{1}{2}\sum_{\alpha=1}^{n}\dot{\tau}_\alpha^2\mu_\alpha && (b) \end{aligned} \tag{9.46}$$

where $\boldsymbol{K}_\alpha$ denotes the "acoustic stiffness" matrix of the α-*th* mode defined by: (the definition of $\boldsymbol{M}_\alpha$ is given in (7.53)):

$$\boldsymbol{K}_\alpha = \omega_\alpha^2 \frac{\boldsymbol{C}_\alpha \boldsymbol{C}_\alpha{}^T}{\mu_\alpha} = \omega_\alpha^2 \boldsymbol{M}_\alpha \tag{9.47}$$

---

[7] $\boldsymbol{F}$ being considered as generalized forces.

The Lagrange equations deduced from (9.46) then yield the following *symmetric* matrix equations in harmonic regime:

$$\begin{aligned}\left(\boldsymbol{K}_E+\boldsymbol{K}^{(c)}+\sum_{\alpha=1}^{n}\boldsymbol{K}_\alpha\right)\boldsymbol{U}-\omega^2\left(\boldsymbol{M}+\sum_{n+1}^{\infty}\boldsymbol{M}_\alpha\right)\boldsymbol{U} &= \sum_{\alpha=1}^{n}\omega_\alpha^2\boldsymbol{C}_\alpha\tau_\alpha+\boldsymbol{F} \quad (a)\\ (-\omega^2+\omega_\alpha^2)\mu_\alpha\tau_\alpha &= \omega_\alpha^2{\boldsymbol{C}_\alpha}^T\boldsymbol{U} \quad (b)\end{aligned} \tag{9.48}$$

where we have used the modal expansion (7.54) of $\boldsymbol{M}^0_{Ac}$, and which may immediately be written in *symmetric* matrix form, with "stiffness coupling".

Superelement of fluid with stiffness coupling — These equations may be interpreted as the assemblage, according to the degrees of freedom $\boldsymbol{U}_\Sigma$:

1. of the **matrix model of the structure** *in vacuo* $(\boldsymbol{K}_E, \boldsymbol{M})$,

2. and of the **"superelement of fluid"** in $(\boldsymbol{U}_\Sigma, \tau_\alpha)$ made up of the following stiffness and mass matrices:

$$\left\{\left[\begin{array}{c|c}\boldsymbol{K}^{(c)}+\sum_{\alpha=1}^{n}\boldsymbol{K}_\alpha & \cdots\ -\omega_\alpha^2\boldsymbol{C}_\alpha\ \cdots\\ \hline \vdots & \ddots\\ -\omega_\alpha^2{\boldsymbol{C}_\alpha}^T & \omega_\alpha^2\mu_\alpha\\ \vdots & \ddots\end{array}\right], \left[\begin{array}{c|c}\sum_{n+1}^{\infty}\boldsymbol{M}_\alpha & \cdots\ 0\ \cdots\\ \hline \vdots & \ddots\\ 0 & \mu_\alpha\\ \vdots & \ddots\end{array}\right]\right\} \tag{9.49}$$

Direct derivation of $(\boldsymbol{u}, \tau_\alpha)$ stiffness coupling formulation

We seek φ in the form (7.33):

$$\boxed{\varphi=\sum_{\alpha}^{n}\tau_\alpha\varphi_\alpha} \tag{9.50}$$

Substituting (9.50) into(9.34ab), then putting successively $\delta\varphi = \varphi_1, \varphi_2, \ldots$, and using the orthogonality relations (7.26cd), we obtain, $\forall \delta u \in \mathcal{C}_u$, and for $1 \le \alpha \le n$:

$$\begin{aligned}k^0(u,\delta u)-\omega^2\int_{\Omega_S}\rho_S u\cdot\delta u\,dx-\omega^2\sum_{\alpha}^{n}\tau_\alpha\int_\Sigma\rho_F\varphi_\alpha\delta u_N\,d\sigma=\int_{\Sigma_d}F\cdot\delta u\,d\sigma \quad (a)\\ (-\omega^2+\omega_\alpha^2)\mu_\alpha\tau_\alpha=\omega_\alpha^2\int_\Sigma\rho_F u_N\varphi_\alpha\,d\sigma \quad (b)\end{aligned} \tag{9.51}$$

With the notations (9.39), equations (9.51) in discretized form are then written, for $1 \le \alpha \le n$:

$$\begin{aligned}\left(\boldsymbol{K}_E+\boldsymbol{K}^{(c)}\right)\boldsymbol{U}-\omega^2\boldsymbol{M}\boldsymbol{U}=\omega^2\sum_{\alpha=1}^{n}\boldsymbol{C}_\alpha\tau_\alpha+\boldsymbol{F} \quad (a)\\ (-\omega^2+\omega_\alpha^2)\mu_\alpha\tau_\alpha=\omega_\alpha^2{\boldsymbol{C}_\alpha}^T\boldsymbol{U} \quad (b)\end{aligned} \tag{9.52}$$

Symmetric matrix equations in U, τ_α — Equations (9.52) may be written in symmetric matrix form as follows: in the second member of (9.52a), $\omega^2\tau_\alpha$ is replaced by its expression $\omega_\alpha^2\tau_\alpha - (\omega_\alpha^2/\mu_\alpha)\boldsymbol{C}_\alpha{}^T\boldsymbol{U}$ taken from (9.52b). Equations (9.52) are then written in the following form, with $1 \le \alpha \le n$:

$$\begin{aligned}\left(\boldsymbol{K}_E + \boldsymbol{K}^{(c)} + \sum_{\alpha=1}^{n}\boldsymbol{K}_\alpha\right)\boldsymbol{U} - \omega^2\boldsymbol{M}\boldsymbol{U} &= \sum_{\alpha=1}^{n}\omega_\alpha^2\boldsymbol{C}_\alpha\tau_\alpha + \boldsymbol{F} \quad (a)\\ (-\omega^2+\omega_\alpha^2)\mu_\alpha\tau_\alpha &= \omega_\alpha^2\boldsymbol{C}_\alpha{}^T\boldsymbol{U} \quad (b)\end{aligned} \tag{9.53}$$

which may immediately be written in *symmetric* matrix form.

These latter equations may be interpreted as the assemblage, following the degrees of freedom $\boldsymbol{U}_\Sigma$, of the *matrix model of the structure* $(\boldsymbol{K}_E, \boldsymbol{M})$, and if a "superelement of fluid", whose stiffness matrix can be verified to coincide with that of (9.49), and whose mass matrix differs from that of (9.49) in the replacement of the expression $\sum_{n+1}^{\infty}\boldsymbol{M}_\alpha$ by 0.

Effects of truncation

We know that in the limit case $\omega = 0$, the "exact" behaviour of the fluid is described by the potential φ^0 introduced in chapter 7. We now analyse the effects of modal truncation, introduced in the preceding formulations, on the representation of the dynamic behaviour of the fluid, for $\omega \longrightarrow 0$ and for $\omega \longrightarrow \infty$. We begin by recalling the modal expansion (7.55) of φ^0 which, for discretized u, is written:

$$\varphi^0 = \sum_{\alpha=1}^{\infty}\frac{\boldsymbol{C}_\alpha{}^T\boldsymbol{U}}{\mu_\alpha}\varphi_\alpha \tag{9.54}$$

Symmetric matrix formulation with mass coupling

- For $\omega \longrightarrow 0$, from(9.40b), κ_α tends to 0 and from (9.37), φ tends to the exact value φ^0.

- For $\omega \longrightarrow \infty$, $\kappa_\alpha \longrightarrow -\frac{\boldsymbol{C}_\alpha{}^T\boldsymbol{U}}{\mu_\alpha}$. From (9.37), $\varphi \longrightarrow \varphi^0 - \sum_{\alpha=1}^{n}\frac{\boldsymbol{C}_\alpha{}^T\boldsymbol{U}}{\mu_\alpha}\varphi_\alpha$. Referring to (9.54), we then see that $\varphi \longrightarrow \sum_{n+1}^{\infty}\frac{\boldsymbol{C}_\alpha{}^T\boldsymbol{U}}{\mu_\alpha}\varphi_\alpha$ (which tends to 0 for $n \longrightarrow \infty$).

Symmetric matrix formulations with stiffness coupling

We begin by noting that the above conclusions apply to the symmetric matrix formulation with stiffness coupling (9.48), since it is equivalent, in construction, to the formulation with mass coupling (9.40).
Formulation (9.53), however obtained by direct modal analysis of (9.34) by means of (9.50), has the following particularities:

- For $\omega \longrightarrow 0$, from (9.53b) $\tau_\alpha \longrightarrow \frac{\boldsymbol{C}_\alpha{}^T\boldsymbol{U}}{\mu_\alpha}$. Consequently, from (9.50) and (9.54) $\varphi \longrightarrow \varphi^0 - \sum_{n+1}^{\infty} \frac{\boldsymbol{C}_\alpha{}^T\boldsymbol{U}}{\mu_\alpha}\varphi_\alpha$.
- For $\omega \longrightarrow \infty$, from (9.53b), τ_α tends to 0, and therefore φ tends to the exact value 0.

In conclusion, when the behaviour at the limit for $\omega \longrightarrow 0$ must be correctly restored, we have a choice between the symmetric formulations with mass coupling (9.41) or the equivalent formulation with stiffness coupling (9.49).

Remark: use of the supplementary variable p^0

Equations (9.42) involve the matrix $\boldsymbol{K}^{(c)}$ which is calculated as follows. This matrix may be written $\frac{\rho_F c^2}{Vol(\Omega_F)}\boldsymbol{C}_1\boldsymbol{C}_1{}^T$, where $\boldsymbol{C}_1$ is a vector discretizing the linear form $\int_\Sigma u_N \, d\sigma$. The calculation is equivalent to applying the condensation algorithm (1.65) with $\alpha = \frac{Vol(\Omega_F)}{\rho_F c^2}$ (*cf.* the analogous calculation § 6.3).
One variation consists, in the description of the fluid, of introducing the variable $p^0 = -\frac{\rho_F c^2}{Vol(\Omega_F)}\int_\Sigma u_N \, d\sigma$ as additional unknown (e.g. for formulation by mass coupling, this is equivalent to applying the technique of modal projection of the fluid starting from (8.54), and for formulation with stiffness coupling *cf.* Ohayon [173] and Kehr-Candille & Ohayon [108]).

Calculation of $\boldsymbol{M}^0_{Ac}$

- The direct calculation of $\boldsymbol{M}^0_{Ac}$ was discussed in § 7.5.
- $\boldsymbol{M}^0_{Ac}$ may also be calculated by means of the following summation rule (*cf.* (7.54) and (9.39d)):

$$\boldsymbol{M}^0_{Ac} = \sum_{\alpha=1}^{\infty} \boldsymbol{M}_\alpha \quad \text{with } \boldsymbol{M}_\alpha = \frac{\boldsymbol{C}_\alpha \boldsymbol{C}_\alpha{}^T}{\mu_\alpha} \tag{9.55}$$

  which involves the discretized form $\boldsymbol{M}_\alpha$ of the modal acoustic mass operators M_α (*cf.* (7.53)).

Complete dynamic substructuring

In the previous section we developed a "partial" dynamic substructuring technique based on a hybrid representation of the problem of fluid-structure interaction in terms of the "physical coordinates" $\boldsymbol{U}$, and the "generalized coordinates" κ_α or τ_α relating to the acoustic modes.
We develop here a "total" dynamic substructuring technique for the coupled system which, while keeping the description in $\boldsymbol{\kappa} = \{\kappa_\alpha\}$ or $\boldsymbol{\tau} = \{\tau_\alpha\}$ of the fluid, consists of substituting for the variables $\boldsymbol{U}$, n_S generalized coordinates q_σ as follows:

$$\boxed{\boldsymbol{U} = \sum_{\sigma=1}^{n_S} q_\sigma \boldsymbol{U}_\sigma} \tag{9.56}$$

where $\{\boldsymbol{U}_\sigma\}$ is a basis of projection chosen to diagonalize the matrix relating to the degrees of freedom $\boldsymbol{U}$ involved in equations (9.40) and (9.48).
The projection technique consists of applying rule (1.20) to the matrix equations in $(\boldsymbol{U}, \boldsymbol{\kappa})$ or $(\boldsymbol{U}, \boldsymbol{\tau})$, with the projection matrix $\boldsymbol{H}$ defined by:

$$\boldsymbol{H} = \left[\begin{array}{c|c} \boldsymbol{U}_1 \cdots \boldsymbol{U}_{ns} & 0 \\ \hline 0 & \boldsymbol{I} \end{array}\right] \tag{9.57}$$

where $\boldsymbol{I}$ is the unity matrix of rank n.

Diagonalization leads to taking, for $\boldsymbol{U}_\sigma$, the eigenvectors which are solutions to the two problems with distinct eigenvalues which we shall analyze successively.

Coupling by mass matrix — Diagonalization of the matrix relating to $\boldsymbol{U}$ in (9.40), involves seeking eigenvectors $\{\boldsymbol{U}_\sigma\}$ which are solutions of: [8]

$$\left(\boldsymbol{K}_E + \boldsymbol{K}^{(c)}\right)\boldsymbol{U} = \lambda(\boldsymbol{M}+\boldsymbol{M}^0_{Ac})\boldsymbol{U} \tag{9.58}$$

We denote by $\boldsymbol{U}_\sigma$ the eigenvectors of (9.58) which are assumed to be normalized to 1 in generalized mass, i.e. verifying the orthogonality relations:

$$\begin{array}{rcll} \boldsymbol{U}_\sigma{}^T(\boldsymbol{M} + \boldsymbol{M}^0_{Ac})\boldsymbol{U}_{\sigma'} & = & \delta_{\sigma\sigma'} & (a) \\ \boldsymbol{U}_\sigma{}^T(\boldsymbol{K}_E + \boldsymbol{K}^{(c)})\boldsymbol{U}_{\sigma'} & = & \omega_\sigma^2\delta_{\sigma\sigma'} & (b) \end{array} \tag{9.59}$$

Taking for $\boldsymbol{U}_\sigma$ in (9.57), the eigenvectors (9.58) with conditions (9.59), projection of equations (9.40) yields:

$$\boxed{\begin{bmatrix} \textbf{Diag}\,\omega_\sigma^2 & 0 \\ 0 & \textbf{Diag}\,\omega_\alpha^2 \end{bmatrix}\begin{bmatrix} \boldsymbol{q} \\ \boldsymbol{\kappa} \end{bmatrix} - \omega^2\begin{bmatrix} \boldsymbol{I}_{ns} & [C_{\sigma\alpha}] \\ [C_{\alpha\sigma}] & \boldsymbol{I}_n \end{bmatrix}\begin{bmatrix} \boldsymbol{q} \\ \boldsymbol{\kappa} \end{bmatrix} = \begin{bmatrix} \boldsymbol{f} \\ 0 \end{bmatrix}} \tag{9.60}$$

where $\boldsymbol{I}_{ns}$ and $\boldsymbol{I}_n$ are the unity matrices of respective dimensions n_S and n, where $\boldsymbol{f}$ denotes the vector of components $f_\sigma = \boldsymbol{U}_\sigma{}^T\boldsymbol{F}$, and where we have let:

$$C_{\sigma\alpha} = \boldsymbol{C}_\alpha{}^T\boldsymbol{U}_\sigma \tag{9.61}$$

- We note that the problem in $\boldsymbol{U}$ (*cf.* 9.58) takes into account the non-resonant behaviour of the fluid through $\boldsymbol{K}^{(c)}$ and $\boldsymbol{M}^0_{Ac}$, *independently of the modal truncation* introduced for the description of the fluid. [9]
- An alternative solution consists of taking as basis of projection $\{\boldsymbol{U}_\sigma\}$, the modes of the structure *in vacuo*, which are solutions to $\boldsymbol{K}_E\boldsymbol{U} = \lambda\boldsymbol{M}\boldsymbol{U}$. In this case, in the stiffness matrix of (9.60), $\omega_\sigma^2\delta_{\sigma\sigma'}$ is replaced by $(\omega_\sigma^2\delta_{\sigma\sigma'} + \boldsymbol{U}_\sigma{}^T\boldsymbol{K}^{(c)}\boldsymbol{U}_{\sigma'})$, and in the mass matrix, $\delta_{\sigma\sigma'}$ is replaced by $(\delta_{\sigma\sigma'} + \boldsymbol{U}_\sigma{}^T\boldsymbol{M}^0_{Ac}\boldsymbol{U}_{\sigma'})$. We then arrive at a system of matrix equations, where the generalized coordinates $\{q_\sigma\}$ are coupled to each other.

[8] $\boldsymbol{K}_E + \boldsymbol{K}^{(c)}$ is a positive matrix, $\boldsymbol{M} + \boldsymbol{M}^0_{Ac}$ is positive, definite.

[9] This type of approximation, neglecting $\boldsymbol{M}^0_{Ac}$, is made in order to take into account the additional rigidity caused by the presence of small gaseous volumes above the liquid in liquid reservoirs. In this case, we have to use the matrix $\boldsymbol{K}^{(c)}$ acting on the degrees of freedom of the surface, which is made up of the structure-gas contact surface (of normal displacement u_N), and the free surface Γ of the liquid (of normal displacement η).

- We note the similarity which exists between the two equations (9.60), and equations (1.85) describing the modal coupling of two elastic substructures. Discussion of the "strong" and "weak" interactions are transposed to the case of coupling of *structure* modes and *acoustic* modes. (*cf.* § 1.11).

- Starting with the unsymmetric (u, p) formulation (8.5), we arrive naturally, after modal projection, at an equally non-symmetric matrix formulation (similar to equations (9.52)) which has been used by Craggs & Stead [47], Ramakrishnan & Koval [189]), and for applications in the automobile industry, Sung & Nefske [204].

  However – through a substitution analogous to that used in going from (9.52) à (9.53) — formulation (8.5), after modal projection, yields the symmetric formulation (9.60) (*cf.* Morand [150], Morand & Ritchie [152]).

- We can also set up the modal scheme (9.60) starting with the "condensed" symmetric formulation in (u, p) (8.34), with mass coupling (*cf.* Morand & Ohayon [138]).

Coupling by the stiffness matrix — Diagonalization of the matrix relating to $\boldsymbol{U}$ in (9.48), leads to seeking the eigenvectors $\{\boldsymbol{U}_\sigma\}$ which are solutions of: [10]

$$\left(\boldsymbol{K}_E + \boldsymbol{K}^{(c)} + \sum_{\alpha=1}^{n} \boldsymbol{K}_\alpha\right)\boldsymbol{U} = \lambda\left(\boldsymbol{M} + \sum_{n+1}^{\infty} \boldsymbol{M}_\alpha\right)\boldsymbol{U} \tag{9.62}$$

We denote by $\boldsymbol{U}_\sigma$ the eigenvectors of (9.62) which are assumed to be normalized to 1 in generalized mass, i.e. which verify the orthogonality relations:

$$\boldsymbol{U}_\sigma{}^T\left(\boldsymbol{M} + \sum_{n+1}^{\infty} \boldsymbol{M}_\alpha\right)\boldsymbol{U}_{\sigma'} = \delta_{\sigma\sigma'}\,,\; \boldsymbol{U}_\sigma{}^T\left(\boldsymbol{K}_E + \boldsymbol{K}^{(c)} + \sum_{\alpha=1}^{n} \boldsymbol{K}_\alpha\right)\boldsymbol{U}_{\sigma'} = \omega_\sigma^2\delta_{\sigma\sigma'} \tag{9.63}$$

Taking for $\boldsymbol{U}_\sigma$ in (9.57), the eigenvectors of (9.62) with conditions (9.63), projection of equations (9.48) yields:

$$\boxed{\begin{bmatrix} \mathbf{Diag}\,\omega_\sigma^2 & -[\omega_\alpha^2 C_{\sigma\alpha}] \\ -[C_{\alpha\sigma}\omega_\alpha^2] & \mathbf{Diag}\,\omega_\alpha^2 \end{bmatrix}\begin{bmatrix} \boldsymbol{q} \\ \boldsymbol{\tau} \end{bmatrix} - \omega^2\begin{bmatrix} \boldsymbol{I}_{ns} & 0 \\ 0 & \boldsymbol{I}_n \end{bmatrix}\begin{bmatrix} \boldsymbol{q} \\ \boldsymbol{\tau} \end{bmatrix} = \begin{bmatrix} \boldsymbol{f} \\ 0 \end{bmatrix}} \tag{9.64}$$

with the preceding definitions of $\boldsymbol{I}_{ns}, \boldsymbol{I}_n, \boldsymbol{f}$ and $C_{\sigma\alpha}$ (*cf.* (9.61)).

Remarks

- We note that the problem (9.62) in $\boldsymbol{U}$, involves the non-resonant behaviour of the fluid *independently of the modal truncation* n relating to the fluid, due to the presence of the terms $\sum_{\alpha=1}^{n} \boldsymbol{K}_\alpha$ and $\sum_{n+1}^{\infty} \boldsymbol{M}_\alpha$.

[10] $\boldsymbol{K}_E + \boldsymbol{K}^{(c)} + \sum_{\alpha=1}^{n} \boldsymbol{K}_\alpha$ and $\boldsymbol{M} + \sum_{n+1}^{\infty} \boldsymbol{M}_\alpha$ are positive matrices.

- An alternative solution which eliminates the practical disadvantage of using a projection basis $\{\boldsymbol{U}_\sigma\}$ dependent on n, consists of using the basis of the structure modes *in vacuo*. We then arrive at a system of matrix equations where, in contrast, the generalized coordinates $\{q_\sigma\}$ are coupled with each other (because of $\boldsymbol{K}^{(c)}$).

- A variant of the formulations (9.60) and (9.64) consists of introducing the supplementary variable p^0 for the description of the fluid. As we have seen, this technique avoids direct calculation of $\boldsymbol{K}^{(c)}$. Under these conditions, modal projection of this formulation in $(\boldsymbol{U}, p^0, \boldsymbol{\kappa})$ yields matrix equations in $(\boldsymbol{q}, p^0, \boldsymbol{\kappa})$.

- Various derivations and applications in the aerospace field of the modal scheme (9.64) can be found in Morand [150] and, with an interface damping taken into account, from a variant of the symmetric formulation in u, p, φ with stiffness coupling using the auxiliary variable p^0, in Ohayon [173] and Kehr-Candille & Ohayon [108].

Case of a free surface

We consider here application of the preceding techniques of modal reduction to a study of the vibrations of an elastic reservoir containing a compressible liquid with a free surface, to incorporate, in a simple way, the effects of compressibility in a hydroelastic matrix model (chapter 5), through the allocation of generalized coordinates corresponding to acoustic modes of the liquid with a free surface (chapter 7).

Variational formulation — Modeling of an incompressible liquid with a free surface in the absence of gravity has been described in § 7.6.
We denote by Γ the free surface and by Σ the liquid-structure interaction surface. We recall that in this case, the free surface condition is written $\varphi = 0$ on Γ, and that the relation between the pressure p and the displacement potential φ is written $p = \rho_F \omega^2 \varphi$.
The variational formulation in (u, φ) of this problem is then readily deduced from (9.34), through the following modifications:

1. k^0 is replaced by $\int_{\Omega_S} \sigma_{ij}(u)\epsilon_{ij}(\delta u)\, dx$,

2. and the admissible space C^*_φ becomes $C^*_\varphi = \{\varphi \mid \varphi = 0 \text{ on } \Gamma\}$.

We then obtain: for given ω and F find $(u, \varphi) \in C_u \times C^*_\varphi$ such that $\forall(\delta u, \delta\varphi) \in C_u \times C^*_\varphi$:

$$\boxed{\begin{aligned}
&\int_{\Omega_S} \sigma_{ij}(u)\epsilon_{ij}(\delta u)\, dx - \omega^2 \int_{\Omega_S} \rho_S u \cdot \delta u\, dx - \omega^2 \int_\Sigma \rho_F \varphi \delta u_N\, d\sigma = \int_{\Sigma^d} F \cdot \delta u\, d\sigma \quad (a)\\
&\int_{\Omega_F} \rho_F \nabla\varphi \cdot \nabla\delta\varphi\, dx - \omega^2 \int_{\Omega_F} \frac{\rho_F}{c^2} \varphi \delta\varphi\, dx - \int_\Sigma \rho_F u_N \delta\varphi\, d\sigma = 0 \quad (b)
\end{aligned}} \tag{9.65}$$

Coupling by the mass matrix

Using the modal expansion (9.37) of φ in (9.65), and recalling that, from (7.65), que $\boldsymbol{M}^0_{Ac} = \boldsymbol{M}_A$, we arrive at the following matrix equations:

$$\begin{aligned} \boldsymbol{K}_E\boldsymbol{U} - \omega^2(\boldsymbol{M} + \boldsymbol{M}_A)\boldsymbol{U} &= \omega^2 \sum_{\alpha=1}^{n} \boldsymbol{C}_\alpha \kappa_\alpha + \boldsymbol{F} \quad (a) \\ (-\omega^2 + \omega_\alpha^2)\mu_\alpha \kappa_\alpha &= \omega^2 \boldsymbol{C}_\alpha{}^T \boldsymbol{U} \quad (b) \end{aligned} \tag{9.66}$$

These equations may be interpreted as the assemblage, according to the degrees of freedom $\boldsymbol{U}_\Sigma$,

1. of the **the hydroelastic matrix model** $(\boldsymbol{K}_E, \boldsymbol{M} + \boldsymbol{M}_A)$

2. and of **the superelement of the liquid** given by:

$$\left\{ \left[\begin{array}{c|ccc} 0 & \cdots & 0 & \cdots \\ \hline \vdots & \ddots & & \\ 0 & & \omega_\alpha^2 \mu_\alpha & \\ \vdots & & & \ddots \end{array} \right] , \left[\begin{array}{c|ccc} 0 & \cdots & \boldsymbol{C}_\alpha & \cdots \\ \hline \vdots & \ddots & & \\ \boldsymbol{C}_\alpha{}^T & & \mu_\alpha & \\ \vdots & & & \ddots \end{array} \right] \right\} \tag{9.67}$$

Complete dynamic substructuring

We proceed as before using the expansion (9.56) in (9.66) and choosing for $\boldsymbol{U}_\sigma$ the eigenvectors of the hydroelastic problem $\boldsymbol{K}_E\boldsymbol{U} = \lambda(\boldsymbol{M} + \boldsymbol{M}_A)\boldsymbol{U}$, assumed to be normalized to 1 in generalized mass, i.e. to verify the orthogonality relations:

$$\boldsymbol{U}_\sigma{}^T(\boldsymbol{M} + \boldsymbol{M}_A)\boldsymbol{U}_{\sigma'} = \omega_\sigma^2 \delta_{\sigma\sigma'} \quad , \quad \boldsymbol{U}_\sigma{}^T \boldsymbol{K}_E \boldsymbol{U}_{\sigma'} = \delta_{\sigma\sigma'} \tag{9.68}$$

Taking the vectors $\boldsymbol{U}_\sigma$ in (9.57), the projection of equations (9.66 yields:

$$\boxed{\begin{bmatrix} \mathbf{Diag}\,\omega_\sigma^2 & 0 \\ 0 & \mathbf{Diag}\,\omega_\alpha^2 \end{bmatrix} \begin{bmatrix} \boldsymbol{q} \\ \boldsymbol{\kappa} \end{bmatrix} - \omega^2 \begin{bmatrix} \boldsymbol{I}_{ns} & [C_{\sigma\alpha}] \\ [C_{\alpha\sigma}] & \boldsymbol{I}_n \end{bmatrix} \begin{bmatrix} \boldsymbol{q} \\ \boldsymbol{\kappa} \end{bmatrix} = \begin{bmatrix} \boldsymbol{f} \\ 0 \end{bmatrix}} \tag{9.69}$$

with the preceding definitions $\boldsymbol{I}_{n_S}$, $\boldsymbol{I}_n$, $\boldsymbol{f}$ and $C_{\sigma\alpha}$.

9.4 Conclusion and open problems

We have shown that a study of the vibratory behaviour of an elastic structure containing an *incompressible fluid with a free surface and gravity effects* , or a *compressible weightless* fluid, may be reduced to a problem of interaction between

1. structure modes comprising two "extreme" non-resonant behaviours of the liquid – respectively "at zero frequency" and "at infinite frequency" — corresponding to the generalized coordinates $\boldsymbol{q}$,

2. and modes of vibration of the fluid in a fixed cavity, corresponding to the generalized coordinates $\boldsymbol{\kappa}$ ou $\boldsymbol{\tau}$.

We have distinguished two main schemes, according to whether the variables $\boldsymbol{q}$ are coupled to the variables $\boldsymbol{\kappa}$ by the mass matrix of the coupled system, or to the variables $\boldsymbol{\tau}$ by the stiffness matrix of the coupled system.
We have shown that these schemes are based *a posteriori* on an energy interpretation in terms of the discrete variables $(\boldsymbol{q}, \boldsymbol{\kappa})$ ou de $(\boldsymbol{q}, \boldsymbol{\tau})$; the symmetric matrix equations that we obtain can be considered as the Lagrange equations associated with the potential and kinetic energies in terms of generalized coordinates.
These schemes have been derived from a special formulation of the set problem – in this case the (u, φ) formulation. The energy interpretation we give them however, gives them an intrinsic character. In other words, these schemes are independent of the chosen initial variational formulation.
In the case of linear vibrations of structures coupled to liquids, we emphasize the key role of the hydroelastic modes, taking account both of sloshing effects and of compressibility effects.
The modal vibroacoustic models derived in this chapter are recommended for the study of low-frequency environments.
Problems currently being investigated include:

- Numerical analysis of the convergence of the modal methods studied in this chapter, together with the use of modal bases calculated with "incompatible" meshes (which do not coincide on the fluid-structure interaction surface; for example in the case of coupled elastic structures *cf.* Farhat & Géradin [68]).

- In situations of high modal density, extension of the methods of modal coupling presented here, in conjunction with procedures for projection by frequency band (*cf.* for example Morand [151]).

- In the nonlinear domain, modeling of sloshing-structure coupling phenomena which appear when the amplitudes of vibration of the free surface and of the structure cannot be considered as infinitely small of the same order (*cf.* Chu & Kana [34], Donea [59], Kana [106]). For a study of the dynamic buckling of coupled fluid-structure systems in relationship with the prestresses in the structure, *cf.* Paidoussis *et al* [178], [179], Liu *et al* [122].

- One important problem concerns low frequency noise reduction in aircraft:

  – by means of "passive" techniques requiring modelling of the dissipation afforded by damping materials or by systems of resonators located at the structure-fluid interface; [11]

[11] independently of coupling with the problem of radiation in an external medium.

– by means of active control techniques using for example adaptive structures — intelligent systems with so-called smart materials — (*cf.* for example Wada, Fanson & Crawley [210], Kohudic [105], Tzou & Tseng [207], Destuynder & Legrain [58], Fuller *et al* [75]). More generally, see the application of new methods of control for distributed systems in Lions [120], Lagnese & Lions [110] and Glowinski, Li & Lions [87].

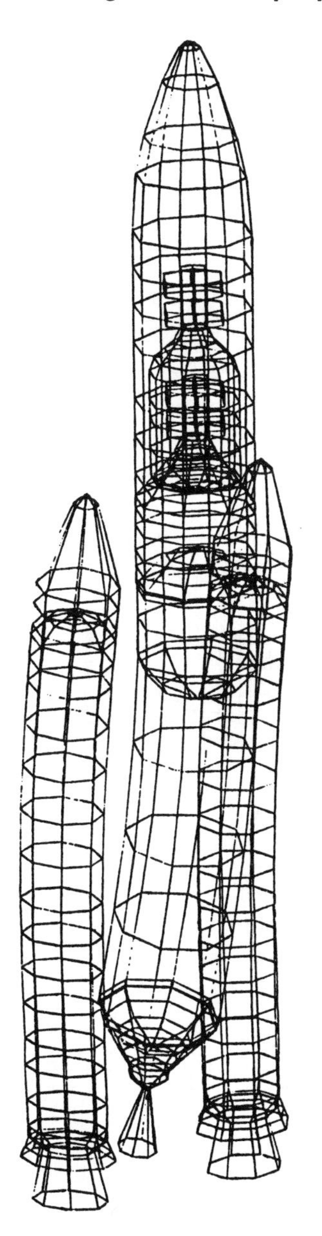

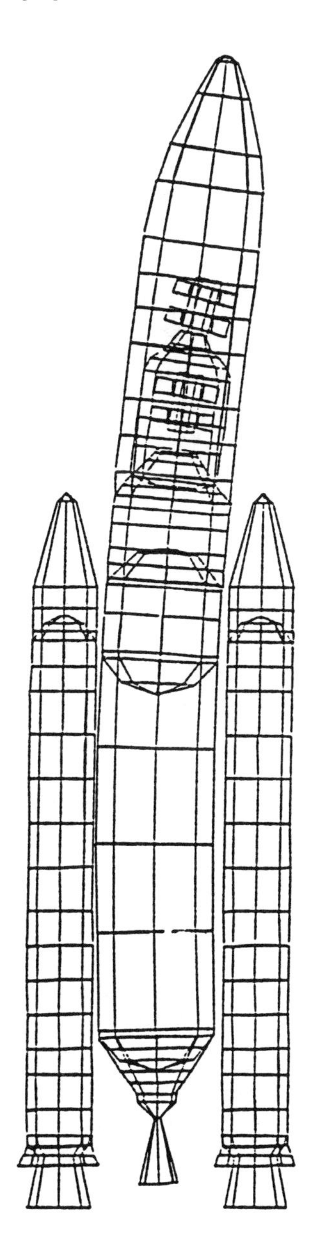

Example of vibration eigenmodes of the launcher Ariane 5 obtained by dynamic substructuring (in a satellite launching configuration).

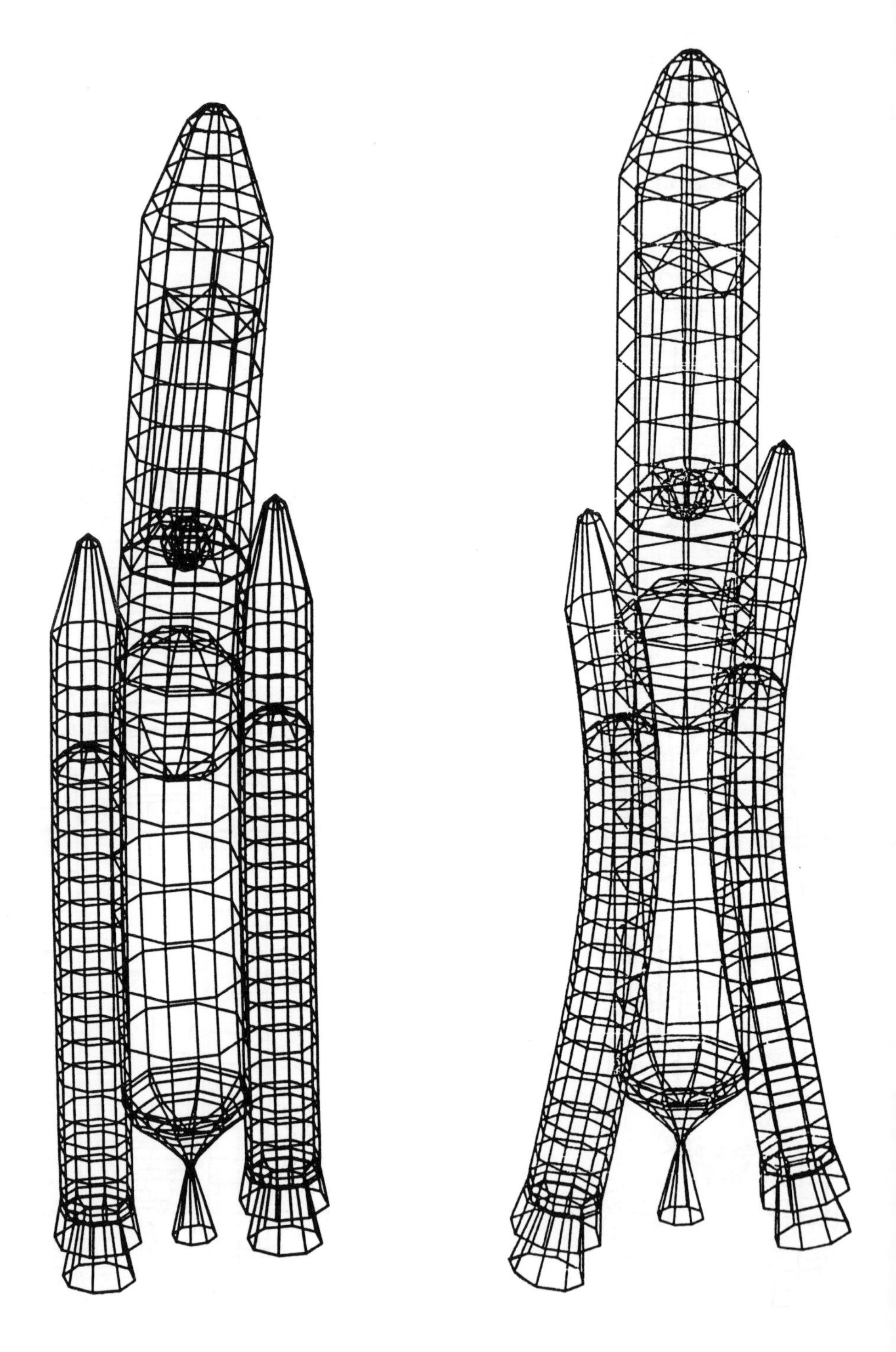

Example of vibration eigenmodes of the launcher Ariane 5 obtained by dynamic substructuring (in a satellite launching configuration).

Bibliography

[1] H.N. ABRAMSON. *The Dynamic Behavior of Liquids in Moving Containers.* NASA SP-106, 1966.

[2] I. AGANOVIC. On a spectral problem of hydroelasticity. *J. de Mécanique* **20** (3), 1981.

[3] S. AMINI, P.J. HARRIS, and D.T. WILTON. *Coupled Boundary and Finite Element Methods for the Solution of the Dynamic Fluid-Structure Interaction Problem.* Lecture Notes in Eng., Vol. 77, Springer, 1992.

[4] I. ANTONIADIS and A. KANARACHOS. Decoupling procedures for fluid-structure interaction problems. *Comp. Meth. in Appl. Mech. and Eng.* **70**, 1–25, 1988.

[5] J. ARGYRIS and H-P. MLEJNEK. *Dynamics of Structures.* Texts on Computational Mechanics, Vol. 5, North-Holland, 1991.

[6] V. ARNOLD. *Mathematical Methods of Classical Mechanics.* Springer, New York, 1978.

[7] F. AXISA and R.F. GIBERT. Non-linear analysis of fluid-structure coupled transients in piping systems. *Proc. ASME-PVP Conf.*, Vol. 63, 151–165, 1982.

[8] I. BABUSKA and J. OSBORN. *Eigenvalue Problems.* Handbook of Numerical Analysis, Vol. 2, P.G. Ciarlet and J.L. Lions (eds), North-Holland, 1991.

[9] K.J. BATHE. *Finite Element Procedures in Engineering Analysis.* Prentice-Hall, 1982.

[10] J-L. BATOZ, G. DHATT. *Modélisation des structures par éléments finis.* Vol.2-3, Editions Hermès, 1991.

[11] H. BAUER. Axisymmetric natural frequencies and response of a spinning liquid column under strong surface tension. *Acta Mechanica* **90**, 21–35, 1991.

[12] T. BELYTSCHKO and T.L. GEERS (eds). Computational methods for fluid-structure interaction. *ASME-AMD*, Vol. 26, Atlanta, 1977.

[13] T. BELYTSCHKO and T.J.R. HUGHES (eds). *Computational Methods for Transient Analysis.* North-Holland, 1983.

[14] W.A. BENFIELD and R.F. HRUDA. Vibration analysis of structures by component mode substitution. *AIAA Journal* **9** (7), 1255–1261, 1971.

[15] T.B. BENJAMIN and J.C. SCOTT. Gravity-capillarity waves with edge constraint. *J. Fluid Mech.* **92**, 241–267, 1979.

[16] A. BENSOUSSAN, J.L. LIONS, and G. PAPANICOLAOU. *Asymptotic Analysis for Periodic Structures.* North-Holland, 1987.

[17] H. BERGER, J. BOUJOT, and R. OHAYON. Computation of elastic tanks partially filled with liquids. *J. Math. Anal. Appl.* **51**, 272–298, 1975.

[18] M. BERGER and B. GOSTIAUX. *Differential Geometry: Manifolds, Curves, and Surfaces.* Springer-Verlag, 1987.

[19] A. BERMUDEZ and R. RODRIGUEZ. Finite element computation of the vibration modes of a fluid-solid system. *Publ. Dept. Mat. Apl.*, no. 11, Univ. Santiago de Compostela, Spain, 1993. *Comp. Meth. Appl. Mech. Eng.* **119**, 355–370, 1994.

[20] P. BETTESS. *Infinite elements.* Penshaw Press, Sunderland, U.K., 1993.

[21] G.W. BLUMAN and S. KUMEI. *Symmetries and Differential Equations.* Applied Math. Sciences. Vol. 81. Springer-Verlag, 1989.

[22] A. BOSSAVIT. Symmetry, groups and boundary value problems. *Comp. Meth. Appl. Mech. Eng.* **56**, 167–215, 1986.

[23] J. BOUJOT. Sur le problème spectral associé aux vibrations d'un réservoir déformable. *C.R. Acad. Sc. Paris, t. 277, série A*, 1973.

[24] J. BOUJOT. Un problème spectral en élasto-acoustique. *C.R. Acad. Sc. Paris, t. 281, série A*, 1975.

[25] J. BOUJOT. Mathematical formulation of fluid-structure interaction problems. *Mod. Math. Anal. Num.* **21**, 239–260, 1987.

[26] A. BOURGINE. Sur une approche statistique de la dynamique vibratoire des structures. *Thèse Doct. Univ. Orsay*, 1973.

[27] F. BOURQUIN and F. D'HENNEZEL. Numerical study of an intrinsic component mode synthesis method. *Comp. Meth. Appl. Mech. Eng.* **97**, 49–76, 1992.

[28] H. CABANNES. *General Mechanics.* Blaisdell Publ., 1968.

[29] J. O' CALLAHAN. Comparison of reduced model concepts. *Proc. of 8th Int. Modal Analysis Conf.*, Orlando, Florida, 422–430, 1990.

[30] H. CARTAN. *Formes différentielles.* 2ème Ed., Hermann, Collection Méthodes, 1977.

[31] H.C. CHEN and R.L. TAYLOR. Vibration analysis of fluid-solid systems using a finite element displacement formulation. *Int. J. Num. Meth. Eng.* ***29***, 683–698, 1990.

[32] P.T. CHEN and J.H. GINSBERG. On the relationship between veering of eigenvalue loci and parameter sensitivity of eigenfunctions. *Journal of Vibration and Acoustics* **114**, 141–148,1992.

[33] B. CHEMOUL and H.J-P. MORAND. Fourier type dynamic superelements for the analysis of the Ariane launchers. *Proc. European Conf. on New Advances in Comp. Struct. Mech.*, O.C. Zienkiewicz P. Ladevèze, (eds), April 1991.

[34] W.H. CHU and D.D. KANA. A theory for nonlinear transverse vibrations of a partially filled elastic tank. *AIAA 5th Aerospace Sciences Meeting*, Paper 67-74, New York, 1967.

[35] P.G. CIARLET. *The Finite Element Method for Elliptic Problems.* North-Holland, 1979.

[36] P.G. CIARLET. *Introduction à l'analyse numérique matricielle et à l'optimisation.* Masson, 1982.

[37] P.G. CIARLET. *Elasticité tridimensionnelle.* R.M.A. [1], Masson, 1986.

[38] P.G. CIARLET. *Mathematical Elasticity, Vol.I: Three-Dimensional Elasticity.* North-Holland, Amsterdam, 1988.

[39] P.G. CIARLET. *Plates and Junctions in Elastic Multi-Structures. An Asymptotic Analysis.* R.M.A. [14], Masson, 1990.

[40] R.W. CLOUGH and J. PENZIEN. *Dynamics of Structures.* Mac Graw-Hill, 1975.

[41] C. CONCA, J. PLANCHARD, B. THOMAS, and M. VANNINATHAN. *Problèmes mathématiques en couplage fluide-structure.* Collection EDF, Vol. 85, Eyrolles, 1994.

[42] P. CONCUS and R. FINN. Exotic containers for capillary surfaces. *J. Fluid Mech.* **224**, 383–394, 1991.

[43] R.N. Coppolino. A numerically efficient finite element hydroelastic analysis. *Proc. AIAA/ASME/SAE 17th SDM Conf.*, 298–312, 1976.

[44] L. Coquart, A. Depeursinge, A. Curnier, and R. Ohayon. Fluid structure interaction problem in biomechanics: prestressed vibrations of the eye by the finite element methods. *J. Biomechanics* **25** (10), 1105-1118, 1992.

[45] R. Courant and D. Hilbert. *Methods of Mathematical Physics.* Vol.1, Intersc., 1962.

[46] A. Craggs. The transient response of a coupled plate-acoustic system using plate acoustic finite elements. *J. Sound Vib.* **15**, 509–528, 1971.

[47] A. Craggs and G. Stead. Sound transmission between enclosures. A study using plate and acoustic finite elements. *Acustica* **35**, 89–98, 1976.

[48] R. R. Craig, Jr. and M.C.C. Bampton. Coupling of substructures for dynamic analysis. *AIAA Journal* **6**, July 1968.

[49] R.R. Craig, Jr. A review of time domain and frequency domain component mode synthesis method. *In Combined Experimental-Analytical Modeling of Dynamic Structural Systems*, ASME-AMD, Vol. 67, D.R. Martinez, A.K. Miller (eds), 1985.

[50] J.M. Crolet and R. Ohayon (eds). *Computational Methods for Fluid-Structure Interaction.* Longman Scientific and Technical, 1994.

[51] F.A. Dahlen. Elastic dislocation theory for a self-gravitating elastic configuration with an initial static stress field. *Geophys. J. R. Astr. Soc.* **28**, 357–383, 1972.

[52] W.J.T. Daniel. Modal methods in finite element fluid-structure eigenvalue problems. *Int. J. Num. Meth. Eng.* ***15***, 1161–1175, 1980.

[53] W.T.J. Daniel. Performance of reduction methods for fluid-structure and acoustic eigenvalue problems. *Int. J. Num. Meth. Eng.* ***15***, 1585–1594, 1980.

[54] R. Dautray and J.L. Lions. *Mathematical Analysis and Numerical Methods for Science and Technology.* Vol. 1-9, Springer, 1990.

[55] J.F. Debongnie. On a purely lagrangian formulation of sloshing and fluid-induced vibrations of tanks eigenvalue problems. *Comp. Meth. Appl. Mech. Eng.* ***58***, 1–18, 1986.

[56] A.C. Deneuvy. Optimization of a coupled fluid-structure problem for maximum difference between adjacent eigenfrequencies. *Doct. Diss., Ecole Centrale Lyon*, 1986.

[57] P. Destuynder. Remarks on dynamic substructuring. *European. J. Mech. A. Solids* **8**, 201–218, 1989.

[58] P. Destuynder and I. Legrain. Piezoelectric devices for vibration control on flexible structures in structural vibration and acoustics. *Proc. ASME-DE Conf.*, Vol. 34, 43–56, 1991.

[59] J. Donea. An arbitrary Lagrangian-Eulerian finite element method for transient fluid-structure interactions. *Comp. Meth. Appl. Mech. Eng.*, **33**, 689–723, 1982.

[60] E.H. Dowell, G.F. Gorman, and D.A. Smith. Acoustoelasticity general theory, acoustic natural modes and forced response to sinusoidal excitation, including comparison with experiments. *J. of Sound and Vib.* **52**, 519–542, 1977.

[61] G. Duvaut and J.L. Lions. *Inequalities in Mechanics and Physics.* Springer, 1976.

[62] I. Ekeland and R. Temam. *Convex Analysis and Variational Problems.* North-Holland, 1976.

[63] M. El-Raheb and P. Wagner. Vibration of a liquid with a free surface in a spinning tank. *J. Sound Vib.* **76** (1), 83–93, 1981.

[64] G.C. EVERSTINE. A symmetric potential formulation for fluid-structure interaction. *J. Sound and Vib.* **79** (1), 157–160, 1981.

[65] G.C. EVERSTINE and M.K. YANG (eds). Advances in fluid-structure interaction. *Pressure Vessel and Piping Conf.*, ASME, PVP-Vol. 78, San Antonio, 1984.

[66] D.J. EWINS. *Modal Testing: Theory and Practice.* Wiley, 1984.

[67] F. FAHY. *Sound and Structural Vibration.* Academic Press, 1987.

[68] C. FARHAT and M. GÉRADIN. A hybrid formulation of a component mode synthesis method. *33rd AIAA/ASME/AHS/ASC Structural Dynamics Conference*, AIAA-92-2383, Dallas, 1992.

[69] C.A. FELIPPA. Symmetrization of the contained compressible-fluid vibration eigenproblem. *Comm. in Appl. Num. Meth.* **1**, 241–247, 1985.

[70] C.A. FELIPPA. Symmetrization of coupled eigenproblems by eigenvector augmentation. *Comm. in Appl. Num. Meth.* **4**, 561–563, 1988.

[71] C.A. FELIPPA and R. OHAYON. Mixed variational formulation of finite element analysis of acoustoelastic/slosh fluid-structure interaction. *Int. J. of Fluids and Structures* **4**, 35–57, 1990.

[72] R. FINN. *Equilibrium Capillary Surfaces.* Springer-Verlag, 1986.

[73] D.W. FOX and J.R. KUTTLER. Sloshing frequencies. *J. Appl. Math. Phys. (ZAMP)* **34**, 668–696, 1983.

[74] B. FRAEIJS DE VEUBEKE, M. GÉRADIN, A. HUCK, and M.A. HOGGE. *Structural Dynamics. Heat Conduction.* International Center for Mechanical Sciences, Udine. CISM Lecture Notes, no. 126. Springer-Verlag, 1972.

[75] C.R. FULLER, S.D. SNYDER, C.H. HANSEN, and R.J. SILCOX. Active control of interior noise in model aircraft fuselages using piezoceramic actuators. *AIAA J.*, Vol. 30, No. 1, 2613–2617, 1992.

[76] Y.C. FUNG. *Foundations of Solid Mechanics.* Prentice Hall, 1968.

[77] R.D. FU. Finite element analysis of lateral sloshing response in axisymmetric tanks with triangular elements. *Computational Mechanics* **12**, 51–58, 1993.

[78] T.L. GEERS and C.A. FELIPPA. Doubly asymptotic approximations for vibration analysis of submerged structures. *J. Acoust. Soc. Am., Vol. 73*, 1152–1159, 1983.

[79] T.L. GEERS and P. ZHANG. Doubly asymptotic approximations for internal and external acoustic domains. *Proc. Symp. on Response of Structures to High-Energy Excitations*, ASME/WAM, AM-3D, Atlanta, 1991.

[80] M. GÉRADIN and D. RIXEN. *Mechanical Vibrations.* Wiley, 1994.

[81] M. GÉRADIN, G. ROBERTS, and J. HUCK. Eigenvalue analysis and transient response of fluid-structure problems. *Eng. Comp. J.* **1**, 152–160, 1984.

[82] P. GERMAIN. *Cours de mécanique des milieux continus.* Masson, 1973.

[83] P. GERMAIN. *Mécanique.* Tomes 1 et 2. Ellipses, Paris, 1986.

[84] R.J. GIBERT. *Vibrations des structures. Interactions avec les fluides. Sources d'excitation aléatoires.* Collection EDF, Vol. 69, Eyrolles, 1986.

[85] V. GIRAULT and P.A. RAVIART. *Finite Element Approximation of the Navier-Stokes Equations.* Lecture Notes in Mathematics, Vol. 749, Springer-Verlag, 1979.

[86] G.M.L. GLADWELL. A variational formulation of damped acousto-structural vibration problems. *J. of Sound and Vib.* **4**, 172–186, 1966.

[87] R. GLOWINSKI, C.H. LI, and J.L. LIONS. A numerical approach to the exact boundary controllability of the wave equation. Dirichlet controls: description of the numerical methods. *Japan J. Appl. Math.* **7**, 1–76, 1991.

[88] S.H. GOULD. *Variational Methods for Eigenvalue Problems.* Oxford Univ. Press, 1966.

[89] H.P. GREENSPAN. *The Theory of Rotating Fluids.* Cambridge Univ. Press, 1968.

[90] R.K. GUPTA and G.L. HUTCHINSON. Solid-water interaction in liquid storage tanks. *J. of Sound and Vib.* **135** (3), 357–374, 1989.

[91] R.J. GUYAN. Reduction of stiffness and mass matrices. *AIAA Journal*, **3**, 1965.

[92] M.A. HAMDI, Y. OUSSET, and G. VERCHERY. A displacement method for the analysis of vibrations of coupled fluid-structure systems. *Int. J. Num. Meth. Eng.* **13** (1), 1978.

[93] I. HARARI and T.J.R. HUGHES. Studies of domain-based formulations for computing exterior problems of acoustics. *Int. J. Num. Meth. Eng.* **37** (17), 2891–3014, special issue on Comp. Struc. Acoust., P. Pinsky (ed), 1994.

[94] E.J. HAUG, K.K. CHOI, and V. KOMKOV. *Design Sensitivity Analysis of Structural Systems.* Academic Press, 1986.

[95] M. H. HOLMES. A spectral problem in hydroelasticity. *J. Diff. Eq.* **32**, 388–397, 1979.

[96] M. H. HOLMES. A mathematical model of the dynamic of the inner ear. *J. Fluid Mech.* **116**, 59–75, 1982.

[97] P. HOLMES. Chaotic motions in a weakly nonlinear model for surface waves. *J. Fluid Mech.* **162**, 365–388, 1986.

[98] P. C. HUGHES. Modal identities for elastic bodies with application to vehicle dynamics and control. *J. Appl. Mech.* **47**, 177-184, 1980.

[99] T.J.R. HUGHES. *The Finite Element Method.* Prentice Hall, 1987.

[100] W.C. HURTY. Dynamic analysis of structural systems using component modes. *AIAA Journal* **3** (4), 678–685, 1965.

[101] K.W. MIN, T. IGUSA, and J.D. ACHENBACH. Frequency window method for strongly coupled and multiply connected structural systems: multiple-mode windows. *Journal of Applied Mechanics* **59** (2), S244–S252, 1992.

[102] B.M. IRONS. Role of part inversion in fluid-structure problems with mixed variables. *AIAA Journal.* **7**, 568, 1970.

[103] B.M. IRONS. Structural eigenvalue problems: elimination of unwanted variables. *AIAA Journal* **3**, May 1965.

[104] M.C. JUNGER and D. FEIT. *Sound, Structures and their Interaction.* MIT, Cambridge, MA 1986.

[105] M.A. KOHUDIC (ed). *Advances in Vibration Control for Intelligent Structures.* Technomic Publ., 1994.

[106] D.D. KANA. Parametric oscillations of a longitudinally excited cylindrical shell containing liquid. *Southwest Research Institute, Report 02-1786 (IR)*, 1967.

[107] A. KANARACHOS and I. ANTONIADIS. Symmetric variational principles and modal methods in fluid- structure interaction problems. *J. Sound Vib.* **121** (1), 77–104, 1988.

[108] V. KEHR-CANDILLE and R. OHAYON. Elasto-acoustic damped vibrations. Finite element and modal reduction methods. *New Advances in Comp. Struct. Mech.*, O.C. Zienkiewicz and P. Ladevèze (eds), Elsevier, 1992.

[109] G.R. KHABBAZ. Dynamic behavior of liquids in elastic tanks. *AIAA Journal* **9**, 1985–1990, 1970.

[110] J.E.LAGNESE, J.L. LIONS. *Modelling Analysis and Control of Thin Plates.* R.M.A. [6], Masson, 1988.

[111] M. LALANNE and G. FERRARIS. *Rotordynamics Prediction in Engineering.* Wiley, 1990.

[112] P. LANCASTER. *Lambda-Matrices and Vibrating Systems.* Pergamon Press, 1966.

[113] C. LANCZOS. *The Variational Principles of Mechanics.* Dover, 1986.

[114] L. LANDAU and E. LIFCHITZ. *Quantum Mechanics.* Pergamon Press, 1992.

[115] L. LANDAU and E. LIFCHITZ. *Fluid Mechanics.* Pergamon Press, 1992.

[116] A.W. LEISSA. *Vibrations of Shells.* NASA SP-288, 1973.

[117] C. LESUEUR (ed). *Rayonnement acoustique des structures.* Collection EDF, Vol. 66, Eyrolles, 1988.

[118] A.Y.T. LEUNG. *Dynamic stiffness and substructures.* Springer, 1993.

[119] J. LIGHTHILL. *Waves in Fluids.* Cambridge University Press, 1978.

[120] J.L. LIONS. *Contrôlabilité exacte, perturbations et stabilisation de systèmes distribués.* R.M.A. [8], [9], Masson, 1988.

[121] W.K. LIU and R.A. URAS. Variational approach to fluid-structure interaction with sloshing. *Nuclear Engineering and Design* **106**, 69–85, 1988.

[122] W.K. LIU, Y.J. CHEN, K. TSUKIMORI, and R.A. URAS. Recent advances in dynamic buckling analysis of liquid-filled shells. *Trans. of the ASME, J. of Pressure Vessel Technology* **113**, 314–320, 1991.

[123] W.K. LIU, Y. ZHANG, and M.R. RAMIREZ. Multiple scale finite element methods. *Int. J. Num. Meth. Eng.* ***32***, 969–990, 1991.

[124] W. LUDWIG and C. FALTER. *Symmetries in Physics: Group theory applied to Physical Problems.* Springer Verlag, 1988.

[125] J.C. LUKE. A variational principle for a fluid with a free surface. *J. Fluid Mech.* **27**, 395–397, 1967.

[126] D.C. MA (ed). Sloshing, Fluid-Structure Interaction and Structural Response due to Shock and Impact Loads. *Proc. ASME-PVP Conf.*, Vol. 272, Chicago, 1994.

[127] L. MALVERN. *Introduction to the Mechanics of a Continuous Medium.* Prentice Hall, 1969.

[128] J. MANDEL. *Cours de mécanique des milieux continus.* Tomes 1 et 2. Gauthier-Villars, Paris, 1966.

[129] J.E. MARSDEN and T.J.R. HUGHES. *Mathematical Foundations of Elasticity.* Prentice Hall, 1983.

[130] D. R. MARTINEZ and A. K. MILLER (eds). Combined experimental/analytical modeling of dynamic structural systems. *ASME-AMD*, Vol. 67, 1985.

[131] L. MEIROVITCH. *Computational Methods in Structural Dynamics.* Martinus Sijthoff and Noordhoff, 1980.

[132] L. MEIROVITCH. *Dynamics and Control of Structures.* Wiley, 1990.

[133] J.S. MESEROLE and A. FORTINI. Slosh dynamics in a toroidal tank. *J. of Spacecraft* **24** (6), 523–531, 1987.

[134] J. MILES and D. HENDERSON. Parametrically forced surface waves. *Annual Rev. Fluid Mech.* **22**, 143–165, 1990.

[135] N.N. MOISEEV and V.V. RUMYANTSEV. *Dynamic Stability of Bodies Containing Fluid.* Springer, 1968.

[136] H. J-P. MORAND. Analyse dynamique de systèmes conservatifs évolutifs. Discussion des croisements de modes. *Assoc. Techn. Mar. et Aéro. (ATMA)*, Paris, 1976.

[137] H.J-P. MORAND and R. OHAYON. Internal pressure effects on the vibration of partially filled elastic tanks. *Proc. World Cong. Finite Element Meth. Struct. Mech.*, Bournemouth, TP ONERA 66, 1975.

[138] H.J-P. MORAND and R. OHAYON. Investigation of variational formulations for the elasto-acoustic problem. *Proc. of the Int. Symp. on Finite Element Method in Flow Problems*, Rappallo, Italy, 1976.

[139] H.J-P. MORAND and R. OHAYON. An efficient variation-iteration procedure applied to a mixed formulation of hydroelastic problem - finite element results. *Proc. Symp. Appl. Comp. Meth. Eng.*, Univ. Southern Cal., 1977.

[140] H. J-P. MORAND and R. OHAYON. Variational formulation of hydrocapillary vibration problems. Numerical finite element results. *Proc. Int. Conf. ESA-CNES*, Publ. SP-129, Toulouse, 1977.

[141] H.J-P. MORAND. Deux théorèmes de congruence relatifs aux vibrations de liquides couplés à des structures. *Proc. Int. Conf. ESA-CNES*, Publ. SP-129, Toulouse, 1977.

[142] H.J-P. MORAND and R. OHAYON. Substructure variational analysis for the vibrations of coupled fluid-structure systems. *Int. J. Num. Meth. Eng.* **14** (5), 741–755, 1979.

[143] H.J-P. MORAND. Règles de sommation en mécanique vibratoire. Application aux problèmes d'interaction fluide-structure. *Proc. Congrès Chocs et Vibrations.* Revue du GAMI, 1979.

[144] H.J-P. MORAND, M. LE GOARANT, P.BODAGALA, B.CHEMOUL. Prediction of dynamic environment of satellites. *Tendances Actuelles en Calcul des Structures*, Pluralis, Paris, 197–214, 1985.

[145] H.J-P. MORAND, M. LE GOARANT, B.CHEMOUL, P. BODAGALA, B.BELON, J-C. AGNÈSE, P. LABOURDETTE. Logiciel du CNES pour l'étude dynamique des lanceurs. *Tendances Actuelles en Calcul des Structures*, Pluralis, Paris, 483–501, 1985.

[146] H.J-P. MORAND. Quand les fusées vibrent: l'effet Pogo. *La Recherche*, no. 165, Vol. 16, 485–495, 1985.

[147] H.J-P. MORAND, B.CHEMOUL, P. BODAGALA. Interaction dynamique lanceur charge utile. *Proc. Spacecraft Structures Int. Conf.*, ESA SP-238 , 101-108, April 1986.

[148] H.J-P. MORAND. Vibrations en vol des lanceurs à liquides: formulation variationnelle symétrique complète. *Calcul des Structures et Intelligence Artificielle*, Vol. 1, J.M. Fouet, P. Ladevèze, R. Ohayon (eds), Pluralis, 1987.

[149] H.J-P. MORAND and R. OHAYON. Finite element method applied to the prediction of the vibrations of liquid-propelled launch vehicles. *Proc. ASME-PVP Conf.*, Vol. 176, Hawaii, 1989.

[150] H.J-P. MORAND. Modal methods for the analysis of the vibrations of structures coupled with fluids. *Proc. ASME-PVP Conf.*, Vol. 176, Hawaii, 1989.

[151] H. J-P. MORAND. A modal hybridization method for the reduction of dynamic models in the medium frequency range. *Proc. European Conf. on New Advances in Comp. Struct. Mech.*, O.C. Zienkiewicz, P. Ladevèze (eds), Elsevier, 1991.

[152] H. J-P. MORAND and B.RITCHIE. Etudes vibroacoustiques des lanceurs. *Proc. Prévision du bruit émis par des structures mécaniques vibrantes* Cetim 26–28 mars 1991, Revue française de mécanique, no. spécial 1991.

[153] H. J-P. MORAND, R. OHAYON. *Interactions fluides-structures.* Coll. Rech. Math. Appl. RMA 23, P.G. Ciarlet et J.L. Lions (eds), Masson, 1992.

[154] A.D. MYSHKIS, V.G. BABSKII, N.D. KOPACHEVSKII, L.A. SLOBOZHANIN, AND A.D.TYUPTSOV. *Low-Gravity Fluid Mechanics.* Springer-Verlag, 1987.

[155] W.C. MÜLLER. Simplified analysis of linear fluid-structure interaction. *Int. J. Num. Meth. Eng.* **17**, 113–121, 1981.

[156] T. NAKAYAMA and K. WASHIZU. Nonlinear analysis of liquid motion in a container subjected to forced pitching oscillation. *Int. J. Num. Meth. Eng..* **15**, 1207–1220, 1980.

[157] R.H. MAC NEAL. A hybrid method of component mode synthesis. *Computers and Structures 1,* 581–601, 1971.

[158] J.C. NÉDÉLEC. Mixed finite elements in $\mathbb{R}^3$. *Report no. 49, Centre Math. Appl., Ecole Polytechnique,* May 1979.

[159] D.J. NEFSKE, JR. WOLF, and L.J. HOWELL. Structural-acoustic finite element analysis of the automobile passenger compartment: a review of current practice. *J. of Sound and Vib.* **80**, 247–266, 1982.

[160] D.J. NEFSKE and S.H. SUNG. Power flow finite element analysis of dynamic systems. *ASME/WAM,* NCA Vol. 3, Boston, 1988.

[161] P. NEZIT. Hydrodynamique. Ondes de gravitation avec tension superficielle. *J. de Mécanique* **16** (1), 39–66, 1977.

[162] C.G. NGAN and E.B. DUSSAN V. On the nature of the dynamic contact angle: an experimental study. *J. Fluid Mech.* **118**, 27–40, 1982.

[163] V.V. NOVOZHILOV. *The Theory of Thin Shells.* P. Noordhoff, 1959.

[164] J.T. ODEN and J.N. REDDY. *Variational Methods in Theoretical Mechanics.* Universitext, Springer-Verlag, 1983.

[165] R. OHAYON. Symmetric variational formulation of harmonic vibrations problem by coupling primal and dual principles. Application to fluid-structure coupled systems. *La Recherche Aérospatiale* **3**, 69–77, 1979.

[166] R. OHAYON and B. NICOLAS-VULLIERME. An efficient shell element for the computation of the vibrations of fluid-structure systems of revolution. *Proc. SMIRT-6 Conf.,* North-Holland, 1981.

[167] R. OHAYON, R. VALID. Principes variationnels symétriques couplés de type primal-dual en élastodynamique linéaire. *C.R. Acad. Sc.*, Série II, t. 297, 1983.

[168] R. OHAYON and R. VALID. True symmetric variational formulation for fluid-structure interaction in bounded media. Finite element results. *In Numerical Methods in Coupled Systems,* chapter 10, R.W. Lewis, P. Bettess and E. Hinton (eds), Wiley, 1984.

[169] R. OHAYON. Transient and modal analysis of bounded medium fluid-structure problems. *Proc. Int. Conf. on Num. Meth. for Transient and Coupled Problems,* R.W. Lewis, P. Bettess, E. Hinton (eds), Pineridge Press, 1984.

[170] R. OHAYON. Interaction fluide-structure en milieu borné. Nouvelles formulations symétriques. *Tendances Actuelles en Calcul des Structures,* Pluralis, 435–448, 1985.

[171] R. OHAYON. Variational analysis of a slender fluid-structure system : the elastic-acoustic beam. A new symmetric formulation. *Int. J. for Num. Meth. in Eng.,* Vol. 22, 1986.

[172] R. OHAYON, N. MEIDINGER, and H. BERGER. Symmetric variational formulations for the vibration of damped structural-acoustic systems. Aerospace applications. *Proc. AIAA/ASME 28th Structures, Structural Dynamics Conf.,* 1987.

[173] R. Ohayon. Fluid-structure modal analysis. New symmetric continuum-based formulations. Finite element applications. *Proc. Int. Conf. NUMETA 87*, G.N. Pande, J. Middleton (eds), Martinus Nijhoff Publ., 1987.

[174] R. Ohayon and C.A. Felippa. The effect of wall motion on the governing equations of contained fluids. *Journal of Applied Mechanics*, Vol. 57, 783–785, 1990.

[175] R. Ohayon. Alternative variational formulations for static and modal analysis of structures containing fluids. *Proc. ASME-PVP Conf.*, Vol. 176, Hawaii, 1989.

[176] L. Olson and Th. Vandini. Eigenproblems from finite element analysis of fluid-structure interactions. *Computers and Structures* **33** (3), 679–687, 1989.

[177] E. Onate, J. Périaux, and A. Samuelsson (eds). *The Finite Element Method in the 1990's.* CIMNE, Barcelona, Springer-Verlag, 1991.

[178] M.P. Paidoussis, T.P. Luu, and B. E. Laithier. Dynamics of finite-length tubular beams conveying fluid. *J. Sound Vib.* **106**, 311–331, 1986.

[179] M.P. Paidoussis, T. Akylas, and P.B. Abraham (eds). 3rd Int. Symp. on Flow-Induced Vibration and Noise - Vol. 7: Fundamental Aspects of Fluid-Structure Interactions. *Proc. ASME/WAM*, AMD-Vol. 151/PVP-Vol. 247, Anaheim, 1992.

[180] B.N. Parlett. *The Symmetric Eigenvalue Problem.* Prentice Hall, 1980.

[181] N.C. Perkins and Jr. C.D. Mote. Comments on a curve veering in eigenvalue problems. *J. Sound Vib.* **106** (3), 451–463, 1986.

[182] M. Petyt and S.P. Lim. Finite element analysis of the noise inside a mechanically excited cylinder. *Int. J. Num. Meth. Eng.* **13**, 109–122, 1978.

[183] M. Petyt. *Introduction to Finite Element Vibration Analysis.* Cambridge Univ. Press, 1990.

[184] N. Piet-Lahanier and R. Ohayon. Finite element analysis of a slender fluid-structure system. *Int. J. for Fluids and Structures* **4**, 631–645, 1990.

[185] C. Pierre. Mode localization and eigenvalue loci veering phenomena in disordered structures. *J. Sound Vib.* **126**, 485–502, 1988.

[186] P.M. Pinsky and N.N. Abboud. Two mixed variational principles for exterior fluid-structure interaction problems. *Computers and Structures*, Vol. 33, no. 3, 621–635, 1989.

[187] O. Pironneau. *Finite Element Methods for Fluids.* Wiley, 1989.

[188] D. Poelaert. Some properties of the spectral expansion of dynamic impedances and their use in the Distel program. *Proc. ESA*, WPP-09, 1989.

[189] J.V. Ramakrishnan and L.R. Koval. A finite element model for sound transmission through laminated composite plates. *J. Sound Vib.* **112** (3), 433–446, 1987.

[190] I.M. Rapoport. *Dynamics of Elastic Containers Partially Filled with Liquids.* Springer-Verlag, 1968.

[191] P.A. Raviart, J.M. Thomas. *Introduction à l'analyse numérique des équations aux dérivées partielles.* Masson, 1982.

[192] M. Roseau. *Vibrations in Mechanical Systems.* Springer, 1980.

[193] S. Rubin. Improved component mode representation for structural dynamic analysis. *AIAA Journal* **18** (8), 995–1006, 1975.

[194] J. Salençon. *Mécanique des milieux continus.* Tomes 1 et 2, Ellipses, Paris, 1988.

[195] J. Sanchez-Hubert and E. Sanchez-Palencia. *Vibration and Coupling of Continuous Systems. Asymptotic Methods.* Springer, 1989.

[196] G. SANDBERG and P. GÖRANSSON. A symmetric finite element formulation for acoustic fluid-structure interaction analysis. *J. Sound Vib.* **123** (3), 507–515, 1988.

[197] H.M. SATTERLEE and W.C. REYNOLDS. The dynamics of the free surface in cylindrical containers under strong capillary and weak gravity conditions. *T.R. no. LG-2*, Stanford University, Palo-Alto, Cal., 1964.

[198] R.M. SCHULKES and C. CUVELIER. On the computation of normal modes of a rotating, viscous incompressible fluid with a capillary free boundary. *Comp. Meth. in Appl. Mech. and Eng.* **92** (1), 97–120, 1991.

[199] M.J. SEWELL. *Maximum and Minimum Principles.* Cambridge Univ. Press, 1987.

[200] J. SIEKMANN and U. SCHILLING. Calculation of the free oscillations of a liquid in axisymmetric motionless containers of arbitrary shape. *Zeit. Flug.*, Vol. 5, 1974.

[201] W. SOEDEL. *Vibrations of Shells and Plates.* Marcel Dekker, Inc., 1993.

[202] C. SOIZE, P.M. HUTIN, A. DESANTI, J.M DAVID, and F. CHABAS. Linear dynamic analysis of mechanical systems in the medium frequency range. *Computers and Structures* **23**, 605–637, 1986.

[203] C. SOIZE, A. DESANTI, and J.M. DAVID. Numerical methods in elastoacousticity for low and medium frequency ranges. *La Recherche Aérospatiale* **5**, 24–44, 1992.

[204] S.H. SUNG and D.J. NEFSKE. Component mode synthesis of a vehicle structural-acoustic system model. *AIAA Journal* **24** (6), 1021–1026, 1986.

[205] B. TABARROK. Dual formulations for acousto-structural vibrations. *Int. J. Num. Meth. Eng.* **13** (1), 197–291, 1978.

[206] P. TONG. Liquid sloshing in an elastic container. *Ph.D. thesis*, Cal.Tech., Pasadena, Cal., AFOSR-66-0943, 1966.

[207] H.S. TZOU and C.I. TSENG. Distributed modal identification and vibration control of continua: Piezoelectric finite element formulation and analysis. *J. of Dynamic Systems, Measurement and Control, Trans. of the ASME* **113**, 500–505, 1991.

[208] B. VALETTE. About the influence of pre-stress upon adiabatic perturbations of the Earth. *Geophys. J. R. Astr. Soc.* **85**, 179–208, 1986.

[209] A.E.P. VELDMAN and M.E.S. VOGELS. Axisymmetric liquid sloshing under low gravity conditions. *Acta Astronom.* **11**, 641-649, 1984.

[210] B.K. WADA, J.L. FANSON, and E.F. CRAWLEY. Adaptive structures. *Journal of Intelligent Material Systems and Structures*, **1**(1), 157–174, 1990.

[211] H.F. WEINBERGER. *Variational Methods for Eigenvalue Approximation.* Regional Conf. Series in Applied Mathematics. SIAM, Vol. 15, 1974.

[212] A. WEINSTEIN and W. STENGER. *Methods for Intermediate Problems for Eigenvalues.* Academic Press, 1972.

[213] G.B. WHITHAM. *Linear and Nonlinear Waves.* Wiley, 1974.

[214] C.H. WILCOX. *Sound Propagation in Stratified Fluids.* Applied Math. Sciences, Vol. 50, Springer, 1984.

[215] O.C. ZIENKIEWICZ and R.E. NEWTON. Coupled vibrations of a structure submerged in a compressible fluid. *Int. Symp. Finite Element Techn.*, Stuttgart, 1969.

[216] O.C. ZIENKIEWICZ and P. BETTESS. Fluid-structure interaction and wave forces. An introduction to numerical treatment. *Int. J. Num. Meth. Eng.* **13** (1), 1–17, 1978.

[217] O.C. ZIENKIEWICZ and R.L. TAYLOR. *The Finite Element Method.* Fourth Edition, Mc Graw-Hill, Vol. 1, 1989, Vol. 2, 1991.

Index

MASSON Éditeur
120, boulevard Saint-Germain
75280 Paris Cedex 06
Dépôt légal : avril 1995

Normandie Roto Impression s.a.
61250 Lonrai
N° d'imprimeur : I5-0358
Dépôt légal : mars 1995